THE GEOGRAPHY OF URBAN TRANSPORTATION

The Geography of Urban Transportation

FOURTH EDITION

Edited by

GENEVIEVE GIULIANO
SUSAN HANSON

THE GUILFORD PRESS
New York   London

Library of Congress Cataloging-in-Publication Data

Names: Giuliano, Genevieve, editor. | Hanson, Susan, 1943– editor.
Title: The geography of urban transportation / edited by Genevieve Giuliano,
    Susan Hanson.
Description: Fourth edition. | New York : Guilford Press, [2017] | Includes
    bibliographical references and index.
Identifiers: LCCN 2016036819 | ISBN 9781462529650 (hardcover : acid-free
    paper)
Subjects: LCSH: Urban transportation.
Classification: LCC HE305 .G46 2017 | DDC 388.4—dc23
LC record available at *https://lccn.loc.gov/2016036819*

To our children and grandchildren:

Dante, Tammy, Josie, and Allie
Marcello, Christine, Enzo, and Vincent

Kristin and Bill
Erik, Meg, Will, and Luke

And to the memory of Vincent

Preface

Transportation is arguably the lifeblood of urban existence. Without transportation, activities in cities grind to a halt, but it is also the source of many seemingly intractable urban problems such as congestion, pollution, inequality, and reliance on fossil fuels. This fourth edition of *The Geography of Urban Transportation* sustains the fundamental line of argument that informed the book's previous incarnations: how citizens and policymakers conceptualize a problem informs how they go about studying and analyzing it; analysis, in turn, informs policy formulation, decision making, and ultimately the shape of the urban transportation system itself.

The book encourages students to see the links among problem formulation, research design, analytical approach, and planning decisions. We hope that students can appreciate how the current geography of urban transportation can be understood in large part as the outcome of policy choices, themselves a result of how planners, citizens, business and labor interests, and elected officials have conceptualized problems, envisioned solutions, and taken action. And we hope that understanding will enable students to imagine—and actively work for—new transportation geographies. The book is appropriate for advanced undergraduates and beginning graduate students. It also serves as a comprehensive overview of contemporary urban transportation for the professional community.

WHY THE NEED FOR A FOURTH EDITION?

For many years the urban transportation problem was equated with congestion, and the analytical structure devised to address the problem (the four-stage urban transportation model system) aimed to guide the building of capacity-increasing new infrastructures, most often highways. Growing concerns about air pollution and other environmental damage, mobility problems of those without access to a private vehicle, and the long-term consequences of an urban transport system almost entirely dependent on the private vehicle brought pressure for policy change. The passage of the Intermodal Surface Transportation Efficiency Act (ISTEA) in 1991 was the culmination of these forces and symbolized a fundamental change in perspective. The urban transportation problem was no longer conceived of simply as congestion; questions of environmental management, historic preservation, and citizen participation, among others, were placed firmly on the mainstream transportation agenda.

Since the sea change in thinking embodied in ISTEA a quarter century ago, many new technologies, trends, and concerns have emerged and continue to alter the urban transportation landscape. Among these are the growing importance of planning for freight movement, the increased interest in nonmotorized travel modes, the shifts in transportation funding sources, the changes underway related to mobile technologies, and the need to understand the transportation-related ramifications of global climate change. In this fourth edition we retain the overall approach and philosophy of the previous editions while thoroughly updating content in light of these ongoing developments.

ORGANIZATION OF THE BOOK

The fourth edition retains the basic three-part structure of the previous editions: Part I asks how stakeholders have *conceptualized* urban transportation; Part II asks how scholars and planners have gone about *analyzing* urban transportation; and Part III asks how data and analysis might help *expand understanding and resolution* of major policy issues in urban transportation.

The four chapters in Part I set the scene by explaining core concepts, providing overviews of current trends in passenger and freight movements, describing the historical and contemporary role of transportation in urban development, and assessing the impacts of information technologies on travel patterns and urban form. The three chapters in Part II introduce students to the urban transportation planning process and contemporary trends in this process, emphasizing the political context of the planning process and the differences between aggregate and disaggregate approaches in transport analysis and planning. Each of the seven chapters in Part III takes up a pressing policy issue: public transit, land use, finance, environment, energy, equity, and the future. Across all three parts we emphasize the importance of attention to geographic scale and the links among conceptualization, analytical approach, and policymaking.

By design, each of the three parts builds on what's come before. We therefore recommend that instructors adhere to the sequential order of the three parts. Within each part, however, the order of the chapters is flexible, although Chapter 1 is intended as the introductory chapter and Chapter 14 is intended as the concluding chapter.

New to the Fourth Edition

In addition to updated content in every chapter, nine of the 14 chapters have new authors and therefore have completely new content. In Part I, Chapter 2 for the first time focuses on urban freight, and Chapter 4 is an entirely new treatment of telecommunications and travel. Part II departs from previous editions by having an overview chapter of the urban transportation planning process, followed by a chapter on regional transportation planning focused on the urban region. The final chapter in Part II focuses on neighborhood-level analysis. In Part III, new authors present new perspectives on public transportation, environmental impacts, and social and environmental justice. In every chapter authors recognize new technologies, trends, and concerns that were not relevant when the third edition was written.

Pedagogical Features

Edited volumes face the challenge of presenting a single voice. This book provides consistency through its organization in three linked parts. Each chapter is organized to provide an opening overview (including concepts and theories), present evidence and analysis, and close with a discussion of future issues. The first chapter introduces the core concepts to be discussed throughout the book, and the last chapter summarizes the main points of the book while taking a look to the future.

ACKNOWLEDGMENTS

Thanks are due to many. First and foremost, we thank the contributors, whose research, ideas, and insights are the heart of this book. We appreciate the patience and cooperation of the chapter authors as we worked through often many drafts of each chapter in an effort to bring coherence and consistency to a volume with multiple authors. Second, we thank our colleagues and users of the previous edition for providing valuable advice on how the book could be improved. Third, we thank graduate classes at the University of Southern California for providing input on draft versions of the chapters. Finally, we thank the editors at The Guilford Press for persuading us to produce a fourth edition and for their excellent support and assistance in generating the final product.

Contents

THE GEOGRAPHY OF URBAN TRANSPORTATION

PART I

SETTING THE SCENE

CHAPTER 1

Introducing Urban Transportation

SUSAN HANSON

Many trace the dawn of the modern civil rights movement in the United States to events on a city bus in Montgomery, Alabama, on December 1, 1955, when Rosa Parks refused an order from a municipal bus driver to give up her seat to a white man. Her arrest and the subsequent Montgomery bus boycott (1955–1959), in which blacks refused to patronize the segregated city bus system, proved the power of collective action and brought Martin Luther King, Jr., to prominence. That the civil rights movement should have been born on a city bus is just one measure of how urban transportation is woven into the fabric of U.S. life.

Can you imagine what life would be like without the ease of movement that we now take for granted? The blizzards that periodically envelop major cities give individuals a fleeting taste of what it is like to be held captive (quite literally) in one's own home (or some other place) for several days. With roads buried under 6 feet of packed snow, you cannot obtain food, earn a living, get medical care for a sick child, or visit friends. As floods and earthquakes occasionally remind us, the collapse of a single bridge or destruction of a small segment of roadway can disrupt the daily lives of tens of thousands of people and hundreds of businesses.

Transportation is vital to urban life around the world; without transport, the food and other goods that come from distant places and sustain life in cities would not appear in city markets. Because cities consist of spatially separated, highly specialized land uses—food stores, law firms, banks, hospitals, libraries, schools, and so on—obtaining necessary goods and services involves travel. Moreover, home and work are in the same location for only a small percentage of the workforce (less than 4% of the U.S. workforce in 2009), so that to earn an income as well as to spend it one must travel.

Although people do sometimes engage in travel entirely for its own sake (as in taking a family bike ride), most urban travel occurs as a by-product of some other, nontravel activity such as work, shopping, or seeing the dentist. In this sense, the demand for urban transportation is referred to as a *derived demand* because it is derived from the need or desire to do something else. A trade-off always exists between doing an activity at home (such as eating a meal or watching a movie) or paying the costs of movement to accomplish that activity somewhere else, such as at a restaurant or a movie theater.

All movement incurs a cost of some sort, which is usually measured in time or money. Some kinds of travel, such as that made by automobile, bus, or train, incur both time and monetary costs; other trips, such as those made on foot, involve an outlay almost exclusively of time. In deciding which mode(s) to use on a given trip (e.g., car or bus), travelers often trade off time against money costs, as the more costly travel modes are usually the faster ones. A trade-off is also involved in the decision to make a trip: the traveler weighs the expected benefits to be gained at the destination against the expected costs of getting there. Each trip represents a triumph of such anticipated benefits over costs, although for the many trips that are made out of habit this intricate weighing of costs and benefits does not occur before each and every trip.

Although transportation studies have emphasized the costs of travel, recent research suggests that for many people daily mobility can also be a source of pleasure and is not simply a hardship to be endured in order to accomplish a necessary activity, like going to work. Some people, for example, actually enjoy the time they spend alone in the car on the commute, saying it's the only time during the day they have to themselves. Contrary to most transportation theory, these people don't seek to minimize the time or distance traveled on the journey to work or other trips (Mokhtarian, Solomon, & Redmond, 2001). In this case, the demand for travel is not entirely "derived" from the demand to accomplish other activities, but something undertaken at least to some extent for its own sake.

This chapter introduces some key concepts in urban transportation and sets the stage for the chapters that follow. In particular, I describe (1) the concepts of accessibility, mobility, equity, and externalities; (2) certain aspects of the urban context within which travel takes place; (3) recent trends in U.S. travel patterns; and (4) the policy context within which transportation analysis and planning in the United States are set. The overall goal of this book is to help you understand the central role of transportation and transportation planning in shaping urban places and urban life. While many concepts have broad applicability and international comparisons enrich many of the book's chapters, our primary focus is on the United States.

CORE CONCEPTS

Accessibility and Mobility

Two concepts that are central to understanding transportation are accessibility and mobility. *Accessibility* refers to the ease of reaching potential destinations, also called "opportunities" or "activity sites"; it depends on the number of opportunities available within a certain distance or travel time, and on *mobility,* which refers to the ability to move between different activity sites (e.g., from home to grocery store). As the distances between activity sites have become longer (because of lower density settlement patterns), accessibility has come to depend increasingly on mobility, particularly in privately owned vehicles.

Accessibility and Land Use Patterns

Let me give an example from my neighborhood in inner-city Worcester, Massachusetts. About 50 years ago, many kinds of activities were located within three blocks of my house: schools, churches, parks, and many kinds of retail stores and services. In addition, several large manufacturing employers (a steel plant, a carpet-making firm, a textile machine manufacturer) were located close to the residential neighborhood. Anyone who could walk had excellent accessibility to goods and services as well as to employment. Access depended on pedestrian mobility rather than vehicular mobility. Since then, many of these places have closed, including the manufacturing companies and the supermarket; food stores across the metropolitan United States have become significantly larger and simultaneously fewer and farther apart. Access to most goods and services now requires mobility by bus, car, or taxi. The successful creation of ever larger retail establishments *depends* on ever-escalating levels of mobility, made possible because we can now travel much farther by car in about the same amount of time it took us to get somewhere on foot.

This example illustrates how the need for mobility can be seen as the *consequence* of the spatial separation between different types of land uses in the city, but enhanced mobility can also be seen as *contributing* to increased separation of land uses. Because improved transportation facilities enable people to travel farther in a given amount of time than they could previously, transportation improvements contribute to the growing spatial separation between activity sites (especially between home and work) in urban areas. As you will learn in the following chapters, the major goal of transportation planning has been to increase people's mobility as *the* way to increase accessibility. Planners and policymakers now recognize, however, that increased accessibility can also be achieved through attention to land use planning, that is, by creating high-density urban neighborhoods much like many urban neighborhoods of yore (see Chapter 7).

This symbiotic relationship between transportation and land use is one reason geographers are interested in urban transportation. One could never hope to understand the spatial structure of the metropolis or to grasp how it is changing without knowledge of the movement patterns of people and goods. The accessibility of places has a major impact upon their land values, and hence on how the land is used. The location of a place within the transportation network determines its accessibility. Thus, in the long run, the transportation system (and the travel on it) shapes the land use pattern, In Chapter 2, Laetitia Dablanc and Jean-Paul Rodrigue introduce the role of freight transport in shaping urban landscapes, and, in Chapter 3, Peter O. Muller provides numerous historical examples of the interaction between transportation innovation and urban land development. In the short run, however, the existing land use configuration helps to shape travel patterns. The intimate relationship between transportation and land use is explicitly acknowledged by the fact that at the heart of every city's transportation plan is a land use forecast. In Chapter 4, Giovanni Circella and Patricia L. Mokhtarian explore the fascinating question of how information and communication technologies such as the Internet and mobile devices are changing the relationship between distance and accessibility, and therefore the relationship between accessibility and land use.

Measuring Accessibility

We can talk about the accessibility of *places* (i.e., how easily certain places can be reached) or of *people* (i.e., how easily a person or a group of people can reach activity sites). As we saw in the example above, an individual's level of accessibility will depend largely on where activity sites are located vis-à-vis the person's home and the transportation network, but it will also be affected by when such sites are open and even by how much time someone can spare for making trips. Urban planners and scholars have long argued that the ease with which people can get where they want to go—in other words, accessibility—should be considered in any assessment of the health of a city or any measure of the quality of life (see, e.g., Chapin, 1974; Scott, 2000; Wachs & Kumagi, 1973). Measuring accessibility in a meaningful way can be difficult, however.

Personal accessibility is usually measured by counting the number of activity sites (also called "opportunities") available at a given distance from the person's home and "discounting" that number by the intervening distance. Often accessibility measures are calculated for specific types of opportunities, such as shops, employment places, or medical facilities. One measure of accessibility is presented in the following equation:

$$A_i = \sum_j O_j d_{ij}^{-b}$$

where A_i is the accessibility of person i, O_j is the number of opportunities at distance j from person i's home, d_{ij} is some measure of the separation between i and j (this could be travel time, travel costs, or simple distance), and b is a measure of how quickly accessibility declines with increasing distance. Such an accessibility index is a measure of the number of potential destinations available to a person and how easily they can be reached. Accessibility is usually assessed in relation to the person's home because that is the base from which most trips originate; personal accessibility indices could (and perhaps should) also be computed around other important bases, such as the workplace.

The accessibility of a place to other places in the city can be measured by the same equation, with A_i now the accessibility of zone i, and O_j the number of opportunities in zone j.

Although we can use the same equation, the difference between measuring the accessibility of individuals and that of places (or zones) within a city is important. When we measure accessibility at the level of places, the access measure treats all those living in zone i as if they have the same level of accessibility to activity sites in the city; it does not distinguish among different types of people within a zone, such as those with or without a car.

Both these measures of accessibility are highly simplified representations; neither really addresses mobility nor includes dimensions such as the ability to visit places at different times of day. A third measure—that of space–time autonomy—takes both accessibility and mobility into consideration; it is a more satisfying measure conceptually than measure (1) but far more difficult operationally. The concept of space–time autonomy has been developed in the context of time geography and focuses on the constraints that impinge on a person's freedom of movement (Hägerstrand, 1970). These constraints include:

Capability constraints—the limited ability to perform certain tasks within a given transportation technology and the fact that we can be in only one place at a time; for example, if the only means of transport available to you are walking and biking, the number of activity sites you can visit in, say, half an hour is lower than it would be if you had access to a car.

Coupling constraints—the need to undertake certain activities at certain places with other people; for instance, that lunch meeting with your boss can only be scheduled when you both can be in the same place at the same time.

Authority constraints—the social, political, and legal restrictions on access—for example, you can only see your dentist or go to the movies during the hours they are open, and certain locations are off-limits to people without access permits.

Your access to places and activities is restricted by these constraints.

A measure of an individual's space–time autonomy is the *space–time prism*, a visual representation of the possibilities in space and time that are open to a person, given certain constraints (see Figure 1.1). The larger the prism, shown in each frame of Figure 1.1 as a parallelogram, the greater the individual's space–time autonomy in a specific situation.

Figure 1.1a, for example, shows the space–time autonomy for a person who is currently (at 5:00 P.M.) at work and who must arrive at the childcare center no later than 6:00 P.M. to pick up his daughter; the distance between these two locations is shown on the "space" axis. Somewhere in between he must stop at a food store to buy salad greens and tofu. In addition to these location and time constraints, the father in this example must conduct all travel either on foot or by bicycle. The slope of the lines in Figure 1.1 shows the maximum speed (in 1.1a, presumably by bicycle) that he can travel. The prism outlines the envelope within which lies the set of all places that are accessible to him given these constraints. If no food store selling what he needs exists between x and y (shown on the "space" axis), then he lacks accessibility in this instance.

The concept of a space–time prism also illustrates how changes in constraints can affect accessibility. If, in this example, the childcare center were to extend its hours to 6:30 P.M., the prism defining the set of possibilities would be enlarged (see shaded area in Figure 1.1b), and this man's space–time autonomy would be increased. Or suppose he traveled by car: he could then travel farther in the same amount of time, and the prism would therefore be larger. Notice that this greater speed is shown by the slope of the lines in Figure 1.1c, which is not as steep as in 1.1a and 1.1b, where he

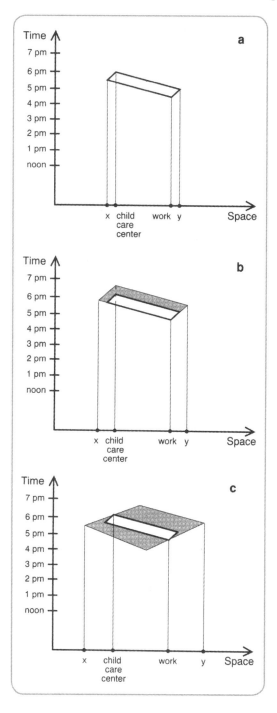

FIGURE 1.1. One measure of space–time autonomy is the space–time prism. (a) The prism defines the set of possibilities that are open to this father who must travel on foot or by bike from his place of work, where he is at 5:00 P.M., to the child care center, which closes at 6:00 P.M. (b) Effects of extended hours; the shading shows the increases in space–time autonomy if the child care center were to extend its hours from 6:00 P.M. to 6:30 P.M. (c) Effects of car availability; the shading shows the increase in space–time autonomy if a car is available, thereby permitting higher speed travel.

is assumed to be traveling by bike. The shading in Figure 1.1c indicates the increase in space–time autonomy that would result from the availability of a car. Notice that the outer spatial limits of possibilities, shown in each case by x and y on the space axis, shift outward as constraints are eased. In general, the prisms show the relationship between time and space, and you can see that as the time constraints facing this father are reduced, the greater the space within which he can move.

Many factors can, then, affect space–time autonomy. For example, flexible work schedules, longer store hours, and purchasing an additional family car all enhance space–time autonomy by adding margins to the space–time prism. Lower speed limits, rigid school hours, and traffic congestion all constrain choice. Large families impose coupling constraints, which often affect women more than men. Babysitters, day care centers, and children's growing up all reintroduce issues of space–time autonomy for parents. You can see that measuring space–time autonomy by including all of these relevant factors would be complicated; nevertheless, the concept has been influential in thinking about transportation planning (see Chapter 5).

Increasing people's space–time autonomy seems desirable in that it implies a greater accessibility to places and more discretion for spending one's time. We might question, however, the need for ever-increasing space–time autonomy and ever-increasing personal mobility. Transportation geographers among others have begun to ponder whether or not there is such a thing as too much mobility.

Equity

As we can see from the concept of space–time autonomy, someone's ability to reach places depends only in part on the relative location of those places; it also depends on *mobility,* the ability to move to activity sites, which in many places requires an automobile. We have seen how the spatial organization of contemporary society demands—indeed assumes—mobility; yet not all urban residents enjoy the high level of mobility that the contemporary city requires for the conduct of daily life. Assessing the equity of a transportation system or a transportation policy requires that we consider who gains accessibility and who loses it as a result of how that system or policy is designed; it requires that we consider to what degree people's travel patterns are the outcomes of choice or constraints. How are the costs and benefits of transportation systems distributed across different groups of people?

At the time of the bus boycott in Montgomery, Alabama, in the late 1950s, a disproportionate share of people with fewer economic resources relied on buses for transportation (as they do today). At the start of the boycott in 1955, blacks comprised 45% of Montgomery's population but 75% of the city's bus ridership, and the majority of bus riders were women (Garrow, 1988; Powledge, 1992). People without access to cars are especially likely to lack the mobility necessary to reach job locations or other activity sites. In fact, lower-income people travel significantly less (they "consume" less transportation) than do higher-income people. In 2009, among all households in the U.S. National Household Travel Survey with valid data on vehicle miles traveled (VMT), households with incomes under $25,000 were 14% of the total, but accounted for only 7% of all vehicle miles traveled.[1] Equity issues are so important in transportation that we devote a chapter to this topic (see Chapter 13).

Externalities

By its very nature, transportation creates externalities: unpriced costs and benefits. Externalities are "unpriced" in the sense that those who produce unwanted effects or who enjoy certain benefits from elements of the transportation system did not pay for them. Examples of the many *negative externalities* (external costs) stemming from transportation include the costs associated with (1) air pollution from auto and truck emissions, (2) the neighborhood disruption and safety hazards of major urban arterials, and (3) increased travel times from congestion. In each of these cases those who contribute to creating the problem have not paid the full costs of the resulting health problems, injuries and deaths, or lost incomes associated with these examples, respectively. Examples of the many *positive externalities* (also called "merit goods") linked to transportation include the benefits associated with (1) increased access and mobility from autos and urban arterials, (2) improved health and safety from a pedestrian or bicycle path, and (3) decreased travel times for autos using corridors also served by high-speed bus or rail, which divert traffic from roadways. In each of these cases, those who benefit are not paying the full costs of these advantages. If, for example, the accessibility benefits of living near a transit stop are fully reflected in higher housing prices, or if the full health care costs of air pollution were included in the price of gasoline, then these benefits and costs would no longer be externalities.

As these examples illustrate, the impact of an externality, whether negative or positive, is almost always place-specific; moreover, the same facility or service can produce positive *and* negative externalities. Households living close to a major arterial will enjoy better access to places they want to go but also will experience worse air quality than will those living farther from the arterial. The geographic specificity of externalities often motivates place-based groups either to oppose or to support infrastructure projects, depending on whether such groups see mainly the negative or the positive externalities of a project. The impacts of externalities depend in large part on *geographic context,* meaning the nature of the places in which they occur. The next section of this chapter describes the overall urban context of travel and a few of the many, diverse urban contexts in which U.S. travel takes place.

THE CHANGING URBAN CONTEXT

How have U.S. cities been changing in recent decades? In particular, how have residential and employment patterns been changing? In addition to looking at patterns for U.S. cities as a whole, we focus on two medium-sized metro areas, one from the Rust Belt—Worcester, Massachusetts—and one from the Sun Belt—Modesto, California—to examine intraurban patterns and trends.

Residential Patterns

Table 1.1 presents data on some important demographic trends from 1970 to 2010 for U.S. metropolitan areas as a whole, and Table 1.2 contains data for the Worcester and Modesto metropolitan areas for the same years. Although both metro areas had

TABLE 1.1. Demographic Trends, Metropolitan Areas in the United States, 1970–2010

	1970[b]	2010
Population of MSAs[a]	139,418,811	262,348,562
Number of households in MSAs	43,862,993	96,674,419
Percentage of households in MSAs that are single person	18.1	33.7
Percentage of MSA population living in		
City	45.8	38.6
Suburbs	54.2	61.4
Percentage of households with no vehicle		
MSA	18.6	9.7
City	28.4	15.7
Suburbs	9.2	5.8
Percentage of households with more than one vehicle		
MSA	35.6	56.0
City	26.2	45.3
Suburbs	44.7	63.0
Percentage of population over 65 years of age		
MSA	9.3	12.5
City	10.8	11.5
Suburbs	8.0	13.2
Percentage of families below the poverty level		
MSA	8.5	11.2
City	11.0	15.5
Suburbs	6.3	8.9
Percentage of families headed by women		
MSA	11.5	20.2
City	15.5	26.7
Suburbs	8.3	16.7

[a]MSAs: Metropolitan Statistical Areas, which includes a central city/cities and the surrounding suburbs.

[b]For 1970, figures refer to SMSAs (Standard Metropolitan Statistical Areas) as defined at that time.

Source: Adapted from the Censuses of Population and Housing (U.S. Bureau of the Census, 1970, 2010, and the American Community Survey).

TABLE 1.2. Demographic Trends, Worcester, Massachusetts, and Modesto, California, 1970–2010

	Worcester		Modesto	
	1970	2010	1970	2010
Population of MSA[a]	344,320	548,050	194,506	515,358
Number of households in MSA	104,694	219,625	62,100	201,520
Percentage of households in MSA that are single person	17.6	26.6	18.6	23.8
Percentage of MSA population living in central city	51.3	33.0	31.7	39.1
Percentage of households with no vehicle				
MSA	17.7	8.6	10.7	7.0
City	26.2	16.2	11.2	8.1
Suburbs	7.6	5.0	10.5	6.2
Percentage of households with more than one vehicle				
MSA	28.6	57.2	41.1	61.7
City	19.4	40.6	43.9	57.9
Suburbs	39.3	65.0	39.7	64.4
Percentage of population over 65 years of age				
MSA	12.0	12.8	10.3	10.6
City	14.7	11.7	9.9	11.6
Suburbs	9.2	13.4	10.5	10.0
Percentage of families below the poverty level				
MSA	5.4	7.2	11.8	17.2
City	7.1	15.4	8.8	17.1
Suburbs	3.7	4.0	9.1	17.3
Percentage of families headed by women				
MSA	11.3	12.3	10.6	20.4
City	15.2	17.0	11.9	24.3
Suburbs	7.2	10.0	10.1	17.8

[a]MSA: Metropolitan Statistical Area, which includes the principal city and the surrounding suburbs. Source: Adapted from the Censuses of Population and Housing (U.S. Bureau of the Census, 1970, 2010).

similar 2010 populations of more than half a million, their different histories reflect their locations in the industrial northeast and the agricultural Central Valley of California, respectively. Located about 50 miles west of Boston, Worcester's once strong manufacturing employment base has been replaced by health care and higher education as major employers. Worcester's low-cost housing relative to Boston's, along with increased job opportunities to the west of Boston, have contributed to the Worcester area's growth in recent years.

With the San Francisco Bay Area about 90 miles to the west, the state capital Sacramento 60 miles to the north, and Fresno 60 miles to the south, Modesto has served as a central place for a large swath of California's Central Valley. It also has served as a food-processing center for the agricultural products grown in the surrounding area, although many food processors once located in the center of Modesto have closed. High housing prices in the Bay Area, together with freeway access, have made Modesto attractive as a bedroom community. Clearly, neither Worcester not Modesto exist in isolation from other places; both are linked into—and therefore are in part shaped by—the national and international systems of cities, perhaps most notably via commute flows to the nearby large metro areas of Boston and San Francisco, respectively, but also via freight flows. Nevertheless, the effects of their distinctive regional contexts are also evident.

The census figures in Tables 1.1 and 1.2 disclose a number of trends that hold important implications for travel patterns and for access, mobility, and urban transportation planning. Worcester and Modesto illustrate interesting similarities to and differences from these national trends.

First, while the *populations* of U.S. metro areas as a whole (Table 1.1) and of the two metro areas in Table 1.2 have certainly increased in the 40 years between 1970 and 2010, the number of households and the number of single-person households have grown faster than has the population. The proportion of single-person households increased from 18.1% of all Metropolitan Statistical Area (MSA) households in 1970 to 33.7% of all households in 2010 (Table 1.1). The greater increase in households relative to population has implications for trip making because the number of trips made per person per day generally declines as household size increases. The trend to more households and more single-person households contributes significantly, then, to an overall growth in travel.

A second national trend is that the proportion of the U.S. metropolitan population residing in central cities continues to decline. A larger proportion (61.4% in 2010 vs. only 54.2% in 1970) now lives in the suburbs, which, with their lower density, are more difficult to serve efficiently with public transportation. This trend is clear in Worcester, where the central-city proportion of MSA population fell from 51.3% to 33%, but in the Modesto case the central-city proportion actually increased, from 31.7% to 39.1% (Table 1.2). Why? Whereas in eastern urban areas city boundaries remain fixed as population shifts occur, in the U.S. west, cities often annex land as it becomes developed, thereby extending the boundary of the metro area's central city to encompass the growing population. Modesto illustrates this process. In the early 1980s when Modesto became a bedroom community for the San Francisco Bay Area, the newly developed areas were incorporated into the city. Third, although the proportion of households having no vehicle has dropped in cities and suburbs across the

United States since 1970, the percentage of households without a car has remained higher in cities than in suburbs. This latter point is to be expected, given the higher incidence of low-income households in the central city and the greater availability of public transportation there. Nevertheless, despite fewer U.S. metro households lacking a car now than was the case in 1970 ("only" about 10% in 2010 vs. 18.6% in 1970), many people must still rely for mobility upon the bus, taxis, a bicycle, their own feet, or rides from other people. The much smaller proportion of central-city carless households in Modesto (8.1%) than in Worcester (16.2%) in 2010 reflects in part Worcester's higher density and better public transportation. Fourth, while the proportion of carless households has declined, the proportion of households with more than one vehicle has grown dramatically in both city and suburbs (Table 1.1); note especially the higher proportion of multivehicle households in low-density central-city Modesto (57.9%) than in higher-density central-city Worcester (40.6%) (Table 1.2).

A fifth national trend that is reflected also in the data for Worcester and Modesto is the growth in the numbers and the proportions of two types of households that are likely to have special transportation needs: low-income households and households headed by women. Lack of access to a vehicle is likely to pose mobility problems for low-income households, many of whom must rely on public transportation, a problem that can be especially acute in suburbs where public transportation is limited or entirely absent. The travel problems of single-parent households, headed mostly by women, stem from the difficulty of running a household single-handedly; earning an income, shopping, obtaining medical care and childcare all must be done by the one adult in the household, sometimes without the aid of an automobile.

Employment Patterns

Since the 1960s, jobs have been decentralizing from the central city to the suburbs. Traditionally, especially from the standpoint of transportation planning, the suburbs were viewed as bedrooms for the central-city workforce. Radial transportation systems, focused on the urban core, were organized in large part around moving workers from the suburbs to the central city in the morning and back to the suburbs again in the evening. But this simple pattern now describes only a small portion of current reality. In Worcester in 1960, for example, 42% of suburban workers had jobs in the central city; by 2010 only 20% of employed people living in the suburbs worked in the central city, the same proportion as for all metro areas in 2010. Similarly, the proportion of the metropolitan labor force that works in the City of Worcester as opposed to surrounding suburbs has declined from more than two-thirds in 1960 to less than one-third in 2010.[2]

In an iconic case study, Hughes (1991) documented the extent to which employment moved from central-city Newark, New Jersey, and into the surrounding region in the 30 years after 1960. Although the Newark region as a whole experienced considerable job growth during this period, the spatial distribution of employment shifted dramatically within the region, from the central city to the suburbs. Central-city job loss coupled with suburban job growth makes access to employment extremely difficult for people who live in the central city but do not have a car. Relatively few suburban jobs in the Newark region could be reached by carless people living in central

Newark; if they took a commuter train, how could they reach the employment site from a suburban train station?

Hughes links this decentralization of employment over the past few decades to the increase in poverty in inner-city Newark. Clearly, as we saw in the case of Worcester, large numbers of residences as well as jobs have been moving to the suburbs in the past four decades. But because of the unequal access of different groups of people to suburban housing, not all social groups have been able to decentralize to the same degree. In particular, low incomes, racial discrimination in the housing market, and people's preferences for living with others who are like themselves have hindered many people, especially those from minority groups, from moving to the suburbs. Hughes's analysis, as well as work by other scholars (e.g., Wilson, 1987), underlines how the reality of residential segregation in U.S. cities, together with changes in job location, has important implications for people's access—or lack of access—to employment opportunities. The term *spatial mismatch* refers to this "mismatch" between inner-city residential location and suburban job location, without the automobility needed to "connect the dots" (for reviews of the spatial mismatch literature, see Holzer, 1991; Mouw, 2000).

In a detailed study of the Boston metropolitan area, Shen (2001) extends and deepens understanding of the spatial access of low-skilled job seekers to employment. In particular, Shen argues that analysts should focus on the location of job *openings* rather than on the location of *employment* as Hughes (1991) did, and he shows that preexisting employment, concentrated in the central city, is the main source of job openings. Shen's analysis also demonstrates that residential location (e.g., city vs. suburb) is not as important as transportation mode is in accounting for differences in job seekers' access to jobs. That is, job seekers who travel by car will have higher than average accessibility to job openings from just about any residential location, whereas job seekers who depend on public transit will have substantially lower than average accessibility from most residential locations (Shen, 2001, p. 65). Evelyn Blumenberg (Chapter 13, this volume) takes up these issues of equity in access in greater detail.

THE ISSUE OF SCALE

Our discussions of residential and employment location patterns provide a useful snapshot of some important urban processes that have transportation implications: the decentralization of population and employment and the concentrations of low-income, carless, and female-headed households in the central city. But the spatial resolution of the information discussed thus far is generalized to large areas, in that we've emphasized distinctions no finer than that between central city and suburbs. For understanding some problems, such general data are sufficient, but if transportation policies and facilities are to be tailored to the specific needs of different kinds of people such as those who lack access to autos, then it is important to know as precisely as possible where, *within* the suburbs and *within* the central city, members of a target group live.

Maps at the level of the census tract (an area comprising 4,000–5,000 people on average) or the census block group (an area within a census tract, encompassing about 1,000 people) reveal the degree to which people and households with certain characteristics are clustered in certain areas within the city or within the suburbs. Of particular interest are maps showing the residential locations of people who are likely to have special transportation needs. Census tract maps for the City of Worcester, Massachusetts, provide examples. Compare Figures 1.2, 1.3, and 1.4, which show the distributions of households in poverty, households without a car, and female-headed households in 2010, respectively. First, look carefully at the mapped categories for each of these variables; these show, for example, that in at least one census tract in Figure 1.2 (poverty) more than 64% of the households had annual incomes in 2010 that fell below the federal poverty threshold of $22,050 for a family of four, while in many tracts less than 4% of households had incomes this low.

Second, the high level of spatial clustering of households with low incomes (Figure 1.2), without a car (Figure 1.3), and headed by women (Figure 1.4) within certain tracts is clear. The same level of clustering is not evident in the suburban tracts (Figure 1.5, suburban carless households, and Figure 1.6, suburban female-headed households), where the majority of tracts have relatively low incidences of these types

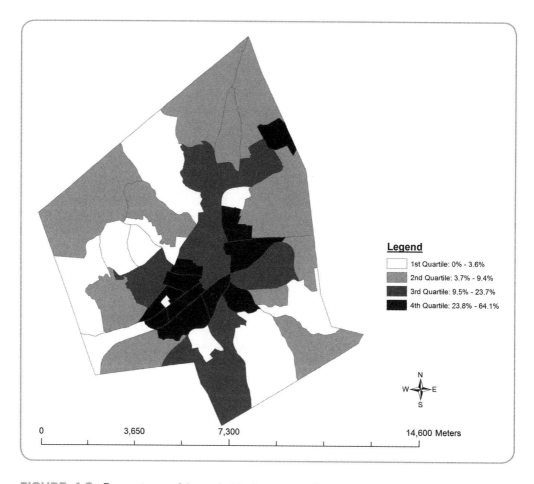

FIGURE 1.2. Percentage of households in poverty by census tract, 2010, Worcester, Massachusetts. Source: U.S. Bureau of the Census, American Community Survey 5-year estimates (2007–2011).

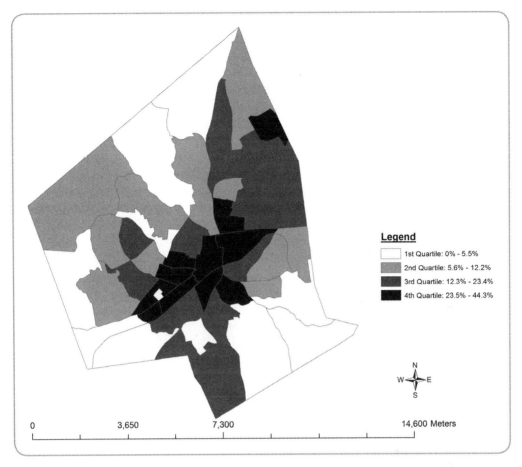

FIGURE 1.3. Percentage of households without a vehicle by census tract, 2010, Worcester, Massachusetts. Source: U.S. Bureau of the Census, American Community Survey 5-year estimates (2007–2011).

of households and the tracts with the highest proportions of carless or female-headed households are dispersed.

Third, note the spatial coincidence of female-headed households, the carless, and the poor within the City of Worcester; that is, the same areas tend to have high proportions of female-headed households, households in poverty, and households without an automobile. These spatial correlations are not as strong for suburban tracts, which show little overlap in the locations of carless (Figure 1.5) and female-headed (Figure 1.6) households. Finally, note the high levels of variation among suburban tracts in the percentages of households without a vehicle (Figure 1.5) and households headed by women (Figure 1.6), demonstrating the folly of generalizing about "the suburbs" as if they were a homogeneous region, even within one MSA.

Policies aimed at providing mobility for low-income carless people might effectively be focused on the census tracts that have the largest percentages of households with these characteristics. You can see that such policies would be far easier to implement in the city, where target tracts are clustered together, than in the suburbs, where

FIGURE 1.4. Percentage of households headed by women in each census tract, 2010, in Worcester, Massachusetts. Source: U.S. Bureau of the Census, American Community Survey 5-year estimates (2007–2011).

they are widely dispersed. Even when an area-targeted policy can be implemented, it provides services to many households who live in the targeted tracts but *do* have a car or are not in poverty, and it would miss the many carless households that do *not* live in the target census tracts. Also, numerous *individuals* (rather than households) are carless for much of the day—people, for example, who remain at home while someone else takes the household's one car to work. The census tract maps are little help in locating these people.

What these maps show is the familiar pattern of people with similar characteristics clustering together in space. What they do not show is the extent to which different types of people live within each census tract or the extent to which certain variables that covary at the area (tract) level also covary at the individual level. For instance, what percentage of female-headed households within a tract are below the poverty line or do not have access to a car?

Consider the three hypothetical census tracts in Figure 1.7. All have an average household income of $30,000, and in this fictitious example we have information not only for the tract but also for households within the tract. In the tract in Figure 1.7a, every household's income is identical, exactly $30,000, so the zonal average income is an accurate measure of the individual household incomes within the zone. In Figure 1.7b, however, the zonal average masks two distinct subareas within the zone. In one part, every household's income is $35,000, and in the other, every household's income is $22,000. In Figure 1.7c the $35,000 households are interspersed with the $22,000 households.

The complete zonal homogeneity depicted in Figure 1.7a simply does not occur in the real world; data for areas (or zones) smooth out whatever internal heterogeneity exists. People in $35,000 households are likely to have quite different travel patterns from members of $22,000 households, but the zonal data will portray only an "average" behavior for the people of the zone.

The more homogeneous an area is, the closer the zonal data will come to approximating the characteristics of the individuals living within that zone. Census tract boundaries or the boundaries of traffic zones (areal units often used in transportation studies) sometimes split relatively homogeneous areas, adding heterogeneity to the resulting zones. In general, the larger a tract or zone, the less likely it is that all the households living there will share similar characteristics.

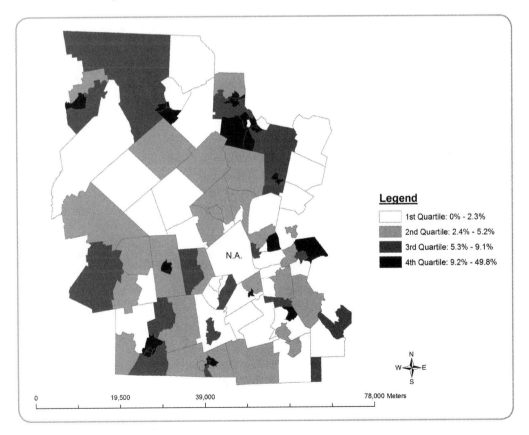

FIGURE 1.5. Percentage of households without a vehicle in each census tract, 2010, Worcester, Massachusetts, suburbs. Source: U.S. Bureau of the Census, American Community Survey 5-year estimates (2007–2011).

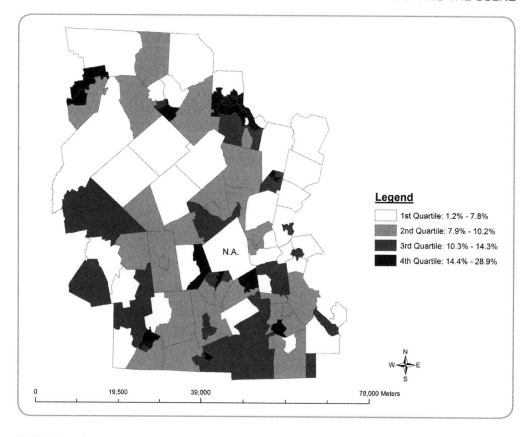

FIGURE 1.6. Percentage of households headed by women in each census tract, 2010, in Worcester, Massachusetts, suburbs. Source: U.S. Bureau of the Census, American Community Survey 5-year estimates (2007–2011).

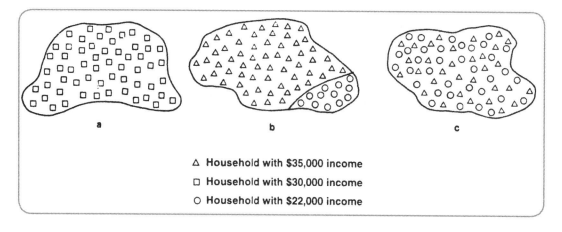

FIGURE 1.7. Hypothetical distributions of households with different income levels.

Maps of data for areas are useful for providing an overview of population distributions and employment locations within an MSA, for showing where certain population characteristics coincide in space, and for suggesting where certain transportation policies might best be deployed. They are not particularly useful for indicating what characteristics occur together at the *household* or *individual* level or for investigating how and why people make travel decisions or how they might respond to a particular transportation policy such as increased headways on a bus route (i.e., longer times between buses) or the installation of a bicycle lane on a certain route. Such questions require data for individuals rather than for areas.

UNDERSTANDING URBAN TRAVEL: AGGREGATE AND DISAGGREGATE APPROACHES

Transportation analysts use both area (*aggregate*) and individual-level (*disaggregate*) data in studying movement patterns in cities. Studies taking an aggregate approach use data for areal units called "traffic zones" and group separate trips together according to their zone of origin and their zone of destination (see Figure 1.8). Data are usually collected at the individual or household level (e.g., by asking people about their daily trip making), but, in the aggregate approach, for analytical purposes such data are aggregated into zones, as shown in Figure 1.8 as well as in the census tract maps of Worcester. In aggregate transportation studies, the focus is on the flows between zones: how many trips does a particular zone "produce" (in other words, how many trips leave zone *i*) or "attract" (how many of those trips end in zone *j*)? Disaggregate travel analyses use information on individuals and households—not zones—and usually use more finely grained spatial codes as well such as street addresses instead of zones. The conceptual base of the disaggregate approach is the person's daily travel activity pattern, rather than flows between zones. Figure 1.9 shows a schematic,

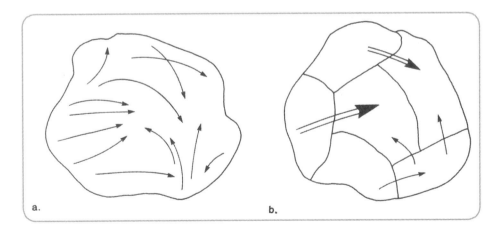

FIGURE 1.8. (a) Individual trips, showing points of origin and destination. (b) Individual trips aggregated by origin and destination zones. Thickness of arrow indicates volume of flow between zones.

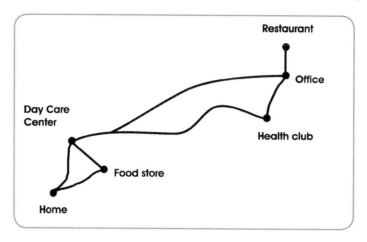

FIGURE 1.9. One person's hypothetical daily travel pattern.

bird's-eye-view of a hypothetical daily travel pattern; you could try mapping your own travel behavior like this over the course of several days. You can also represent your daily travel as a three-dimensional space-time path (see Figure 5.2 in Chapter 5).

Studies of person travel usually focus on the number of trips made and the times or distances traveled for various purposes (e.g., work, shopping, recreation), time of day of travel, locations of destinations, and modes of travel by trip purpose. Transportation analysts view person travel as a function of the characteristics of the traveler (e.g., household size, income, and auto ownership; employment status; gender) and the nature of the travel environment (e.g., available travel modes; density and diversity of, and distances to, potential destinations). For example, people tend to make more trips if they have higher incomes, an automobile for their own use, or are part of a smaller household. Men generally travel longer distances than do women. People tend to make a higher proportion of their trips by transit if they live near a transit stop and a higher proportion of their trips on foot if they live in a dense urban environment instead of a low-density suburb. However, people who want to walk and use transit may selectively choose to live in transit- and walk-friendly places, so it is difficult to determine exactly how much influence the local urban environment has on travel behavior (see Chapter 7).

To understand patterns of aggregate flows, analysts look at the characteristics of trip origin zones i and destination zones j that might account for the volume of flows leaving from i and arriving at j. Such characteristics might be median household size, income, and car ownership in origin zones—all measures of the propensity of people living in that zone to make trips—and the nature of shopping and employment in destination zones—measures of the attractiveness of different zones as trip destinations. Also important in understanding the size of the flow from i to j is the distance between the zones: more trips are made between proximate than between distant zones. Aggregate modeling approaches using data for zones are routinely used at the metropolitan-wide scale to answer transportation planning questions such as: Where within the metro region are new transportation investments (such as a new transit line or a new bridge) most needed? Where should new infrastructure be built in order to accommodate the mobility needs of the metro population 20 years hence? In Chapter 5, Harvey J. Miller describes the aggregate models used to help answer questions like these, and, in Chapter 6, Gian-Claudia Sciara and Susan Handy discuss transportation planning at the metropolitan-wide scale.

In disaggregate studies, data are not smoothed out into zonal averages, and different kinds of questions as well as questions about subareas within the metro area can be posed: What factors affect why a person selects one destination or mode rather than another? What proportion of those who live in the suburbs and work in the central city will shift from commuting by drive-alone auto to a car pool or van pool if a high-speed, high-occupancy vehicle (HOV) lane is installed on their journey-to-work route? The authors of Chapter 5 (Harvey J. Miller) and 7 (Marlon G. Boarnet) discuss the nature of the disaggregate data and the models used to answer such questions.

The scale distinction—between aggregate and disaggregate approaches—threads throughout many of the chapters in this book, particularly those in Part II, which focuses on the ways planners analyze movement patterns in order to design and implement changes to the urban transportation system. It is important to understand at the outset the close interdependencies among the scale at which you collect data, the types of models you can build (i.e., how you can simplify and make more comprehensible some of the overwhelming complexities that characterize flow patterns), and the kinds of policy analysis you can carry out. Always ask, "At what scale is this transportation issue or problem being conceptualized?"

TRENDS IN U.S. TRAVEL PATTERNS

Americans have more mobility, particularly the kind that is provided by motorized vehicles, than people anywhere else on earth. Figure 1.10, which compares the United States with several other countries in terms of automobile travel, vividly illustrates this point. In 2011, Americans logged 4.3 trillion passenger miles of travel by all motorized modes (excluding air travel), and 98% of those miles were by private vehicle (a passenger mile is one person traveling 1 mile) (U.S. Department of Transportation, Bureau of Transportation Statistics, 2012). After decades of steadily increasing to a high of 5 trillion miles in 2007, this measure of American person mobility began to decline slightly, reflecting in part the economic downturn that began in 2008 as well as other recent trends discussed in subsequent chapters of this book.

A similar trend is evident in vehicle miles traveled (VMT; 1 VMT is 1 mile traveled by a vehicle; if a vehicle has four passengers, then 1 VMT would equal 4 PMTs [passenger miles traveled]). Annual VMT per person in the United States rose steadily from about 3,000 miles in 1950 to a peak of more than 10,000 miles in 2004; since then it has declined—to 9,360 miles in 2012 (U.S. Department of Transportation, Federal Highway Administration, 2012). The long, steady increase in VMT reflects long-term growth in household income and auto ownership, increased labor force participation by women, shifts from walking and transit to autos as well as lower levels of vehicle occupancy (Pisarski, 2005); average vehicle occupancy for all trips fell from 1.9 persons per vehicle in 1977 to 1.6 in 1990 (Pisarski, 1992, p. 12) and

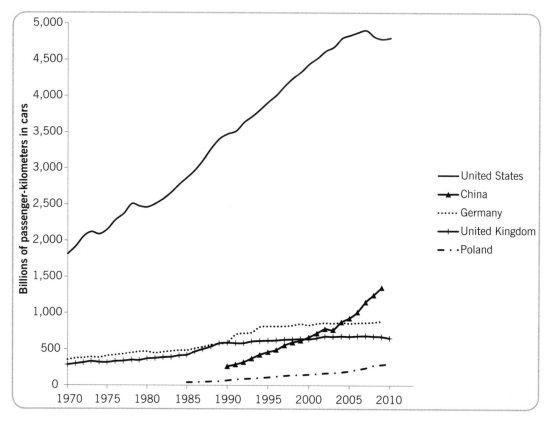

FIGURE 1.10. Trends since 1970 in passenger kilometers traveled in cars, selected countries (Organisation for Economic Co-operation and Development, 2013).

has stayed at that level since (U.S. Department of Transportation, Federal Highway Administration, 2011). Compared to several decades ago, fewer people are now passengers and more are drivers, so that, increasingly, more cars are needed to serve the same number of riders, driving the upward trend in per person VMT. The decline in VMT since 2004 is less well understood: Does it signal the beginning of a long-term trend? Is it due to durable changes in people's preferences for living in denser urban environments and using other travel modes, such as walking, biking, and transit? Marlon G. Boarnet in Chapter 7 explores these questions and examines the potential impacts of neighborhood-scale land use patterns on people's travel patterns.

Urban transportation planning has for decades focused largely on the work trip. This overarching concern with the journey to work reflects several factors. First, of all the purposes for which people travel (including work, socializing, recreation, shopping, and personal business), work used to account for the largest proportion of trips. Second, work trips are associated with the morning and evening "peaking problem"; because most people have to be at work between 7:00 A.M. and 9:00 A.M. and leave 8 hours later, work trips have been concentrated in time. The peak load associated with the work trip has placed the greatest demands on the transportation system. As you will see in the chapters describing the urban transportation planning process (Chapters 5, 6, and 7), urban transportation planners have traditionally aimed to provide a transportation system with enough capacity to handle the work trip, under the assumption that such a system can then easily accommodate travel for other purposes. (Obviously, this assumption will be erroneous if nonwork trips have a markedly different spatial configuration from work trips.) A final reason for transportation planners to focus on the work trip is that people tend to travel longer distances for work than for other purposes. The average work trip length in 2009 (13.7 miles) was more than double the average distance traveled for shopping (6 miles), for example (calculated from the 2009 National Household Travel Survey [NHTS]).

In recent years, however the proportion of travel for nonwork purposes (e.g., socializing, recreation, personal business) has increased significantly. Whereas in 1969, work and work-related travel accounted for more than 41% of all local trips, by 2009, it accounted for only about 15% (NHTS, 2009; U.S. Department of Transportation, 2003). Although travel for all purposes has grown substantially, nonwork travel has increased at a faster rate than work travel has. This increase in nonwork travel can be traced to increases in the number of affluent households and two-earner households, which spur more trips to childcare centers, restaurants, shops, fitness centers, and the like. Another reason is the decline in household size (and therefore a greater number of households for a given population) because "it is the care and upkeep of households, almost independent of the number of persons in the household that frequently governs trip making (U.S. Department of Transportation, Bureau of Transportation Statistics, 1994, p. 54). Commuting costs are now a smaller proportion of the average household's total transportation cost than in the past.

In part because of this increase in nonwork travel and because most nonwork trips begin between late morning and midafternoon (U.S. Department of Transportation, Federal Highway Administration, 2011, Figure 12), congestion that used to be confined to the morning and evening peaks has spread to encompass larger portions of the day (from 4.5 hours in 1982 to 7 hours in 2001) and a greater portion of all travel (from 33% of trips in 1982 to 67% in 2001) (U.S. Department of Transportation, Federal Highway Administration, 2013; data are for the 75 largest U.S. metro areas). The blame for traffic congestion can no longer be placed solely on the work trip.

Not only do people travel longer distances to work than they do for other purposes, work travel *distances* have been increasing, whereas *travel time* to work has been holding fairly steady. In 1975, average travel distance to work was about 9 miles and the average travel time was approximately 20 minutes (U.S. Bureau of the Census, 1979). In 2009, the average work trip covered 13.7 miles and took 23 minutes (NHTS, 2009). As described in greater detail in the chapters that follow, these national averages mask a great deal of variability, of course, among different places (region of the country, metro area size, and central city vs. suburb all affect commute distances and times) and among different groups of people, defined, for example, by age, gender, and travel mode.

Other important characteristics of work travel are the changing spatial pattern of commute trips and the modes of transportation used. The traditional suburb-to-central-city commute has not been the dominant work trip type since at least as long ago as 1970 (Plane, 1981). If we exclude work trips made within and between nonmetropolitan areas and look only at trips made within metropolitan areas, the national pattern of commuting flows looks quite intricate (see Table 1.3). By 2009 the within-suburb commute (which includes suburb-to-suburb) clearly dominated, accounting for 36.8% of all metropolitan work trips. The "traditional" commute (suburb to central city) accounted for only 20% of all journeys to work, and the reverse commute (central city to suburb) was 7.6% of commuting flows.

Given the complexity of the flow patterns depicted in Table 1.3, it is perhaps not surprising that the proportion of work trips made by auto has consistently increased while the proportion made on public transit (bus, commuter rail, subway) has remained fairly stable. In 2009, only 4.5% of work trips in the United States were made on transit, a figure that masks a great deal of place-to-place variability (U.S. Bureau of the Census, 2009). (For a thorough discussion of public transit, see Chapter 8.) By 2009, the proportion of people driving alone to work had increased to nearly 80% (up from 64.4% in 1980; U.S. Bureau of the Census, 1980), while the proportion carpooling had decreased from roughly 20% in 1980 to 9%. Those commuting by private vehicle in 2009 accounted for close to 88% of all work trips (U.S. Bureau of the Census, 2009).

Taken together, these trends—more vehicles on the road, increasing VMT, longer trips in terms of distance—add up not only to more mobility, which people clearly value, but also to many of the problems associated with transportation and primarily with the car. Among these problems, whose impacts are widely felt within and well beyond the United States, are the following: traffic congestion; air, water, and noise pollution; energy consumption; greenhouse gas emissions; urban sprawl; traffic accidents; and health problems. As mentioned earlier, in the section on externalities, a substantial portion of the cost of automobile travel is borne not by the user, but by government and by society, including future generations. Many people perceive transit as being more heavily subsidized than the auto, in that a large portion of the costs of transit are not paid directly by the user but via government support of transit agencies. Transportation analysts have argued that the automobile is also heavily subsidized, but that these subsidies tend to be less visible, more complex, and more difficult to quantify and thus not as much a part of public discourse as are

TABLE 1.3. Commuting Flows in U.S. Metropolitan Areas, 2010	
Suburbs to central city	20.0%
Within suburbs	36.8%
From suburbs to outside home MSA	4.8%
Central city to suburbs	7.6%
Within central city	29.1%
From central city to outside home MSA	1.8%
Source: Data from National Household Travel Survey (2009).	

transit subsidies (Delucchi & McCubbin, 2011; Miller & Moffet, 1993). Delucchi and McCubbin (2011) provide quantitative estimates of the external costs (costs not borne directly by those incurring them) of transport in the United States (see Table 1.4). The ranges provided for each type of cost vary substantially depending on a number of factors such as type of vehicle, road type (urban vs. rural), and density of urban area. The total external costs (in 2006 U.S. cents) per passenger mile, which range from 2.6 to 37.8 cents, can be put in the context of the direct costs of driving (costs borne by the driver), which range from 37.6 to 72.9 cents per passenger mile depending on vehicle size (American Automobile Association, 2006). Clearly the proportion of the total cost of driving (direct plus external) that is not borne by the driver—and therefore might be considered a subsidy—varies a great deal depending on circumstances, but is not trivial.

Although increases in automobile ownership and in VMT by car are evident in most countries, the patterns described in this section for the United States are not replicated everywhere; in fact, high levels of economic efficiency and of personal mobility are possible without the extreme automobile dependency that characterizes the U.S. transportation system. Despite European trends that mimic those in the United States (increases in car usage, reductions in walk, bike, and transit trips), public transport is still far more widely used in Western Europe than in the United States, accounting for about 10% of total urban travel in Europe, but only 2% in the United States (Transportation Research Board, 2001); similarly, Europeans are more likely to get places via walking or biking (they make between 25% and 35% of their trips by these modes) than are Americans, who make only 12% of their trips via walking or bicycle (Buehler & Pucher, 2012). As many of the chapters in this book make clear, public policy plays a vital role in shaping international differences in land use and transportation patterns. Within the United States, policy shifts over the past 25 years demonstrate a more comprehensive approach to conceptualizing transportation issues.

TABLE 1.4. External Costs of Transport in the United States

Costs in 2006 cents per passenger mile	Range
Congestion	0.88–7.5
Accidents	1.4–14.4
Air pollution	0.09–6.7
Climate change	0.06–4.8
Noise	0.00–3.5
Water pollution	0.01–0.05
Energy security	0.2–0.84
Total	**2.56–37.79**

Source: Delucchi and McCubbin (2011).

THE POLICY CONTEXT

In the early 1990s the policy context for transportation planning in the United States changed dramatically with the passage of two key pieces of federal legislation: the Clean Air Act Amendments (CAAA; passed in 1990) and the Intermodal Surface Transportation Efficiency Act (ISTEA; passed in 1991). The Clean Air Act of 1970 identified the automobile as a major contributor to the nation's air pollution problems and explicitly enlisted transportation planners in the effort to meet air quality goals. The 1990 CAAA required that the transportation planning process be broadened to integrate clean air planning and transportation planning at the regional level. Specifically, the CAAA set out goals for cleaner vehicles, for cleaner fuels, and for transportation programs to meet air quality standards.

ISTEA allocated funding support and set out institutional processes to meet these goals. As Howe put it, ISTEA embodied "a whole new attitude toward transportation planning" (1994, p. 11). ISTEA stated, "It is the policy of the United States to develop a National Intermodal Transportation System that is economically efficient and environmentally sound, provides the foundation for the Nation to compete in the global economy, and will move people and goods in an energy-efficient manner." As you can see from this policy statement, ISTEA construed the transportation problem far more broadly than had previous policies—to include energy consumption, air pollution, and economic competitiveness as goals in addition to increasing mobility. In 1998 Congress passed the Transportation Equity Act for the 21st Century (TEA-21), legislation that continued the transportation planning and funding philosophy embodied in ISTEA. TEA-21 was followed in 2005 by SAFETEA-LU (Safe, Accountable, Flexible, Efficient Transportation Equity Act: A Legacy for Users) and in 2012 by MAP-21 (Moving Ahead for Progress in the 21st Century). The latest multiyear transportation funding bill, known as the FAST (Fixing America's Surface Transportation) Act, became law in December 2015.

These federal surface transportation authorization bills (ISTEA and its successors) have increased the flexibility of the regional agencies responsible for transportation planning, known as Metropolitan Planning Organizations (MPOs), in their approaches to solving transportation problems. Funds that earlier had been reserved for highway projects can now be used for all surface modes of transportation, including walking, bicycling, and public transit, which federal transportation funding bills had neglected in the past. Significantly, ISTEA encouraged the building of bicycle and pedestrian facilities and gave priority to managing the existing transportation system more efficiently rather than increasing supply (i.e., building more roads). Beginning with ISTEA, regional planning agencies have enhanced power in the transportation planning arena, and public participation (the involvement of the users of the transportations system) is a mandated, integral part of the planning process. Other goals of the federal transportation spending bills passed since 1991 include protecting environmental quality, preserving the integrity of communities, and providing increased mobility for the elderly, the disabled, and the economically disadvantaged.

All this is a far cry from the days, not so long ago, when transportation planning meant highway building. Throughout the remaining chapters in this book you will see how, together, the CAAA and the ISTEA and its successors—all federal transportation funding bills—have had a significant impact on the way planners conceptualize and try to solve urban transportation planning problems. Keep in mind, however, the close association between transportation and land use (see Chapter 9, in particular); whereas MPOs address transportation issues at the regional scale, local jurisdictions retain control over land use, thereby creating a scale mismatch between planning for transportation and land use.

What are the issues, the problems, the questions that transportation analysts seek to understand and to remedy? Some are evident from the above discussions of recent trends in travel and the contemporary urban context within which travel and transportation planning take place. The increasing separation between home and work and between activity sites in general—together with the growth in population, in households, in the civilian labor force, and in consumption—mean not only that more travel is undertaken for each individual to carry out his or her round of daily activities but also that more and more people are traveling more and more miles. Congestion has long been viewed as the main urban transportation problem to be "solved," mainly by constructing more and more highways with ever greater capacity. Since the 1950s, however, we have learned the ironic lesson that increased highway capacity generally cannot keep pace with the increased travel demand that is attracted by faster movement and lower-cost travel; as a result, even with more highway capacity roads remain congested.

Legislation passed over the past 25 years has articulated a range of transportation-related policy concerns—other than traffic congestion—and a number of these are addressed in Part III of this book. Not every major transportation-related problem is accorded a separate chapter in Part III. One example is health. The growing distance between activity sites along with the overwhelming automobile orientation of U.S. society makes travel on foot or by bicycle difficult and often dangerous. In 2009 pedestrian and bicycle travel accounted for only 0.48% and less than 0.02%, respectively, of all person miles traveled but fully 14% of all traffic fatalities (U.S. Department of Transportation, 2009). One might argue, therefore, that part of the urban transportation problem is the threat to health and safety posed by the monopoly that motorized vehicles seem to have in urban travel. Air pollution, water pollution, and traffic accidents (some 34,000 traffic deaths per year in the United States) are all health problems that can be related to the current configuration of urban transportation. There is also the question as to whether the current U.S. transportation system discourages physical activity and encourages a sedentary lifestyle; how would you go about investigating that question?

The policy concerns that *are* addressed in Part III reflect the range of questions that transportation geographers and planners are grappling with: transit, land use change, transportation finance, environmental impacts, energy, and equity issues. Politics surrounds decision making in all of these policy arenas: careful analysis by transportation planners may conclude that a particular plan or policy would best serve the transportation needs of a community, but whether that plan or policy gets implemented is the result of a political process. Because every transportation-related decision will benefit some people more than others—and because who the "winners" and "losers" will be is often defined by *where they live*—the politics of urban transportation often has a distinct geographic dimension, which is evident in the chapters in Part III.

A topic of much public debate is the appropriate role for public transport in U.S. cities. In the 1960s and the early 1970s planners (and the public) looked to transit to reduce air pollution, energy consumption, and congestion, as well as to revitalize downtown areas and to promote mobility for the carless. It is now clear that, although public transportation is not a panacea for all these urban problems, it does fill an important niche in many, if not all, U.S. cities. What are the reasons behind the precarious finances of transit companies in U.S. cities? What is an appropriate role for public transportation in a country as devoted to the private automobile as is the United States? In Chapter 8, Lisa Schweitzer covers the complex policy issues surrounding public transportation.

The intimate relationship between transportation and land use was highlighted at the outset of this chapter, but what are the policy implications of this close relationship? To what extent are transportation projects responsible for increasing urban land values and for generating urban development? Can urban sprawl be attributed to large-scale transportation improvements? Are certain transportation investments, such as light-rail rapid transit lines, an effective means of changing urban land use patterns (e.g., intensifying urban land use or revitalizing certain parts of the city)? Genevieve Giuliano and Ajay Agarwal take up these and other questions about the land use/transportation relationship in Chapter 9.

Transportation investments involve huge amounts of money. What is the economic rationale for investing public funds in transportation systems? How should public monies for transportation be raised and how should they be allocated? What determines how and where that public money gets spent? How can we assess whether or not transportation funds are being allocated equitably across geographic areas and various social groups? In Chapter 10, Brian D. Taylor delves into these and other complexities of transportation finance.

Because most travel in the United States is conducted in motor vehicles, another dimension of the urban transportation problem is the set of environmental impacts stemming from facility construction and from the use of motor vehicles. Although the amount of air pollution generated per automobile has declined significantly in the past 20 years, increases in VMT mean that transportation sources remain a primary contributor to air quality problems. For example, transportation accounts for 86.5% of carbon dioxide; 45.5% of volatile organic compounds, which contribute to ground-level ozone formation; and 61.9% of nitrogen dioxide released into the air (U.S. Environmental Protection Agency, 2010). Transportation planners are now federally mandated to play a key role in maintaining air quality standards. How can transportation investments be made so as to minimize these and other adverse environmental impacts such as noise and water pollution and wildlife habitat fragmentation? In Chapter 11, Scott DeVine and Martin Lee-Gosselin focus on the environmental impacts of transportation.

Transportation is a major consumer of energy, especially energy from petroleum, accounting for 28% of the energy used in the United States but fully 70% of the petroleum consumed (Bureau of Transportation Statistics, 2014; Knittel, 2012). Although the United States has less than 5% of the world's population, it consumes 30% of the transportation energy used worldwide (Davis, Diegel, & Bundy, 2013). In the 1970s the price of energy rose substantially, and the reality of petroleum shortages—and of

U.S. reliance on petroleum imports—forced its way into the American consciousness. What impact have these earlier changes in energy price and availability had upon American energy consumption? What are the policy options for reducing the consumption of fossil fuels in transportation? In Chapter 12, David L. Greene analyzes the many key issues related to transportation and energy.

Because social status in the U.S. city is closely related to location, as is illustrated in this chapter in the maps of Worcester, Massachusetts, the placement of transportation projects will affect various social groups differently. One dimension of the urban transportation problem is, then, who pays for and who benefits from any given transportation investment. Are public transportation costs and benefits distributed evenly among transit users? How can transportation services be provided in an equitable manner? Similarly, are various social groups equally or differentially exposed to the environmental costs associated with urban mobility (e.g., noise, air pollution, traffic accidents)? In Chapter 13, Evelyn Blumenberg explores these and other questions associated with equity in transportation.

Because transportation is so completely intertwined with all aspects of urban life, questions of policy are closely linked to questions of sustainability, and sustainable transportation has to be at the core of any effort to promote sustainable development. While difficult to define, *sustainable development* involves meeting current needs in ways that improve economic, environmental, and social conditions while not jeopardizing the ability of future generations to meet their own needs (Brundtland Commission, 1987). Strategies for sustainable transportation are primarily aimed at reducing fossil fuel consumption, via changing vehicle technologies or altering people's travel patterns (e.g., reducing vehicle trip frequencies and trip distances; promoting walking, bicycling, and use of public transportation). Concerns about social justice and environmental quality are also integral to sustainable transportation strategies. With the U.S. transportation sector the single major source of greenhouse gas emissions in the world (see Chapter 13), current transportation practices in the United States are far from sustainable. Will transportation become more sustainable through reduced vehicle use, through further technology improvement, or through some of each? We invite you to think carefully about how citizens and transportation professionals might improve the sustainability of urban transportation.

Each of the policy chapters examines the evidence that bears upon an issue related to sustainability. An interesting theme that emerges from these chapters is that careful empirical analysis often yields results that challenge long-held ideas. Some of these established, accepted notions emerged from microeconomic theory; others came from earlier, less carefully controlled empirical work. But the message that comes through again and again in Part III is that we cannot assume that an assertion is true simply because it has been accepted and unquestioned for a long time. So, we invite you to read critically and to think about how *you* would go about improving transportation in cities.

ACKNOWLEDGMENT

The author thanks Joseph Danko of Clark University and Mohja Rhoads of the University of Southern California for their research assistance. Scott Le Vine and Martin Lee-Gosselin graciously agreed to move Figure 1.10 from Chapter 11 to this chapter.

NOTES

1. Calculated from the 2009 National Household Travel Survey; for the U.S. population as a whole, 22% have incomes under $25,000 annually.

2. In 1960, 67.2% of Worcester's MSA labor force worked in the City of Worcester, and in 2000 the percentage was 32.2. The MSA boundaries changed over this period as well; in 1960, the MSA included 20 towns, but by 2000 it included 35 towns. Although the number of workers in the city increased slightly in these four decades (from about 81,500 to 82,800), suburban employment grew at a far greater rate.

REFERENCES

American Automobile Association. (2006). *Your driving costs.* Heathrow, FL: AAA Communications.

Brundtland Commission (World Commission on Environment and Development). (1987). *Our common future.* New York: Oxford University Press.

Buehler, R., & Pucher, J. (2012). Walking and cycling in Western Europe and the United States: Trends, policies, and lessons. *Transportation Research News,* No. 280, 34–42.

Chapin Jr., F. S. (1974). *Human activity patterns in the city: Things people do in time and in space.* New York: Wiley.

Davis, S. C., Diegel, S. W., & Bundy, R. G. (2013) *Transportation energy data book, 32nd edition.* Prepared by Oak Ridge National Laboratory for the U.S. Department of Energy. See *http://cta.ornl.gov/data/index.shtml.*

Delucchi, M., & McCubbin, D. (2011). External costs of transport in the U.S. In A. de Palma, R. Lindsey, E. Quintet, & R. Vickerman (Eds.), *Handbook in transport economics* (pp. 341–386). Northampton, MA: Elgar.

Garrow, D. (1988). *Bearing the cross: Martin Luther King Jr. and the Southern Christian Leadership Conference.* New York: Vintage Books.

Hägerstrand, T. (1970). What about people in regional science? *Papers, Regional Science Association, 24,* 7–21.

Holzer, H. (1991). The spatial mismatch hypothesis: What has the evidence shown? *Urban Studies, 28,* 105–122.

Howe, L. (1994). Winging it with ISTEA. *Planning, 60*(1), 11–14.

Hughes, M. (1991). Employment decentralization and accessibility: A strategy for stimulating regional mobility. *Journal of the American Planning Association, 57,* 288–298.

Knittel, C. (2012). Reducing petroleum consumption from transportation. *Journal of Economic Perspectives, 26*(1), 93–118.

Miller, P., & Moffet, J. (1993). *The price of mobility: Uncovering the hidden costs of transportation.* Washington, DC: Natural Resources Defense Council.

Mokhtarian, P., Solomon, I., & Redmond, L. (2001). Understanding the demand for travel: It's not purely "derived." *Innovation, 14*(4), 355–380.

Mouw, T. (2000). Are black workers missing the connection?: The effect of spatial distance and employee referrals on interfirm racial segregation. *Demography, 39,* 507–528.

National Household Travel Survey (NHTS). 2009. Author calculation from NHTS data.

OECD (Organization for Economic Co-operation and Development). (2013). Data extracted from OECD.Stat, April 11, 2013.

Pisarski, A. E. (1992). *Travel behavior issues in the 90s.* Washington, DC: U.S. Department of Transportation.

Pisarski, A. E. (2005). *Transportation trends and smart growth.* Conference Proceedings No. 32, Transportation Research Board, National Research Council, Washington, DC.

Plane, D. A. (1981). The geography of urban commuting fields: Some empirical evidence from New England. *Professional Geographer, 33,* 182–188.

Powledge, F. (1992). *Free at last?: The civil rights movement and the people who made it.* New York: Harper Perennial.

Scott, L. M. (2000). Evaluating intra-metropolitan accessibility in the information age: Operational issues, objectives, and implementation. In D. G. Janelle & D. C. Hodge (Eds.), *Information, place, and cyberspace: Issues in accessibility* (pp. 21–45). Heidelberg, Germany: Springer.

Shen, Q. (2001). A spatial analysis of job openings and access in a U.S. metropolitan area. *Journal of the American Planning Association, 67,* 53–68.

Transportation Research Board. (2001). *Making transit work: Insight from Western Europe, Canada, and the United States* (TRB Special Report 257). Washington, DC: National Academy Press.

U.S. Bureau of the Census. (1970). *1970 Census of population and housing.* Washington, DC: U.S. Department of Commerce.

U.S. Bureau of the Census. (1979). *The journey to work in the United States: 1975.* Washington, DC: U.S. Government Printing Office.

U.S. Bureau of the Census. (1980). *1980 census of the population. General social and economic characteristics: United States summary.* Washington, DC: U.S. Government Printing Office.

U.S. Bureau of the Census. (2009). *American Community Survey Reports: Commuting in the U.S.* Washington, DC: U.S. Department of Commerce.

U.S. Bureau of the Census. (2010). *Census of Population and Housing.* Washington, DC: U.S. Department of Commerce.

U.S. Department of Transportation. (2003). Summary statistics on demographic characteristics and total travel, 2001 National Household Travel Survey. Available at *http://nhts.ornl.gov/2001/html_files/trends_ver6.shtml.*

U.S. Department of Transportation. (2009). *Traffic safety facts.* Washington, DC: National Highway Traffic Safety Administration's National Center for Statistics and Analysis.

U.S. Department of Transportation, Bureau of Transportation Statistics. (1994). *Transportation statistics annual report 1994.* Washington, DC: U.S. Government Printing Office. Available at *www.rita.dot.gov/bts/sites/rita.dot.gov.bts/files/publications/national_transportation_statistics/index.html#chapter_1.*

U.S. Department of Transportation, Bureau of Transportation Statistics. (2003). *Pocket guide to transportation 2003* (BTS03-01). Washington, DC: Author. Available at *www.bts.gov/publications/pocket_guide_to_transportation/2003/index.html.*

U.S. Department of Transportation, Bureau of Transportation Statistics. (2012). *National transportation statistics,* Table 1.40. Washington, DC: Author.

U.S. Department of Transportation, Bureau of Transportation Statistics. (2014). *National transportation statistics 2014.* Washington, DC: U.S. Department of Transportation, Research and Innovative Technology Administration. Available at *www.rita.dot.gov/bts/sites/rita.dot.gov.bts/files/publications/national_transportation_statistics/index.html.*

U.S. Department of Transportation, Federal Highway Administration. (2001). *1995 NPTS databook.* Washington, DC: U.S. Government Printing Office.

U.S. Department of Transportation, Federal Highway Administration. (2011). Summary of travel trends, 2009 National Household Travel Survey. Available at *http://nhts.ornl.gov/2009/pub/stt.pdf.*

U.S. Department of Transportation, Federal Highway Administration. (2012). Travel trends. Available at *www.fhwa.dot.gov/policyinformation/travel_monitoring/tvt.cfm.*

U.S. Department of Transportation, Federal Highway Administration. (2013). Congestion report. Available at *www.ops.fhwa.dot.gov/congestion_report_04/chapter3.htm.*

U.S. Environmental Protection Agency. (2010). *EPA inventory of US greenhouse gas emissions and sinks: 1990–2008.* Washington, DC: Author.

Wachs, M., & Kumagi, T. G. (1973). Physical accessibility as a social indicator. *Socioeconomic Planning Sciences, 7,* 437–456.

Wilson, W. J. (1987). *The truly disadvantaged: The inner city, the underclass, and public policy.* Chicago: University of Chicago Press.

CHAPTER 2

The Geography of Urban Freight

LAETITIA DABLANC
JEAN-PAUL RODRIGUE

The movement of goods is essential in any urban area; goods of every description must find their way to residences, restaurants, offices, factories, or department stores, and the discarded materials remaining from consumption must be hauled away and disposed. Cities are also centers of production, and with that production comes the distribution of raw materials, parts, and finished goods. Cities compete to attract economic activities, and the efficiency of the goods movement system is critical to this competitiveness. At the same time, the environmental and congestion impacts associated with urban freight activities are a factor of growing concern for public and private stakeholders as well as for urban travelers. Thus freight is a very important element of the urban transportation system. Despite its importance, surprisingly little is known about urban freight.

WHAT IS URBAN FREIGHT?

Urban freight is defined as the transportation of goods by or for commercial entities (as opposed to households) taking place in an urban area. This definition includes all goods movements generated by the economic needs of local businesses: warehousing and activities such as deliveries and pick up of supplies, materials, parts, consumables, mail, and refuse. It also includes home deliveries to households, as these are generally done by means of a commercial transaction. Through traffic, such as trucks circulating in a city en route to another destination without serving any business or household of the city, is also part of urban freight. This definition does not include private transportation undertaken by people to acquire goods for themselves (shopping trips), although these activities can represent an important element of the transportation of goods. In large French cities, shopping trips are estimated to account for about half of the vehicle-miles related to the urban movement of goods (Routhier, 2013). No such data are available for the United States.

Urban freight activities represent an important part of urban transportation in many respects (Rodrigue, 2013). In terms of vehicle flows and traffic, the movement of goods represents 10 to 15% of vehicle equivalent miles traveled in city streets.[1] Freight transportation modes and terminals are important components of urban geography. They interact and sometimes conflict with passenger transportation. Three to five percent of urban land is devoted to freight transportation and warehousing, but this figure can go higher in major port cities acting as gateways to large markets or at large inland hubs such as Chicago and Kansas City. On average, urban freight distribution generates one delivery (or pick-up) per week for every job. About 300–400 truck trips per 1,000 urban residents can be counted per day. Such figures obviously vary according to income and the economic function of cities (e.g., manufacturing as opposed to financial centers) (Dablanc, 2009). Urban freight distribution generates a wide array of externalities, namely, congestion, pollution, and community disruptions. It contributes a significant share to total urban transportation air and noise emissions: a quarter of CO_2 emissions, a third of NOx emissions, and half of particulate matter. These large shares reflect the dominance of road transport (more than 90% of the tonnage) as well as the age of commercial fleets: trucks and vans in cities tend to be older than those used in long-distance hauling.

CHANGING URBAN ECONOMIES

Recent trends related to globalization and rising standards of living imply more, and more diverse, consumption (Rodrigue, 2004). To improve productivity, store inventory levels have shrunk, and businesses are increasingly supplied on a just-in-time basis. This enables a higher intensity of usage of valuable retail space. The quantity and variety of retail goods have increased considerably, and inventories change several times a year. With the rise of the service economy, the demand for express transport and courier services has soared. Additionally, e-commerce has increased the demand for home deliveries and new forms of urban distribution. These factors have made urban economies more dependent on freight transportation systems, with more frequent and customized deliveries. All this generates a greater intensity and frequency of urban freight distribution and correspondingly improved forms, organization, and management of urban freight flows.

Urban Freight in a Global Context

Although urban freight appears to be taking place at the local (urban, metropolitan) level, a comprehensive understanding of its drivers and dynamics requires the following considerations:

• *Urban freight distribution in the context of global supply chains.* Global processes are imposing local forms of adaptation to ensure that freight is delivered in a timely and reliable fashion. *Offshoring* (the shift of manufacturing to lower cost production locations outside the United States) and *outsourcing* (the contracting out of manufacturing to lower cost production locations outside the United States) have contributed to the development of global supply chains (see Box 2.1). Local or regional economic structures alone cannot therefore explain urban freight distribution activities that are part of global supply chains (Hesse, 2008). When supply chains serve entire regions or countries or beyond, cities are clearly trade hubs.

• *Cities as global trade nodes.* As global supply chains have expanded, rising international trade has led to growing freight volumes focused on certain cities. Global freight distribution has taken on a new significance, particularly with new large terminal facilities such as ports, airports, rail yards, and distribution centers. These terminals are handling movements originating from, bound to, or simply passing through a metropolitan area. With containerization and scale economies in shipping as tools supporting growth of international trade, intermodal terminals have become a notable element of the urban/suburban landscape. With the growth of valuable cargo carried over long distances, airports are also active nodes in urban freight distribution. Along with their attached freight distribution facilities (e.g., transloading facilities and warehouses) large terminals form a fundamental element of the interface between global distribution and city logistics. These developments explain the phenomenal increase in the number of warehouses[2] and freight centers in large urban areas in recent decades. For example, in the Atlanta metropolitan area alone, the number of freight and logistics facilities tripled from 1998 to 2008 (Dablanc & Ross, 2012).

• *Continuing urbanization and rising per capita income.* Global urbanization is compounding the challenges of urban freight and logistics as the share and the level of concentration of the global population living in cities increase. Cities present a variety of forms and levels of density, each associated with specific urban freight patterns. Socioeconomic factors, such as rising income and consumer preferences, are also important drivers of urban consumption.

BOX 2.1
The Global Supply Chain

A *global supply chain* is an integrated network of production, trade, and service activities that covers all stages in a supply chain, from the transformation of raw materials, through intermediate manufacturing stages, to the delivery of a finished good to a market. The chain is conceptualized as a series of nodes, linked by various types of transactions, such as sales and intrafirm transfers. Each successive node within a commodity chain involves the acquisition or organization of inputs for the purpose of adding value. Because production (supply) and consumption (demand) are never synchronized, there is the need to store raw materials, parts, and finished goods at different stages of the value chain. As most of the final consumption takes place in urban areas, the last stage of a value chain commonly concerns urban supply chains.

The Emergence of City Logistics

As a response to the increasing challenges facing the urban freight system, *city logistics* has emerged as a strategy ensuring efficient freight movements and innovative responses to consumer and business demands (Dablanc, 2009). City logistics covers all the means by which freight distribution can take place in urban areas, as well as the strategies and planning policies that can improve overall efficiency, such as mitigating congestion, energy consumption, and environmental externalities (Giuliano, O'Brien, Dablanc, & Holliday, 2013; Taniguchi & Thompson, 2008). The first applications of city logistics were undertaken primarily in metropolitan areas in Japan and Western Europe, where land availability is highly constrained and strong traditions of urban planning prevail.

City logistics is an emerging field of investigation. Prior to the 21st century the consideration of freight distribution within urban planning remained limited (Lindholm & Behrends, 2012). As clearly demonstrated in this book, the mobility of people has been the dominant focus of researchers and practitioners. Strategies aimed at improving the mobility and sustainability of passenger transport, such as promoting urban transit, imposing restrictions on driving, and increasing development density can even run counter to freight distribution strategies in urban areas that rely upon trucking. Thus freight transport is a critical component of urban mobility

This chapter describes the diversity in which urban freight distribution takes place and provides a review of the current trends in city logistics. Based on the main challenges that have been identified, we examine city logistics strategies and policies that have been implemented with various degrees of success. Because urban freight distribution systems depend on the specific land use forms, patterns, and functions in which they take place, we suggest a typology to classify global cities by their logistics systems and their city logistics initiatives; this classification highlights the need for performance metrics.

THE DIVERSITY OF URBAN FREIGHT DISTRIBUTION

Main City Functions and Urban Distribution

Each city around the world may display different freight transport and logistics activities, levels of intensity, and urban freight problems:

- Paris aims to limit the environmental footprint of freight distribution so that the quality of life of its residents can be maintained and improved. The city's status as one of the world's leading cultural and tourist attractions has a notable impact on the strategies and priorities accorded to urban freight distribution.

- Lima tries to cope with the contradictory demands related to the copresence of modern (motorized) and traditional forms of urban distribution. This duality leads to contradictions in terms of infrastructure provision and regulations that are aimed at promoting mechanized transportation in an environment where nonmechanized modes still have dominance. Modern logistics services are as vital to the urban economies of developing and emerging countries as are more basic freight activities serving street vendors or home-based manufacturing workshops.

• Chicago aims to maintain its role as a major rail hub and freight distribution platform for North America with a concentration of distribution and manufacturing activities. The metropolitan area is the point of convergence of the rail lines of the Class I carriers, but the different terminal facilities are in separate parts of the city and not well connected. Trucks are required to carry containers from one terminal to the other, adding to traffic congestion.

• Los Angeles is facing not only congestion and environmental problems such as noise and air pollution but also conflicts between its function as a major commercial gateway for East Asian trade and other functions linked with economic, tourist, and cultural activities. One response to these conflicts has been the construction of a partially underground rail link between the port terminals and the major rail yards nearby the central business district. This link, known as the Alameda Corridor, effectively separates a share of the port-related intermodal traffic from local circulation. Another response is a truck replacement program that has greatly reduced port-related particulate emissions.

• Shanghai has become the largest cargo port in the world and acts as the main transport hub supporting China's export-oriented strategies. A significant share of the freight circulating within the city is therefore linked with global distribution processes. Rising standards of living imply growing consumption levels and the emergence of city logistics challenges common in advanced economies.

• Istanbul is coping with rapid urbanization and economic growth, along with unique geographical constraints, namely, a scarcity of flat land and the division of the city by the Bosporus strait. Its commercial function is being strengthened by its role as a platform between Middle Eastern, European, and Black Sea commercial interactions. The outcome has been severe constraints for freight circulation in an environment of accelerated economic and urban growth. The city is embarking on large infrastructure projects with a new mega-airport and the relocation of manufacturing activities to exurban locations.

Because of differences in regional context, urban economic mix, and built environments, material flows associated with urban production, distribution, and consumption vary. They are illustrated in Figure 2.1. The role and extent of these functions vary according to the historical and socioeconomic context of each city,

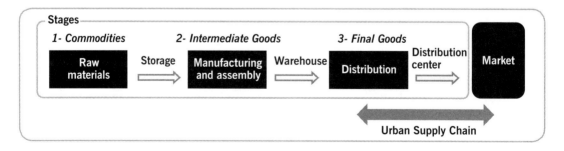

FIGURE 2.1. Synthetic supply chain.

commonly involving a specialization (e.g., financial cities, manufacturing cities), with consumption being, nonetheless, an equalizing factor as it is now a dominant activity everywhere.

Consumption, production, and distribution are each associated with various types of material flows and urban supply chains. For instance, retailing relies on urban deliveries originating from distribution centers, which themselves are likely to have been supplied through terminal haulage. Urban freight distribution is usually most intense around large specialized generators, which come in four major types:

• *Terminals,* such as ports, airports, and railyards, serve as the major connecting nodes for freight movements. Terminals are typically large and generate high traffic levels. As terminals handle a wide variety of freight, the form and means of entry of that freight also vary. Freight may arrive as bulk, containers, full truckloads (TLs), and less than truckloads (LTLs). The market area of a transport terminal is defined as the hinterland, which can involve destinations (logistics zones and manufacturing districts) within the city itself or flows having to transit through urban areas on their way to other destinations. The impact of a transport terminal on city logistics is obviously related to the intensity of the terminal activity, the supply chain it services, and the extent of its hinterland.

• *Logistics zones* include warehouses, distribution centers, and cross-docking facilities, sometimes associated with clustered distribution and light manufacturing activities. Higher consumption levels and global supply chains have been the driving forces in the development and expansion of logistics zones. Facilities are increasingly co-located, yielding more efficient interactions because of proximity and reducing the propensity of freight to enter urban areas. High land prices near terminals and central areas have encouraged the development of logistics zones in peripheral areas (Dablanc, 2014; Dablanc & Ross, 2012).

• *Manufacturing areas.* Many production activities are now related to global processes and to elements of global value chains insofar as such activities produce finished goods (requiring parts coming from all over the world) or, more likely, intermediate goods (e.g., parts). Manufacturing activities are generators of producer-related urban freight movement involving all possible forms of road (and sometimes rail) transport. Manufacturing districts are commonly found in association with transport terminals, particularly for heavy industry. Manufacturing and logistics activities are often mixed; standard manufacturing activities are common in logistics areas. The distinction between a logistics and a manufacturing zone can thus be blurred. For instance, many logistics zones were developed as industrial zones that attracted distribution centers instead.

• *Commercial areas.* Commercial areas are key nodes of high accessibility throughout the urban area (i.e., city centers and suburban subcenters) and are destinations for the bulk of urban passenger flows. They attract consumer-related freight movements mostly through retail activities usually supplied through LTLs (e.g., delivery vans and trucks). In addition, clusters of office towers and large institutions (seats of government, universities, museums, etc.) are large generators of freight demand.

Some central business districts also involve adjacent freight-intensive activities such as rail yards and even port terminals, particularly in older cities or in cities having an important gateway function.

Urban freight generators are commonly interrelated and therefore spatially clustered. For instance, a port district will have maritime terminals as well as nearby distribution centers and industrial activities. The same applies to airport districts, which often have concentrations of distribution centers and commercial activities. Parcel and express deliveries are everywhere, especially since the growth of e-commerce has increased direct goods deliveries to residential areas.

TRANSPORTATION AND URBAN SUPPLY CHAINS

A *supply chain* includes all the activities necessary to satisfy a demand for goods or services such as raw materials, parts, or finished goods (see Box 2.1 above). Each city is supplied by an impressive variety of supply chains servicing a wide array of economic activities such as grocery stores, retail, restaurants, office supplies, raw materials and parts, construction materials and wastes (Rhodes et al., 2012; Routhier, 2013). The level and type of economic development determines the level of urban freight activity because income and consumption levels are interdependent.

Urban supply chains fall into two main functional classes. The first involves consumer-related distribution:

- *Independent retailing.* Urban areas have a notable variety of retailing activities, many of which define the commercial and social character of neighborhoods. Small single-owner stores are located primarily in the densest parts of cities. In developing countries these retailing activities are often complemented by informal street markets and stalls. Supply is often organized by the storeowner (own-account logistics or direct relationship between a specific supplier and the store) and involves mostly small loads and small vehicles.

- *Chain retailing.* In the contemporary commercial landscape, especially in the United States, chain retailing (stores directly affiliated to a common brand or franchised from this brand) have become a dominant element. Like independent retailing, chain retailing covers an extensive array of goods and types of stores. They are supplied by manufacturers that have extensively relied on global sourcing, through a complex chain of warehouses and distribution centers (Figure 2.2). Chain retail outlets are located throughout metropolitan areas, but in different forms, from the classic suburban mall to the multistory city shopping district. Shopping malls are based on the principle of economies of agglomeration and the provision of ample parking space. Chain retailing tends to rely on the expertise of third-party logistics service providers to mitigate urban freight distribution challenges and to organize complex multinational sourcing strategies for mass retailers. Large stores are commonly accessed through dedicated delivery bays where they are resupplied on a daily basis (or more frequently) through their own regional warehousing facilities.

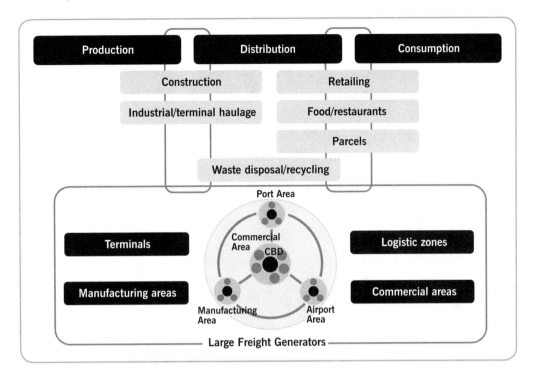

FIGURE 2.2. City functions and urban distribution.

• *Food deliveries.* Because fresh foods are perishable, a specialized form of urban distribution supplies outlets such as grocery stores and restaurants. In the Paris region, the total daily number of deliveries made to cafés, restaurants, and hotels is five times larger than the total number of deliveries made for all large supermarkets.

• *Parcel and home deliveries.* The growth of retail, e-commerce and advanced services such as insurance, finance, or corporate management has led to growth in the movement of parcels and courier services. While some home deliveries are handled by local companies, large parcel carriers have established services covering the majority of the world's main commercial cities. They maintain a network of strategically located distribution hubs where shipments are consolidated or deconsolidated. International shipments are often taken care of by parent companies, namely, air freight integrators. Online shopping (for retail goods) has grown significantly: it represented about 8% of total retail sales in the United States in 2013. With more online shopping comes more home deliveries or deliveries within newly implemented networks of pick-up-points (14,000 Hermes drop-boxes in Germany, 19,000 local stores serving as pick-up-points in France in 2013). These deliveries can happen at unusual times (evening, week-ends) and in unusual neighborhoods (residential) for common carriers accustomed to delivering to businesses.

The second functional class of urban supply chains is related to producer-related distribution:

- *Construction sites.* Urban infrastructures, from roads and residences to office and retail spaces, are constantly being constructed, renovated, and repaired. Such activities are intensive in the use of materials and heavy equipment, which must be supplied on an irregular basis in terms of both the time and the location of deliveries. This specificity makes the planning of such deliveries an ad hoc process that is prone to local disruptions.

- *Waste collection and disposal.* Urban activities generate large quantities of waste. As standards of living are improving across the world, the amount of waste generated by cities has grown accordingly. Waste must be collected and carried to recycling or disposal sites. In particular, recycling has become an important activity taking place in urban areas and involves specialized vehicles and dedicated pick-up tours. This form of urban freight transportation can effectively be organized to minimize disruptions and improve its efficiency.

- *Industrial and terminal haulage.* Large transportation terminals serve as gateways for goods bound elsewhere. From these facilities cargo is moved (drayed) to warehouses and distribution centers where shipments are processed and then redistributed to markets (see Figure 2.2). The decentralization of terminals in far away suburbs increases the net mileage of trucks and vans on metropolitan roads. Transport terminals and logistics zones are therefore generators of goods movements that inevitably affect urban circulation. Gate access at large intermodal terminals such as ports can also lead to congestion (queuing) and local disruptions.

Cities and Logistical Performance

The measurement of urban freight logistical performance requires key performance indicators at the city level. However, such measures do not yet exist. The World Bank has compiled a national measure, the Logistics Performance Index (LPI).[3] A value of less than 3 reflects an array of problems within a nation's freight distribution system causing undue delays and additional costs. While the LPI reflects global trade and supply chains, it can also indicate, to some extent, the logistical capabilities of cities. As urban freight flows are affected by the general level of freight infrastructure in a country, the LPI provides a way to categorize logistics infrastructure quality across countries.

By cross-referencing a dataset composed of the world's 435 cities of more than one million inhabitants (totaling 1,257 million urban dwellers) with their respective national 2010 LPI values, it is possible to provide a basic overview of the national city logistics context (Figure 2.3). Based on such a classification it is noted that 27% of the urban population lived in cities in countries with a low LPI (less than 3) while 47% lived in cities with below average LPI conditions (between 3 and 3.5). Only 26% of the urban population were living in cities with adequate national LPI conditions (more than 3.5). In countries with high LPIs, such as the United States, supply chains tend to be extensive and cover large markets areas, while in countries with low LPIs supply chains tend to be shorter and more unreliable. Cities in high LPI countries also tend to have extensive urban freight distribution systems, whereas such systems in cities in countries with low LPIs are less efficient. Using the LPI as a proxy for

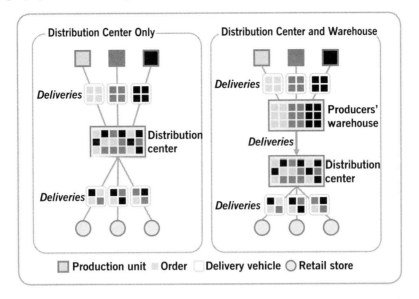

FIGURE 2.3. The role of distribution centers and warehouses.

metropolitan performance should be interpreted with caution, as significant differences may exist between cities of the same nation. For instance, port and airport cities tend to have more capabilities for city logistics because of their infrastructure and distribution capabilities.

There is a level of proportionality ($R^2 = 0.3708$) between the share of the urban population and the LPI; the higher the share of a country's population that is urban, the higher the LPI (Plate 2.1, see color insert1). For instance, China with an urban population standing at 46% of its total population, has 224 million inhabitants living in cities of more than one million with a national LPI of 3.49. The countries above the trend line in Figure 2.4 are highly dependent on international trade and thus have well-developed logistical structures (e.g., Germany, Japan, China, South Korea). Those below the trend line tend to be low-income, less developed countries with deficient transport infrastructure and governance issues and a somewhat more limited participation in international trade (e.g., Brazil, Russia, Nigeria).

As underlined by Figure 2.4, the characteristics of urban freight distribution depend upon local economic, geographic, and cultural features, which can lead to different objectives and preoccupations in urban freight distribution.

THE EXTERNALITIES OF URBAN FREIGHT DISTRIBUTION

Environmental Issues

Modern urban freight distribution systems are generators of environmental and social externalities. In less developed countries, rural migration and population growth have led to rapid urbanization, while the public supply of infrastructure and transport services has lagged behind, impairing the efficiency of urban deliveries. Congestion is an

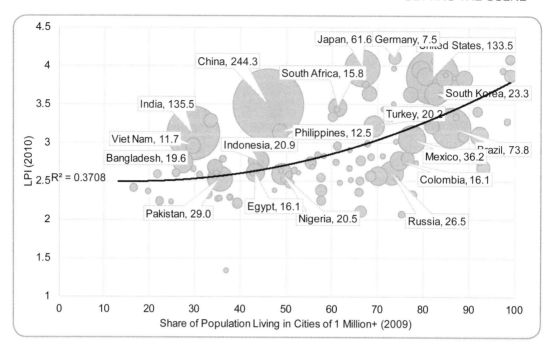

FIGURE 2.4. World's major cities and the Logistics Performance Index, 2010.

important externality of urban freight. The particular conditions of each city influence the nature and intensity of congestion in its urban freight distribution system. Land use patterns, development density, public transit shares, and per capita income levels in cities are factors that contribute to each city's level of congestion. This brings up the question of the specific urban size threshold after which urban freight distribution problems, such as delays and congestion, become prevalent enough to require a concerted approach. Using the United States as evidence (data from Texas A&M Transportation Institute, 2011), it appears that congestion starts to be a serious issue once a threshold of about one million inhabitants is reached. For cities of less than one million, city logistics is less likely to be a problem and may be localized to specific areas such as the downtown, the port, or other terminals areas.

Road transportation is the most polluting surface freight mode per vehicle mile traveled (VMT), but there are limited alternatives to provide for urban deliveries. A positive trend has been the decline of air pollution due to better engine designs and the phasing out of leaded fuel in most countries. Diesel trucks remain significant sources of particulate matter and nitrous oxide (NOx) emissions, a problem compounded by their use as urban delivery vehicles. Urban freight distribution is more polluting than intercity freight transport, particularly because of the following factors.

- *Vehicle age.* On average vehicles used for urban deliveries are older; it is common practice to use trucks at the end of their service life for short distance drayage. This problem is compounded in developing countries where vehicles are even older and thus more prone to higher emissions and accidents.

• *Vehicle size.* The size of vehicles used for urban deliveries is on average smaller, particularly in areas that have high density and limited street parking. Smaller delivery trucks mean more trips and more truck VMT for a given volume of freight compared to large trucks.

• *Operating speeds and idling.* Urban freight deliveries take place on congested street networks characterized by low driving speeds and frequent stops. In some cases, delivery trucks, pedestrians, and urban transit impede one another. The rise in the use of bicycles for passenger mobility in cities such as New York and Paris generates additional conflicts and more safety risk. Driving restrictions, such as one-way streets or truck prohibitions, make routing less efficient. Like passenger cars, trucks consume more fuel and generate more emissions under these conditions. In suburban areas, higher operating speeds, more parking availability, and use of large trucks allow for more efficient deliveries. However, in low-density environments, longer trips contribute to higher overall emissions.

Freight delivery problems are more extensive in developing countries, where motorized and nonmotorized traffic share the same inadequate infrastructures, in typically extreme congestion conditions. Freight deliveries are further complicated by a large informal economy of small-scale vendors.

The "Motor Transition"

The "motor transition" for urban freight refers to the change from predominantly pedestrian or animal-powered transport of goods to motor vehicles, mostly diesel-powered trucks and vans. Figure 2.5 illustrates conceptually this transition in the respective modal share in each of its stages.

In stage 1, the current situation in some cities in the poorest countries, a significant share of the urban movement of goods comes from nonmotorized traditional means of circulating. In stage 2, found in thriving large cities in medium-income countries, diesel trucks and vans are dominant, with nonmotorized traffic still important. Stage 3 sees an automobile-dependent environment, with vans and motorbikes taking over the streets. Stage 4, where green modes of transport dominate (electric, natural gas, or clean diesel vehicles, new cargo-cycles), is the ideal situation envisioned in the transport strategies of many large cities in the world. It has been partially implemented in the Greater London area with a ban on trucks and vans not meeting strict emission standards. In Mexico City, already many trucks run on natural gas, and some electric vehicles are used in the historic center. The urban freight transition from stage 1 to stages 2 and 3 has important impacts, both negative and positive. It makes freight more efficient and provides better service to the urban economy. All the while, however, pollution and energy consumption are dramatically increased because urban fleets of trucks and vans are generally old. In Delhi, one of the most polluted cities in the world, regulations were implemented to promote use of cleaner fuels and conversion to compressed natural gas (CNG) vehicles for passenger as well as freight vehicles. These measures have not met much success because they raise vehicle costs and make it difficult for smaller and less capitalized firms to

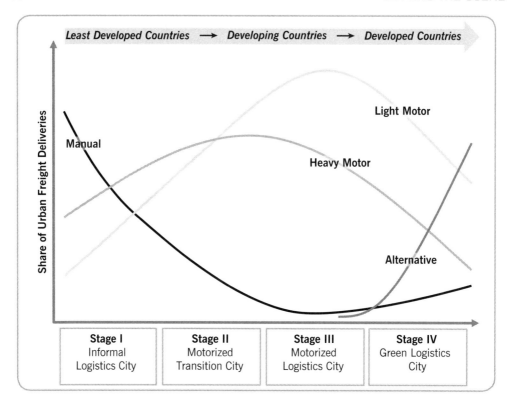

FIGURE 2.5. The "motor transition" for freight distribution in cities.

compete. Stage 3 can have a positive impact on congestion as traffic becomes more homogeneous with the setting of companies specializing in urban deliveries, even though more motorized vehicles are added to the city streets.

KEY LOGISTICAL CHALLENGES

Urban freight distribution has a unique array of challenges, which reflect many dimensions of contemporary logistics such as route and delivery sequence selection and delivery time constraints.

Congestion and Parking

Freight deliveries often take place during peak commuting times. In most developed economies, 8:00–9:00 A.M. is the peak hour for both freight and passenger transport. Passenger cars, transit vehicles, trucks, and vans all share the same road space, and all are competing for that space during peak commute times. An obvious question is why freight deliveries occur during peak periods. It would be cheaper and more efficient for shippers if trucks could avoid the most congested routes and time periods. We will address time shifting of freight deliveries in the next section.

City logistics, like logistics in general, depend on consistent and reliable deliveries. Congestion not only reduces travel time; it also increases travel time variation. In order to fulfill time-definite deliveries, shippers must build extra time into delivery schedules, and this extra time adds to driver and vehicle costs.

High-density areas have limited parking capacity (both on-street and off-street) to accommodate deliveries. Delivery vehicles cope with this challenge by illegally parking in passenger car spots or double parking. Illegal parking reduces the supply of parking for cars, causing more cars to "cruise" in search of a space. Double parking takes away traffic lanes and greatly increases local congestion.

Urban freight distribution is characterized by smaller volumes and more frequent deliveries than is the case in long-haul freight movement. This is driven by two factors: constraints on vehicle size and restocking and inventory practices of urban businesses. When rents are high, businesses minimize space for inventory and rely on frequent deliveries for restocking.

Land Use and Sustainability

Land use patterns determine many features of urban goods movement. The fundamental supply and demand for goods is determined by the spatial distribution of population and employment. The basic form of the distribution network is determined by the scale and location of producers, warehouses, and distribution centers. Because of scale economies in warehousing and distribution, and the extremely large footprint of these facilities, logistics facilities are decentralizing more rapidly than is any other economic activity (Cidell, 2010; Dablanc & Ross, 2012). Termed "logistics sprawl," decentralization pushes distribution farther from the sources of demand and may induce more truck VMT.

Many of these challenges are related to density—the close proximity of people, business activities, and transport, which is at the core of urban vitality. Proximity also generates negative externalities, including noise and pollution exposure. Narrow streets and multistory buildings create pollution and noise hot spots and affect more people. Conflicts between passenger and freight movements become more intense as density increases.

While cities are major consumers of final goods, they are also the site of reverse logistics activities related to the collection of wastes and recycling. Growing environmental concerns and rising consumption increase the pressure to implement urban green logistics strategies.

E-Commerce

Online transactions—e-commerce—such as purchasing by businesses and households is growing rapidly at an annual rate of more than 15% in the United States, where it accounts for about 8% of all retail purchases in 2015. In Europe, online purchases account for 8.4% of all 2015 retail purchases in average, the United Kingdom having the highest rate (13.5%). E-commerce typically involves very small and dispersed deliveries (single packages to households). While e-commerce volumes are still relatively small, its challenges are already evident for urban deliveries. Residential deliveries in cities are growing quickly and account for more than 20% of parcel deliveries in large cities (up to 40% in some cities such as New York City). This adds freight trips to urban areas, in uncommon time windows (evenings, weekends) and locations (residential neighborhoods). Recently, "instant deliveries" (deliveries made to customers' homes in less than 2 hours) have developed quite rapidly in the core of large cities. It is estimated that, by August 2016, 100,000 instant deliveries per week were made in central Paris. Most instant delivery couriers are self-contractors connected to the service through smartphone apps. Legal issues have risen about work contracts and working conditions. Some places have implemented strategies to complement home deliveries with alternate solutions such as pick-up points or automated locker banks (Morganti, Dablanc, & Fortin, 2014). In Japan, pick-up points located in convenience stores have been common for more than two decades. In the United States, Amazon, UPS, and the United States Postal Service offer package locker options in several major cities.

Urban space is thus the site of conflicts among different stakeholders, as high population densities are related to a low tolerance for infringements and disturbances. Urban locations also offer opportunities for collaboration as city logistics open new realms of engagement for urban planning.

URBAN FREIGHT DISTRIBUTION STRATEGIES

From a freight distribution perspective, cities are bottlenecks, where infrastructure resources (namely, road and parking space) are scarce relative to the potential demand and are thus highly valuable. As an optimizing strategy, city logistics can take many forms depending on the type of supply chain (e.g., retailing, parcels, food deliveries) as well as the urban setting in which it takes place. The most commonly considered strategies for addressing the problems outlined above involve changes in how freight is delivered, changes in facilities, or changes in mode.

Optimization of Deliveries

The optimization of deliveries relates to making changes in how freight is delivered (or picked up) so that urban distribution is more productive and costs and externalities, namely, congestion and energy consumption, are minimized. It seeks to make better use of existing assets, particularly vehicles. One of the simplest strategies is to reduce the time constraint on access to specific parts of the city, such as by promoting night-time deliveries in central areas. Distributors can opt for night deliveries or at least extended delivery windows to avoid peak hour traffic, but there are challenges. For receivers, night-time deliveries require additional work shifts, adding to their costs, or risking theft if deliveries are left unattended. Truck drivers may need additional compensation to work a night shift. Night-time deliveries also impose truck noise on nearby residents. To date, the most extensive use of "off-hours deliveries" has taken place in New York City as a test demonstration (Holguin-Veras et al.,

2011). European programs such as the Dutch PIEK program have also been developed to promote night deliveries.

Information technologies are increasingly being used to manage urban freight distribution systems. The most heavily used technologies include global positioning systems that improve vehicle tracking and urban navigation, and load management applications that optimize routes and delivery schedules. Currently these technologies are used within firms (e.g., UPS or Wal-Mart). The City of Lyon in France is working with IBM to develop software for the city's urban carriers that provides live information on traffic coupled with automatic rerouting of delivery tours.

Freight Facilities

Another set of strategies seeks to develop freight distribution infrastructures that are better adapted to the urban context. This approach can involve logistics parks that provide services to tenants (catering for employees, truck stops, security); designated parking areas for deliveries; or the use of urban freight distribution centers and local freight stations. If the opportunity arises, such as the availability of an underused area (e.g., former railyards or some brownfield sites—provided cleanup costs are reasonable) in proximity to the city center, urban logistics zones can be developed to provide a counterweight to logistics zones that have emerged in the periphery of most large urban agglomerations (Cidell, 2010). However, land prices in dense areas can be high. Also, available urban areas fitted for logistics activities tend to be located in poor neighborhoods, generating environmental justice problems related to vehicle noise and emissions.

Using urban consolidation centers (UCC) is a more sophisticated strategy, providing a bundled and coordinated delivery service for a whole neighborhood or a set of products. A UCC is a logistics facility located close to areas with a high density of delivery operations (such as a city center) from which consolidated deliveries are carried out, and which provides a range of other value-added logistics services (Bestufs, 2007). The purpose of a UCC is to reduce total truck trips and VMT within the city by combining deliveries and filling up each truck to capacity. UCCs generate higher shipping costs, because goods must be collected and sorted at the UCC before final delivery and because rental costs in city locations are high. Up to 200 UCCs existed in European cities in the 1990s and early 2000s. Cities provided subsidies to offset the additional handling costs. Most of the UCCs closed down when cities stopped subsidizing them (Dablanc, 2009). Elsewhere, a few projects for UCCs (such as the Motomachi UCC in Japan) (Bossin et al., 2009) have met success.

In some cities, such as Tokyo, logistics facilities are still a feature of dense urban areas (Bossin et al., 2009) thanks to the great care given to minimizing negative impacts on neighboring communities. In Paris, recent examples of urban logistics facilities demonstrate that such facilities can be accommodated within populated areas with minimal impact (Diziain, Ripert, & Dablanc, 2012). These developments, however, are costly. Urban freight facilities are a higher value proposition for very large cities, as their contribution to flow optimization and reduced environmental impacts can outweigh their costs. In smaller cities, such initiatives tend to drive up costs and unreliability.

Alternative Vehicles, Modal Shift

A third strategy relates to using vehicles better adapted to urban freight distribution. Smaller vehicles tend to be better suited for urban deliveries because of their ability to maneuver and the small loads typical of urban deliveries. However, smaller vehicles imply more trips and more VMT. One possible solution is use of medium-size vehicles and regulations on vehicle age to address emissions and noise problems, such as in the London Low Emission Zone. Innovative strategies such as CNG or electric vehicles and even bicycles, tricycles, or "cargo-cycles" have been successfully implemented. They suggest a good potential for modes to adapt to the diversity of the urban landscape, although they tend to remain a niche market (Giuliano et al., 2013). Also, such a shift often brings changes in the capacity and operational characteristics of vehicles and thus imposes a change in urban distribution practices. As their driving range is limited, electric vehicles require nearby freight terminals where delivery tours are prepared. The relative costs and benefits of alternative vehicles and fuels are largely unknown. More research will be required to understand how a variety of vehicles can best be utilized in urban environments.

Barge and rail are minor modes of freight transportation for urban deliveries. Barges do play a role in some river cities for the shipment of building materials. There are also some special cases. For instance, some Dutch cities use their canals for the supply of cafés and restaurants. Many Chinese cities rely extensively on canals for urban freight distribution, but specific figures are not available. Rail freight requires dedicated logistics facilities (tracks, sidings, yards, terminals) that are space-consuming and therefore expensive within central areas. There is often local opposition to railyards and intermodal terminals because of their environmental impacts: old and noisy locomotives, negative aesthetics of industrial areas, and increased truck movements.

In the United States, rail freight volumes have increased in recent decades, and today railroads carry 43% of all ton-miles of domestic freight. Major rail hubs such as Los Angeles, Chicago, or Atlanta have busy intermodal terminals that serve as hubs for national and international freight distribution. They generate economic benefits to the urban areas (jobs and some tax revenues) but also produce significant negative environmental impacts. The rail market is in long-distance, low-value shipments, and hence plays no role in urban freight distribution in the United States. In the largest U.S. cities, competition for rail capacity between passenger and freight services is growing. The freight railroad companies own all the rail infrastructure, and passenger rail (e.g., Amtrak) must pay to use the tracks.

In Europe, growing passenger rail traffic has drastically reduced the available capacity for freight on the rail infrastructure. Several projects to expand urban rail capacity are taking place in response. In Dresden, Germany, a freight light rail service has been in operation since 2000 for parts supplied to a Volkswagen plant. Another major project based on light rail, called Amsterdam City-Cargo, went bankrupt in early 2009. The Monoprix freight train, in Paris, has been in operation since 2007. Monoprix is a chain of supermarkets with 90 stores in Paris. A train enters the city of Paris every evening. Pallets are then transferred to CNG-operated trucks for final deliveries early in the morning. The scheme generates a yearly saving of 10,000 diesel trucks and 280 tons of CO_2. This operation, however, is quite expensive, with an initial additional cost of 25% per pallet compared to the former all-road solution. One of Monoprix's competitors, Franprix, has been operating barges to supply its Paris stores since 2012.

Advantages and Drawbacks of City Logistics Strategies

Policymakers are increasingly interested in innovative city logistics to reduce negative externalities and enhance efficiency. Although each of these city logistics strategies has its own advantages, there are also drawbacks, typically higher distribution costs and additional delays. Few strategies have met obvious success (Giuliano et al., 2013), leaving city authorities reluctant to go further. City logistics strategies are difficult to implement as they often imply higher costs to shippers. An urban freight distribution center linked to a set of suburban distribution centers, each connected to its respective supply chains, could service dense city areas and achieve more efficient distribution within the central city. The total cost of distribution would be higher, however, because of the additional handling and rental costs as discussed above. Thus most UCC experiments have failed. Shifting freight from truck to train or barge or public transit systems also generates higher shipping costs owing to the slower speeds of these modes and the additional handling required. The policy question is whether societal benefits are achieved (Is there a net reduction in emissions or energy consumption?), and whether any proposed strategy offers benefits that are large enough to offset added costs.

The best city logistics strategy entails a framework that ensures coherent management of urban freight distribution, particularly assets such as trucks, logistics terminals, and parking space. Incentives, regulations, and enforcement should aim at promoting better use of existing urban assets: off-peak hour deliveries, available on-site and off-site loading zones, and more full trucks. Policies should promote voluntary changes in urban distribution practices, such as the provision of labels that identify businesses demonstrating optimized and environmentally sound delivery practices (Giuliano et al., 2013). Planning provisions that make the development of logistics parks possible while minimizing their impacts on adjacent communities are also part of a city logistics strategy, which further underlines the need for collaboration between city governments and the freight and logistics industries.

IN THE ERA OF GLOBALIZATION, FOUR CATEGORIES OF LOGISTICS CITIES

Cities remain congested areas where space comes at a premium and where the presence of many stakeholders in close proximity requires special efforts to serve markets in efficient and environmentally friendly ways. A transition toward greener forms of city logistics in response to growing concerns about rising congestion and environmental externalities is underway. Because each city represents a unique setting with its own supply of transport infrastructure and modes, no single encompassing strategy will improve urban freight distribution; rather, a set of strategies, each reflecting the particular circumstances in each city, will be needed. Size and density are major factors in city logistics. Large metropolitan areas (a threshold of about one million inhabitants has been suggested in this chapter) have reached a level of complexity that warrants a concerted city logistics (freight policy) effort. Large cities tend to have frequent transshipment/transloading operations in suburban distribution centers, whereas in smaller cities direct deliveries are more common. Additionally, large cities have experienced rapid growth in the number of logistics centers and warehouses, particularly in suburban areas.

Plate 2.2 and Table 2.1 summarize much of the previous discussion about urban supply chains and freight transport systems by identifying four general models of urban logistics: large metropolitan areas in developed and developing countries; gateway cities (which can also be major metropolitan areas) providing a substantial interface function between national and global freight distribution; and an array of medium-sized cities in developed economies, particularly in Europe, that have implemented city logistics schemes to deal with specific challenges.

The four categories of Global City Logistics can be described as follows.

1. *Large metropolitan areas of developed economies.* The logistics organization in such cities integrates freight and logistics facilities in dense urban environments where mass retailing prevails. The retailing landscape is rapidly changing through e-commerce with parcel transport companies providing finely tuned home delivery services or alternative pick-up-points using information technology tools. In this category of cities, one subcategory deserves special attention: Japanese cities and their *land-efficient urban logistics.* The logistics organization in cities such as Tokyo presents several striking features including the integration of freight and logistics facilities in very dense settings. Responding to the needs of urban consumers, a widespread network of convenience stores is supplied day and night. These advanced logistics strategies are (or are likely to be) adopted by other high-density Asian cities, such as those in South Korea, Taiwan, and coastal China.

2. *Large metropolitan areas in developing countries.* Large cities of fast-growing economies have a two-sided (dual) urban freight system; the logistics needs of a modern economic sector comparable to any city in a developed country coexist with an informal and largely unrecorded system of pick-ups and deliveries for home-based artisans or street vendors. Poorly maintained roads accommodate a diversity of modes ranging from push carts to mopeds, vans, and trucks. Such cities are prone to the "motor transition" described earlier (Figure 2.4).

3. *The gateway city.* Serving as gateways to import-based consumer-oriented economies, these urban regions concentrate the growth of new freight terminals, serving as distribution facilities for important local markets of urban consumers and businesses, as well as being regional hubs for the grouping and redistribution of goods to regional and national markets. A notable characteristic is the growth in number and size of warehouse and distribution facilities located at the metropolitan periphery.

4. *Smart city logistics* (historical centers of middle-sized cities). Innovative schemes of urban deliveries emerge in many European city centers with an emphasis on cleaner and more silent operations and consolidated deliveries.

TABLE 2.1. Proposed Typology of Global City Logistics

Large metropolitan areas of developed economies	Large metropolitan areas of emerging economies	Gateway cities	Medium-sized cities in developed economies
Operations			
• Chain retailing resulting in more optimized urban deliveries. • High share of common carriers, high level of urban delivery subcontracting (Europe). • E-commerce and services activities requiring parcel and express transport.	• Many independent stores and home- and street-based businesses requiring specific patterns of deliveries. • Dual transport and logistics system, prevalence of own-account operations. • Very high diversity of urban supply chains	• Mostly involves inbound (import-based) flows in developed countries. • Mostly involves outbound (export-based) flows in developing countries. • Numerous drayage operations from port (airport, intermodal terminals, large logistics hubs) to region's DCs	• Higher share of direct deliveries: fewer transshipment activities in local distribution centers.
Modes and vehicles			
• Prevalence of vans. • Many old commercial vehicles in European urban areas. • New city logistics schemes (alternative fuel vans, cargo-cycles, barges)	• Huge heterogeneity of modes and types of road uses (from pedestrian carts to two wheelers to trucks), high levels of congestion.	• Additional heavy truck traffic to local freight traffic. • Intermodal traffic.	• Large- and medium-sized trucks still quite visible.
Infrastructure and land			
• Availability of suburban land, generating patterns of logistics sprawl (U.S., Europe). • European attempts at urban consolidation centers. • In Japan: scarcity of land, low differentials in prices between urban and suburban lands, maintaining freight facilities and multistory terminals in dense areas.	• Land generally available but supporting infrastructure often lacking.	• Intermodal terminals, ports, airports, mega-distribution centers serving regional markets.	• Various conditions but land generally available and infrastructures adequate.
Policies			
• European cities involved in new city logistics experiments and environmental zones to reduce the share of old trucks. • U.S. cities lacking data, strategies focused on metropolitan truck traffic, port cities (New York City, Los Angeles, Seattle) more involved in freight issues.	• Freight not yet a prevalent issue despite recent efforts in some cities.	• Issues of infrastructure investments for a better position in global competition (deepening of ports, capacity of airports, renovation of rail infrastructure, dedicated freight corridors, grade crossings, etc.).	• Case-specific, such as access to a congested central area. • Many city logistics initiatives in Europe.

Start-up companies provide a significant share of new delivery services implemented to cope with an increasingly complex setting for city logistics. These experiments still represent a marginal share of total freight activities in metropolitan areas, but they receive a lot of media and decision-maker attention. Large companies in Europe and the United States have started to integrate new urban delivery services (such as Amazon lockers and Amazon Prime Now, or Uber EATS and Uber RUSH). In France, UPS has acquired the pick-up-point company Kiala.

CONCLUSION

Cities are intertwined with their freight transport systems, a dimension that is often neglected when looking at the geography of urban transportation. In this chapter, we provided an international overview of the geography of urban freight. We compared the urban transportation of goods in different cities in the world, in developed as well as developing countries. Large-scale urbanization (particularly in developing economies), globalization, and rising standards of living are factors, among others, that have contributed to the growing volumes of freight circulating through and within cities. Freight in urban areas is highly diversified but involves both consumer-related movements such as retailing and producer-related movements such as waste collection and terminal haulage. These movements are linked with different supply chains, each having their own operational requirements.

Urban freight transport is a generator of an array of externalities. The most salient remain congestion and atmospheric emissions. The urban geography of freight distribution has followed the trend of suburbanization (logistics sprawl), with large distribution centers an important component of today's suburban landscape. New forms of consumption and business transactions are also affecting city logistics, particularly e-commerce and the fast growth of home deliveries. Since cities commonly compete to attract economic activities, whether related to production (manufacturing), distribution (warehouses), or consumption (major retailers), and their logistical performance is a factor of growing concern for public and private actors as it directly influences their competitiveness in the global economy.

The growth in intensity and complexity of urban freight distribution leads to strategies to improve its efficiency and reduce its externalities. The optimization of deliveries aims at better use of existing freight distribution assets by changing delivery patterns, for example, by shifting delivery times away from rush hours or increasing the parking space allocated for deliveries. Specialized urban freight facilities are also considered as they can support the consolidation of deliveries in high-density sections of the city. Vehicles adapted to the urban landscape, such as ones using alternative fuels (electric or compressed natural gas) and even cargo bicycles, represent another mitigation strategy. Still, experience demonstrates the complexity of mitigating the problems associated with urban freight distribution, particularly because of the large number of public and private actors involved.

While specific operational strategies remain a challenge, city logistics has emerged as an active field of research and application in many world regions. We conclude this chapter by emphasizing the need to develop comparisons between cities and to identify city logistics performance metrics. As a first step, we suggested four models of urban freight and logistics situations: large metropolitan areas in both developed and developing economies; gateway cities; and innovative medium-sized cities. All four categories display quite distinctive features of urban freight operations, infrastructures, and strategies, as well as some commonalities. Better specifying the urban freight landscape worldwide and better defining the main patterns between freight activities and urban structures are objectives for further research on the geography of urban freight, in order for this important topic to gain its appropriate place on the urban transportation agenda.

NOTES

1. These data and the following result from a compilation of various sources, mostly European—see Giuliano et al. (2013); Routhier (2013). Few sources are from North America owing to a lack of specifically urban freight data collection.

2. A warehouse is a facility where goods are stored for periods of time, while a distribution center tends to store goods for short periods of time as orders are fulfilled, commonly on a daily basis. However, the word *warehouse* often stands for all sorts of logistics terminals, including distribution centers.

3. The LPI is a composite measure ranging from 1 (worst) to 5 (best) based on six underlying factors of logistics performance: (1) efficiency of the clearance process by customs and other border agencies; (2) quality of transport and information technology infrastructure for logistics; (3) ease and affordability of arranging international shipments; (4) competence and quality of logistics services; (5) ability to track and trace international shipments; and (6) timeliness of shipments in reaching destination.

4. Source: Centre for Retail Research, *www.retailresearch.org/onlineretailing.php*.

REFERENCES

Bestufs. (2007). *Good practice guide on urban freight*. Brussels: European Commission. Retrieved August 30, 2014, from *www.bestufs.net/gp_guide.html*.

Bossin, P., Dablanc, L., Diziain, D., Levifve, H., Ripert, C., & Savy, M. (2009). *City logistics Tokyo visit* (Technical report to INNOFRET-Predit). Paris: Ministry of Transport.

Cidell, J. (2010). Concentration and decentralization: The new geography of freight distribution in US metropolitan areas. *Journal of Transport Geography, 18*, 363–371.

Dablanc, L. (2009). Freight transport: A key for the new urban economy (Report for the World Bank as part of the initiative Freight Transport for Development: A Policy Toolkit). Retrieved August 30, 2014, from *siteresources.worldbank.org/INTTRANSPORT/Resour ces/336291–1239112757744/5997693–1266940498535/urban.pdf*.

Dablanc, L. (2014). Logistics sprawl and urban freight planning issues in a major gateway city—The case of Los Angeles. In J. Gonzalez-Feliu, F. Semet, & J. L. Routhier (Eds.), *Sustainable urban logistics: Concepts, methods and information systems* (pp. 49–69). Berlin/Heidelberg, Germany: Springer-Verlag.

Dablanc, L., & Ross, C. (2012). Atlanta: A mega logistics center in the Piedmont Atlantic megaregion (PAM). *Journal of Transport Geography, 24*, 432–442.

Diziain, D., Ripert, C., & Dablanc, L. (2012). How can we bring logistics back into cities?: The case of Paris metropolitan area. In E. Taniguchi & R. Thomson (Eds.), *Seventh International Conference on City Logistics, Procedia: Social and Behavioral Sciences, 39*, 267–281.

Giuliano, G., O'Brien, T., Dablanc, L., & Holliday, K. (2013). *Synthesis of freight research in urban transportation planning* (National Cooperative Freight Research Program Project 36[05]).

Washington, DC. Retrieved April 4, 2014, from *www.trb.org/Publications/Blurbs/168987. aspx*.

Hesse, M. (2008). *The city as a terminal: The urban context of logistics and freight transport.* Farnham, UK: Ashgate.

Holguín-Veras, J., Ozbay, K., Kornhauser, A., Brom, M. A., Iyer, S., Yushimito, W. F., et al. (2011). Overall impacts of off-hour delivery programs in the New York City metropolitan area. *Transportation Research Record,* No. 2238, 68–76.

Lindholm, M., & Behrends, S. (2012). Challenges in urban freight transport planning: A review in the Baltic Sea region. *Journal of Transport Geography, 22,* 129–136.

Morganti, E., Dablanc, L., & Fortin, F. (2014). Final deliveries for online shopping: The deployment of pickup point networks in urban and suburban areas. *Research in Transportation Business and Management, 11,* 23–31.

Rhodes, S., Berndt, M., Bingham, P., Bryan, J., Cherrett, T., Plumeau, P., et al. (2012). *Guidebook for understanding urban goods movement* (NCFRP 14). Washington, DC: National Academies of Sciences, Transportation Research Board.

Rodrigue, J. P. (2004). Freight, gateways and mega-urban regions: The logistical integration of the BostWash corridor. *Tijdschriftvooreconomischeensocialegeografie, 95*(2), 147–161.

Rodrigue, J. P. (2013). Urban goods transport. In *Planning and design for sustainable urban mobility: Global report on human settlements 2013* (United Nations Human Settlements Programme). London: Earthscan.

Routhier, J. L. (2013). *Urban freight surveys in French cities.* Paper presented at the EU-US Transportation Research Symposium No. 1: City Logistics Research: A Trans-Atlantic Perspective, Transportation Research Board, Washington, DC.

Taniguchi, E., & Thompson, R. (Eds.). (2008). *Innovations in city logistics.* New York: Nova Science.

Texas A&M Transportation Institute (TTI). (2011). Annual urban mobility scorecard. Available from *https://mobility.tamu.edu/ums.*

CHAPTER 3

Transportation and Urban Form

Stages in the Spatial Evolution of the American Metropolis[1]

PETER O. MULLER

As the opening chapter demonstrated, the movement of people, goods, and information within the local metropolitan area is critically important to the functioning of cities. In this chapter, I review the American urban experience of the past two centuries and trace a persistently strong relationship between the intraurban transportation system and the spatial form and organization of the metropolis. Following an overview of the cultural foundations of urbanism in the United States, a four-stage model of intrametropolitan transport eras and associated growth patterns is introduced. Within that framework it will become clear that a distinctive spatial structure dominated each stage of urban transportation development and that geographical reorganization swiftly followed the breakthrough in movement technology that launched the next era of metropolitan expansion.[2] Finally, the contemporary scene is briefly considered, both as an evolutionary composite of the past and as a dynamic arena in which new forces are reshaping the circulatory system of the American metropolis.

CULTURAL FOUNDATIONS OF THE AMERICAN URBAN EXPERIENCE

Americans, by and large, were not urban dwellers by design. The emergence of large cities between the Civil War and World War I was an unintended by-product of the nation's rapid industrialization. Berry (1975), recalling the observations of the eighteenth-century French traveler Hector St. Jean de Crèvecoeur, summarizes the cultural values that have shaped attitudes toward urban living in the United States for the past two centuries:

Foremost . . . was a *love of newness*. Second was the overwhelming desire to be *near to nature*. *Freedom to move* was essential if goals were to be realized, and *individualism* was basic to the self-made man's pursuit of his goals, yet *violence* was the accompaniment if not the condition of success—the competitive urge, the struggle to succeed, the fight to win. Finally, [there is] a great *melting pot* of peoples, and a manifest *sense of destiny.* (p. 175)

As the indigenous culture of the emergent nation took root, its popular Jeffersonian view of democracy nurtured a powerful rural ideal that regarded cities as centers of corruption, social inequalities, and disorder. When mass urbanization became unavoidable as the Industrial Revolution blossomed after 1850, Americans brought their agrarian ideal with them and sought to make their new manufacturing centers noncities. For the affluent, this process began almost as soon as the railroad reached the city in the 1830s; by midcentury, numerous railside residential clusters had materialized just outside the built-up urban area. But middle-income city dwellers could not afford this living pattern because of the extra time and travel costs it demanded. With cities increasingly unlivable as industrialization intensified, pressures mounted after 1850 to improve the urban transportation system to permit the burgeoning middle class to have access to the high-amenity environment of the periurban zone.

The necessary technological breakthrough—in the form of the electric (streetcar) traction motor—was finally achieved in the late 1880s. By the opening of the final decade of the 19th century, the city began to spill over into the much-desired surrounding countryside. By 1900, the decentralization of the middle-income masses was no longer a trickle but a widening migration stream (which has yet to cease its flow) that rapidly spawned the emergence of the full-fledged metropolis, wherein a steadily increasing multitude of urban dwellers shunned the residential life of the industrial city altogether. Hardly had this initial transformation of the American city been completed when the automobile introduced mass private transportation in the 1920s for all but the poorest urban dwellers. As the intrametropolitan highway network expanded in the interwar period, successive rounds of new peripheral residential development were launched, and the urban perimeter was pushed ever farther from the downtown core. But these centrifugal forces still operated at a rather leisurely pace, undoubtedly slowed after 1930 by a decade and a half of economic depression and global war.

Following the conclusion of World War II, however, all constraints were removed, and a massive new wave of deconcentration was triggered. Spurred by a reviving economy, widespread housing demands, federal home-loan policies that favored new urban development, copious highway construction, and more efficient cars, the exodus from the nation's cities reached unprecedented proportions between 1945 and 1970. The proliferation of urban freeways (introduced in Southern California in the late 1930s) heightened the centrifugal drift. With the completion of these high-speed, limited-access, superhighway networks in the 1960s and 1970s came the elimination of the core-city central business district's (CBD's) regionwide centrality advantage, as superior intrametropolitan accessibility became widely available near any expressway-interchange location. As entrepreneurs swiftly realized the consequences of this structural reorganization of the metropolis, nonresidential activities of every variety began their own massive wave of intraurban deconcentration. Manufacturing and retailing led the way.

By 1980, the erstwhile ring of bedroom communities that girdled the aging central city had become transformed into a diversified, expanding *outer city* that was increasingly home to a critical mass (i.e., more than half) of the metropolitan area's industrial, service, and office-based business employers. Moreover, major new multipurpose activity centers have been emerging in the outer city during the past three decades, attracting so many high-order urban functions that residents of surrounding areas have completely reorganized their lives around them. Thus the compact industrial city of the recent past has today turned inside out. The rise of downtown-type centers in the increasingly independent outer city has also forged a decidedly polycentric metropolis, the product of both the cumulative spatial processes outlined above and the emerging forces of a postindustrial society that are shaping new urbanization patterns that represent a clean break with the past.

The legacy of more than two centuries of intraurban transportation innovations and the development patterns they etched on the landscape of metropolitan America is *suburbanization*—the growth of the edges of the urbanized area at a rate faster than that in the already developed interior. Since the spatial extent of the continuously built-up urban area has, throughout history, exhibited a fairly constant time-distance radius of about 45 minutes' travel from the center, each breakthrough in higher-speed transport technology extended that radius into a new outer zone of suburban residential opportunity. In the 19th century, commuter railroads, horse-drawn trolleys, and electric streetcars each created their own suburbs—*and thereby also created the large industrial city,* which could not have been formed without absorbing these new suburbs into the preexisting compact urban center. But the suburbs that developed in the early 20th century began to assert their independence from the enlarged, ever more undesirably perceived central cities. Few significant municipal consolidations occurred after the 1920s, except in postwar Texas and certain other Sunbelt locales (a trend that ended during the 1980s). As the automobile greatly reinforced the intraurban dispersal of population, the distinction between central city and suburban ring grew as well. And as freeways greatly reduced the friction effects of intrametropolitan distance for most urban functions, nonresidential activities deconcentrated to such an extent that by the mid-1970s the emerging outer city became at least the coequal of the neighboring central city that spawned it—making the word *sub*urb an oxymoron.

This urban experience of the United States over the past two centuries is the product of uniquely American cultural values and contrasts sharply with modern urbanization trends in Europe. The European metropolis, though also now experiencing decentralization, retains a more tightly agglomerated spatial structure, and the historic central city continues to dominate its immediate urban region. Sommers (1983) summarizes the concentrative forces that have shaped the cities of postwar Europe:

> Age is a principal factor, but ethnic and environmental differences also play major roles in the appearance of the European city. Politics, war, fire, religion, culture, and economics also have played a role. Land is expensive due to its scarcity, and capital for private

enterprise development has been insufficient, so government-built housing is quite common. Land ownership has been fragmented over the years due to inheritance systems that often split land among sons. Prices for real estate and rent have been government controlled in many countries. Planning and zoning codes as well as the development of utilities are determined by government policies. These are characteristics of a region with a long history, dense population, scarce land, and strong government control of urban land development. (p. 97)

Further evidence of the persistent dominance of the European central city is shown in Figure 3.1. The density gradient pattern of the North American metropolis (Figure 3.1A) is marked by progressive deconcentration whereas the counterpart European pattern (Figure 3.1B) exhibits sustained intraurban centralization. The newest metropolitan trends of the past decade in Europe do show accelerating suburbanization, but the central city remains the intraurban social and economic core.

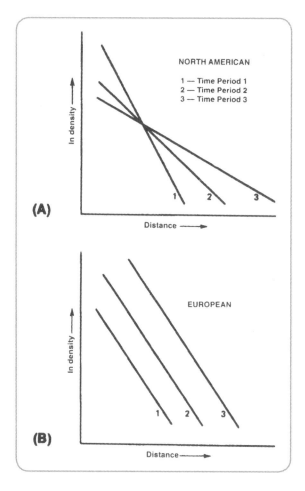

FIGURE 3.1. Density gradients over time in the North American and European metropolis. Source: From *Interpreting the city: An urban geography* (2nd ed., p. 230) by T. A. Hartshorn, 1992, New York: John Wiley & Sons. Copyright 1992 by John Wiley & Sons. Reprinted by permission.

THE FOUR ERAS OF INTRAMETROPOLITAN GROWTH AND TRANSPORT DEVELOPMENT

The evolving form and structure of the American metropolis, briefly outlined in the previous section, may be traced within the framework of four transportation-related eras identified by Adams (1970). Each growth stage is dominated by a particular movement technology and network-expansion process that shaped a distinctive pattern of intraurban spatial organization:

    I. Walking-Horsecar Era (1800–1890)
    II. Electric Streetcar Era (1890–1920)
    III. Recreational Automobile Era (1920–1945)
    IV. Freeway Era (1945–present)

This model is diagrammed in Figure 3.2 and reveals two sharply different morphological properties over time. During Eras I and III uniform transport surface conditions prevailed (as much of the urban region was similarly accessible), permitting directional freedom of movement and a decidedly compact overall development pattern. During Eras II and IV pronounced network biases were dominant, producing an irregularly shaped metropolis in which axial development along radial transport routes overshadowed growth in the far less accessible interstices.

    A generalized model of this kind, while organizationally convenient, risks oversimplification because the building processes of several simultaneously developing cities do not fall into neat time–space compartments (Tarr, 1984, pp. 5–6). An examination of Figure 3.3, which maps Chicago's growth since 1850, reveals numerous

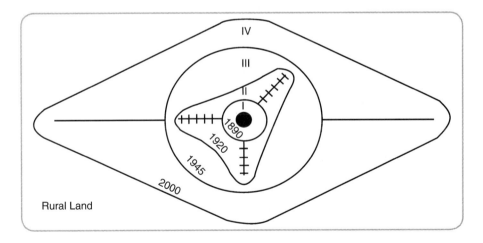

FIGURE 3.2. Intraurban transport eras and metropolitan growth patterns: (I) Walking-Horsecar Era, (II) Electric Streetcar Era, (III) Recreational Auto Era, and (IV) Freeway Era. Source: From "Residential structure of Midwestern cities" by J. S. Adams, 1970, *Annals of the Association of American Geographers, 60,* p. 56. Copyright 1970 by the Association of American Geographers. Reprinted by permission of the Association of American Geographers, *www.aag.org.*

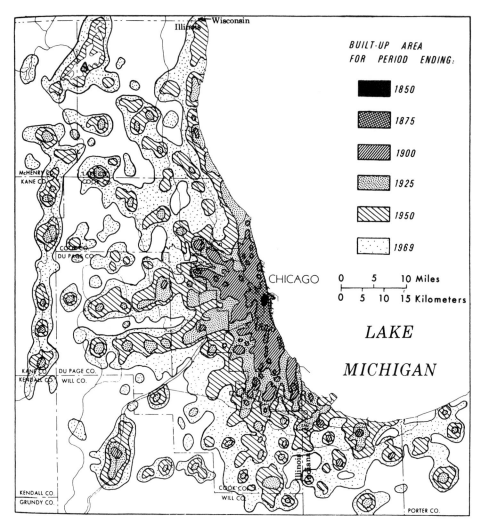

FIGURE 3.3. The suburban expansion of metropolitan Chicago from the mid-nineteenth century through 1970. Source: From *Chicago: Transformations of an urban system* (p. 9) by B. J. L. Berry, 1976, Cambridge, MA: Ballinger Publishing Company. Copyright 1976 by B. J. L. Berry. Reprinted by permission.

empirical irregularities, suggesting that the overall urban growth pattern is somewhat more complex than a simple, continuous, centrifugal thrust. Yet, when developmental ebb-and-flow pulsations, leapfrogging, backfilling, and other departures from the normative scheme are considered, there still remains a reasonably good correspondence between the model and historical–geographical reality. With that in mind, each of the four eras is now examined in detail.

Walking-Horsecar Era (1800–1890)

Prior to the middle of the 19th century, the American city was a highly agglomerated urban settlement in which the dominant means of getting about was on foot

(Figure 3.2, Era I). Thus people and activities were required to cluster within close proximity of one another. Initially, this meant less than a 30-minute walk from the center, later extended to about 45 minutes when the pressures of industrial growth intensified after 1830. Any attempt to deviate from these mobility constraints courted urban failure: Washington, DC, struggled enormously for much of its first century on L'Enfant's 1791 plan, in which blocks and facilities were too widely dispersed for a pedestrian city, prompting Charles Dickens to observe during his 1842 visit that buildings were located "anywhere, but the more entirely out of everyone's way the better" (Schaeffer & Sclar, 1975, p. 12).

Within the walking city, there were recognizable concentrations of activities as well as the beginnings of income-based residential congregations. The latter behavior was clearly evinced by the wealthy who walled themselves off in their larger homes near the city center; they also favored the privacy of horse-drawn carriages to move about town—undoubtedly the earliest American form of wheeled intraurban transportation. The rest of the population resided in tiny overcrowded quarters, of which Philadelphia's now-restored Elfreth's Alley was typical (see Figure 3.4).

The rather crude environment of the compact preindustrial city impelled those of means to seek an escape from its noise as well as the frequent epidemics that resulted from the unsanitary conditions. Horse-and-carriage transportation enabled

FIGURE 3.4. Elfreth's Alley in the heart of downtown Philadelphia, whose restoration provides a good feel for the lack of spaciousness in the revolutionary-era city. Source: Carol M. Highsmith Archive, U.S. Library of Congress, Prints and Photographs Division.

the wealthy to reside in the nearby countryside for the disease-prone summer months. The arrival of the railroad in the early 1830s soon provided the opportunity for year-round daily travel to and from elegant new trackside suburbs. By 1840 hundreds of affluent businessmen in Boston, New York, and Philadelphia were making these round trips every weekday. The "commutation" of their fares to lower prices, when purchasing tickets in monthly quantities, introduced a new word to describe the journey to work: *commuting*. A few years later, these privileges extended to the nouveau-riche professional class (well over 100 trains a day ran between Boston and its suburbs in 1850), and a spate of planned rail suburbs, such as Riverside near Chicago, soon materialized.

As industrialization and its teeming concentrations of modest, working-class housing increasingly engulfed the mid-19th-century city, the worsening physical and social environment heightened the desire of middle-income residents to suburbanize as well. Unable to afford the cost and time of commuting—and with the pedestrian city stretched to its morphological limits—these yearnings intensified the pressure to improve intraurban transport technology.

As early as the 1820s, New York, Philadelphia, and Baltimore had established *omnibus* lines. These intracity adaptations of the stagecoach eventually developed dense networks in and around downtown (see Figure 3.5); other cities experimented with cable-car systems and even the steam railroad, but most efforts proved impractical. With omnibuses unable to carry more than a dozen or so passengers or attain speeds of people on foot, the first meaningful breakthrough toward establishing intracity "mass" transit was finally introduced in New York City in 1852 in the form of the horse-drawn streetcar (see Figure 3.6). Lighter street rails were easy to install, overcame the problems of muddy unpaved roadways, and allowed horsecars to be hauled along them at speeds slightly faster (ca. 5 miles per hour) than those of pedestrians. This modest improvement in mobility allowed a narrow band of land at the city's edge to be opened for new home construction. Middle-income urbanites flocked to these *horsecar suburbs,* which proliferated rapidly after 1860. Radial routes were usually the first to spawn such peripheral development, but the steady demand for housing required the construction of crosstown horsecar lines, thereby filling in the interstices and preserving the generally circular shape of the city.

The nonaffluent remainder of the urban population was confined to the old pedestrian city and its bleak, high-density industrial appendages. With the massive influx of unskilled laborers, increasingly of European origin after the Civil War, huge blue-collar neighborhoods surrounded the factories, often built by the mill owners themselves. Since factory shifts ran 10 or more hours 6 days a week, their modestly paid workers could not afford to commute and were forced to reside within walking distance of the plant. Newcomers to the city, however, were accommodated in this nearby housing quite literally in the order in which they arrived, thereby denying immigrant factory workers even the small luxury of living in the immediate company of their fellow ethnics. Not surprisingly, such heterogeneous residential patterning almost immediately engendered social stresses and episodic conflicts that persisted until the end of the century, when the electric trolley would at last enable the formation of modern ethnic neighborhood communities.

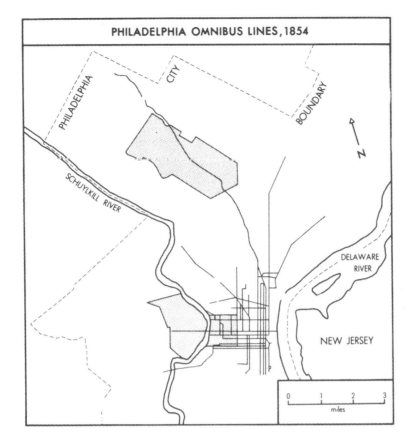

FIGURE 3.5. Philadelphia's omnibus routes in 1854, the year the city annexed all of surrounding Philadelphia County. Source: From "Household activity patterns in nineteenth-century suburbs: A time–geographic exploration" by R. Miller, 1982, *Annals of the Association of American Geographers, 72*, p. 364. Copyright 1982 by the Association of American Geographers. Reprinted by permission of the Association of American Geographers, *www.aag.org*.

   Toward the end of the Walking-Horsecar Era, the scale of the city was slowly but inexorably expanding. One by-product was the emergence of the downtown central business district (CBD). As needs intensified for specialized commercial, retailing, and other services, it was quickly realized that they could best be provided from a single center at the most accessible urban location. With immigrants continuing to pour into the all-but-bursting industrial city in the late 19th century, pressures redoubled to upgrade intraurban transit and open up more of the adjacent countryside.

   In retrospect, horsecars had only been a stopgap measure, relieving overcrowding temporarily but incapable of bringing enough new residential space within the effective commuting range of the burgeoning middle class. The hazards of relying on horses for motive power were also becoming unacceptable. Besides high costs and the sanitation problem, disease was an ever-present threat: for example, thousands of horses succumbed in New York and Philadelphia in 1872 when respiratory illnesses swept through the municipal stables (Schaeffer & Sclar, 1975, p. 22). By the

FIGURE 3.6. The horsecar introduced mass transportation to the teeming American city. This dictograph was taken in downtown Minneapolis in the late 1880s. Source: Minnesota Historical Society.

late 1880s, a desperately needed transit revolution was at last in the making. When it came, it swiftly transformed both city and suburban periphery into the modern metropolis.

The Electric Streetcar Era (1890–1920)

The key to the first urban transport revolution was the invention of the electric traction motor by one of Thomas Edison's technicians, Frank Sprague. This innovation surely must rank among the most important in American history. The first electric trolley line opened in Richmond in 1888, was adopted by two dozen other major cities within a year, and by the early 1890s had already become the dominant mode of intraurban transit. The rapidity of the diffusion of this innovation was enhanced by the immediate recognition of its ability to mitigate the urban transportation problems of the day: motors could be attached to existing horsecars to convert them into self-propelled vehicles, powered via easily constructed overhead wires. Accordingly, the tripling of average speeds (to over 15 miles per hour) now brought a large band of open land beyond the city's perimeter into trolley-commuting range.

The most dramatic impact of the Electric Streetcar Era was the swift residential development of those urban fringes, which expanded the emerging metropolis into a decidedly star-shaped spatial entity (Figure 3.2, Era II). This morphological pattern was produced by radial trolley corridors extending several miles beyond the compact city's limits; with so much new space available for home building within easy walking distance of these trolley lines, there was no need to extend tracks laterally. Consequently, the interstices remained undeveloped.

The typical streetcar suburb around the turn of the 20th century was a continuous corridor whose backbone was the road carrying the trolley tracks (usually lined with stores and other local commercial facilities), from which gridded residential streets fanned out for several blocks on both sides of the tracks. This spatial framework is illustrated in Figure 3.7, whose map reconstructs street and property subdivisioning in a portion of streetcar-era Cambridge, Massachusetts, just outside Boston. By 1900, most of the open spaces between these streets were themselves subdivided into small rectangular lots that contained modest single-family houses.

In general, the quality of the housing and the prosperity of streetcar suburbs increased with distance from the central-city line. As Warner (1962) pointed out in his classic study, however, these continuous developments were home to a highly mobile middle-class population, finely stratified according to a plethora of minor income and status differences. With frequent upward (and local spatial) mobility the norm, community formation became an elusive goal, a process further inhibited by the relentless grid-settlement morphology and the heavy dependence on distant downtown for employment and most shopping. As Warner put it so aptly, this kind of a society generated "not integrated communities arranged about common centers, but a historical and accidental traffic pattern" (p. 158). The desire to exclude the working class also shaped the social transportation geography of suburban streetcar corridors. "Definitional" conflicts usually revolved around the entry of saloons, with middle-income areas voting to remain "dry" while the blue-collar mill towns of lower-status trolley and intercity rail corridors chose to go "wet" (Schwartz, 1976, pp. 13–18).

Within the city, too, the streetcar sparked a spatial transformation. The ubiquity and low fares of the electric trolley now provided every resident access to the intracity circulatory system, thereby introducing truly *mass* transit to urban America in the closing years of the 19th century. For nonresidential activities, this ease of movement among the city's various parts quickly triggered the emergence of specialized land-use

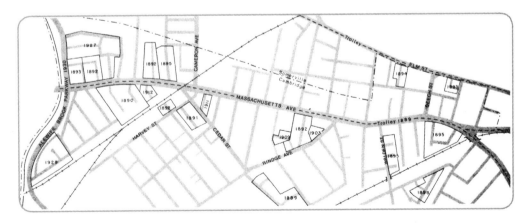

FIGURE 3.7. Streetcar subdivisions outside Boston in North Cambridge, Massachusetts, 1890–1930. Source: From *Northwest Cambridge: Report five, Survey of architectural history in Cambridge* (p. 44) by Cambridge Historical Commission, 1977, Cambridge, MA: MIT Press. Copyright 1977 by Cambridge Historical Commission. Reprinted by permission.

districts for commerce, industry, and transportation as well as the continued growth of the multipurpose CBD—now abetted by the elevator, which permitted the construction of much taller buildings. But the greatest impact of the streetcar was on the central city's social geography, because it made possible the congregation of ethnic groups in their own neighborhoods. No longer were these moderate-income masses forced to reside in the heterogeneous jumble of rowhouses and tenements that ringed the factories. The trolley brought them the opportunity to "live with their own kind," enabling the sorting of discrete groups into their own inner-city social territories within convenient and inexpensive travel distance of the workplace.

The latter years of the Electric Streetcar Era also witnessed additional breakthroughs in public urban rail transportation. The faster electric commuter train superseded steam locomotives in the wealthiest suburban corridors, which had resisted the middle-class incursions of the streetcar sectors because the rich always seek to preserve their social distance from those of lesser socioeconomic status. In some of the newer metropolises that lacked the street-rail legacy, heavier electric railways became the cornerstone of the circulatory system; Los Angeles is the outstanding example, with the interurban routes of the Pacific Electric network (Figure 3.8) spawning a dispersed settlement fabric in preautomobile days, and many lines forging rights-of-way that were later upgraded into major boulevards and even freeways

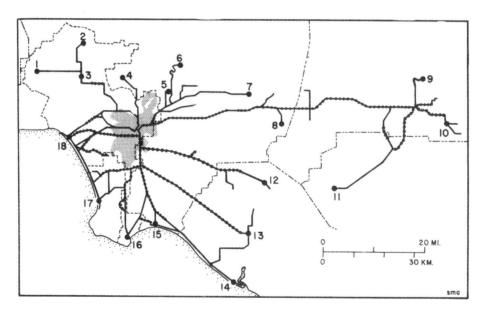

FIGURE 3.8. Interurban railway routes of the Pacific Electric system at their greatest cumulative extent in the mid-1920s. Shading denotes portions of central Los Angeles that were situated within a half-mile of narrow-gauge streetcar lines in the early 1920s. Selected interurban destinations are numbered in clockwise sequence as follows: (1) Canoga Park; (2) San Fernando; (3) Van Nuys; (4) Burbank; (5) Pasadena; (6) Mount Lowe; (7) Glendora; (8) Pomona; (9) Arrowhead Springs; (10) Redlands; (11) Corona; (12) Yorba Linda; (13) Santa Ana; (14) Newport Beach; (15) Long Beach; (16) San Pedro; (17) Redondo Beach; (18) Santa Monica. Source: From *Los Angeles: The centrifugal city* (p. 95) by R. Steiner, 1981, Dubuque, IA: Kendall/Hunt. Copyright 1981 by Kendall/Hunt

(see Banham, 1971, pp. 32–36). Finally, within the city proper elevated and underground rapid transit lines made their appearances, the "el" in New York as early as 1868 (using steam engines—the electric elevated was born in Chicago in 1892) and the subway in Boston in 1898. Such heavy-rail rapid transit was always enormously expensive to build and could be justified only in the largest cities that generated the highest traffic volumes. Therefore, els and subways were restricted to New York, Boston, Chicago, Philadelphia, and Cleveland, and most construction concluded by 1930. (Rapid-transit-system building did not resume until the 1960s with metropolitan San Francisco's Bay Area Rapid Transit network [activated in 1972], followed by the opening of new systems in Washington [1976], Atlanta [1979], Baltimore [1983], Miami [1984], and Los Angeles [1993].)

The Recreational Automobile Era (1920–1945)

By 1920, the electric trolleys, commuter trains, interurbans, and subways had transformed the tracked city into a full-fledged metropolis whose streetcar suburbs and mill-town intercity rail corridors, in the most prominent cases, spread out to encompass an urban complex more than 20 miles in diameter. It was at this time, many geographers and planners would agree, that intrametropolitan transportation achieved its greatest level of efficiency—that the burgeoning city truly "worked." How much closer the American metropolis might have approached optimal workability for all its millions of residents, however, shall never be known because the second urban transportation revolution was already beginning to assert itself through the increasingly popular automobile.

Even though many scholars have vilified the automobile as the destroyer of the city, Americans took to cars as completely and wholeheartedly as they did to anything in the nation's long cultural history. More balanced assessments of the role of the automobile (see, e.g., Bruce-Briggs, 1977) recognize its overwhelming acceptance for what it was: the long-hoped-for attainment of private transportation that offered users almost total freedom to travel whenever and wherever they chose. Cars came to the metropolis in ever greater numbers throughout the interwar period, a union culminating in accelerated deconcentration—through the development of the bypassed, streetcar-era interstices and the pushing of the suburban frontier farther into the countryside—to produce once again a compact, regular-shaped urban entity (Figure 3.2, Era III).

Although it came to have a dramatic impact on the urban fabric by the eve of World War II, the automobile was introduced into the American city in the 1920s and 1930s at a leisurely pace. The first cars had appeared in both Western Europe and the United States in the 1890s, and the wealthy on both sides of the Atlantic quickly took to this innovation because it offered a better means of personal transport. It was Henry Ford, however, with his revolutionary assembly-line manufacturing techniques, who first mass-produced cars; the lower selling prices soon converted them from the playthings of the rich into a transport mode available to a majority of Americans. By 1916, over 2 million autos were on the road, a total that quadrupled by 1920 despite wartime constraints. During the 1920s, the total tripled to 23 million and increased another 4½ million by the end of the depression-plagued 1930s;

passenger car registrations paralleled these increases (Figure 3.9). The earliest flurry of automobile adoptions had been in rural areas, where farmers badly needed better access to local service centers; accordingly, much of the early paved-road construction effort was concentrated in rural America. In the cities, cars were initially used for weekend outings—hence the *Recreational* Auto Era—and some of the first paved roadways built were landscaped parkways that followed scenic waterways (such as New York's Bronx River Parkway, Chicago's Lake Shore Drive, and the East and West River Drives along the Schuylkill in Philadelphia's Fairmount Park).

In the suburbs, however, where the overall growth rate now for the first time exceeded that of the central cities, cars were making a decisive penetration throughout the economically prosperous 1920s. Flink (1975, p. 14) reported that, as early as 1922, 135,000 suburban dwellings in 60 metropolises were completely dependent on motor vehicles. In fact, the subsequent rapid expansion of automobile suburbia by 1930 so adversely affected the metropolitan public transportation system that, through significant diversions of streetcar and commuter-rail passengers, the large cities started to feel the negative effects of the car years before accommodating to its actual arrival. By encouraging the opening of unbuilt areas lying between suburban rail axes, the automobile effectively lured residential developers away from densely populated traction-line corridors into the now-accessible interstices. Thus the suburban home-building industry no longer found it necessary to subsidize privately owned streetcar companies to provide cheap access to their trolley-line housing tracts.

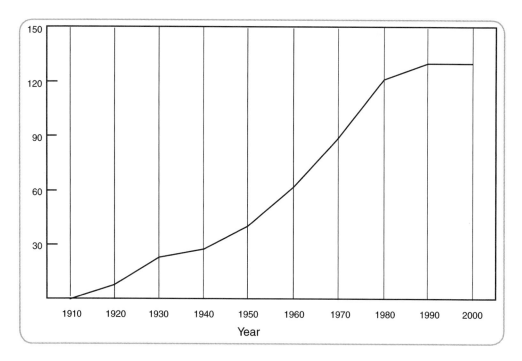

FIGURE 3.9. Passenger car registrations in the United States, 1910–2000. Source: *Statistical Abstract of the United States*, Washington, DC, U.S. Government Printing Office, annual.

Without this financial underpinning, the modern urban transit crisis soon began to unfold. Traction companies, obliged under their charters to provide good-quality service, could not raise fares to the level necessary to earn profits high enough to attract new capital in the highly competitive money markets. As this economic squeeze intensified, particularly during the Great Depression of the 1930s, local governments were forced to intervene with subsidies from public funds; eventually, as transit ridership continued to decline in the postwar era, local governments assumed ownership of transit companies when lines could not be closed down without harming communities.

Several additional factors also combined to accelerate the interwar deterioration of the superlative trolley-era metropolitan transit network: the growing intrasuburban dispersal of population that no longer generated passenger volumes large enough to support new fixed-route public transportation facilities; dispersion of employment sites within the central city, thereby spreading out commuter destinations as well as origins; shortening of the work week from 6 to 5 days; worsening street congestion where trolleys and auto traffic increasingly mixed; and the pronounced distaste for commuting to the city by bus, a more flexibly routed new transit mode that never caught on in the suburbs.

Ironically, recreational motorways helped to accelerate the decentralization of the urban population. Most were radial highways that penetrated deeply into the suburban ring; those connecting to major new bridges and tunnels—such as the Golden Gate and Bay Bridges in San Francisco, the George Washington Bridge and Holland and Lincoln Tunnels in New York—usually served to open empty outer metropolitan sectors. Sunday motorists, therefore, had easy access to this urban countryside and were captivated by what they saw. They responded in steadily increasing numbers to the home-sales pitches of developers who had shrewdly located their new tract housing subdivisions beside the suburban highways. As more and more city dwellers relocated to these automobile suburbs, by the end of the interwar era many recreational parkways were turning into heavily traveled commuter thoroughfares—especially near New York City, where the suburban parkway network devised by planner Robert Moses reached far into Westchester County and Long Island, and in the Los Angeles Basin, where the first freeway (the Arroyo Seco, now called the Pasadena) was opened in 1940.

The residential development of automobile suburbia followed a simple formula that was devised in the prewar years and perfected and greatly magnified in scale in the decade following 1945. The leading motivation was developer profit from the quick turnover of land, which was acquired in large parcels, subdivided, and auctioned off. Accordingly, developers much preferred open areas at the metropolitan fringe where large packages of cheap land could readily be assembled. As the process became more sophisticated in the 1940s, developer-builders came to the forefront and produced huge complexes of inexpensive housing—with William J. Levitt and his Levittowns in the vanguard. Silently approving and underwriting this uncontrolled spread of residential suburbia were public policies at all levels of government that included the financing of highway construction, obligating lending institutions to invest in new home building, insuring individual mortgages, and providing low-interest loans to Federal Housing Administration (FHA) and Veterans Administration (VA) clients.

Whereas the conventional-wisdom view of U.S. suburbanization holds that most of it occurred after World War II, longitudinal demographic data indicate that intrametropolitan population decentralization had achieved sizeable proportions during the interwar era. Table 3.1 reveals that suburban growth rates began to surpass those of the central cities as early as the 1920s, and that after 1930 the outer ring took a commanding lead (which has not ceased widening to this day). With an ever larger segment of the urban population residing in automobile suburbs, their spatial organization was already forming the framework of contemporary metropolitan society. Because automobility removed most of the preexisting movement constraints, suburban social geography now became dominated by locally homogeneous income-group clusters that isolated themselves from dissimilar neighbors. Gone was the highly localized stratification of streetcar suburbia; in its place arose a far more dispersed, increasingly fragmented residential mosaic that builders were only too happy to cater to, helping shape this kaleidoscopic settlement pattern by constructing the most expensive houses that could be sold in each locality.

The long-standing partitioning of suburban social space was further legitimized by the widespread adoption of zoning (legalized in 1916). This regulatory device gave municipalities control of lot and building standards, which, in turn, assured dwelling prices that would only attract newcomers whose incomes at least equaled those of the existing population. For the middle class, especially, such exclusionary economic practices were enthusiastically supported because it now extended to them the capability that upper-income groups had enjoyed to maintain their social and geographic distance from people of lower socioeconomic status.

Nonresidential activities were also suburbanizing at a steadily increasing rate during the Recreational Auto Era. Indeed, many large-scale manufacturers had decentralized during the previous streetcar era, choosing suburban freight-rail locations that rapidly spawned surrounding working-class towns. These industrial suburbs became important satellites of the central city in the emerging metropolitan constellation (see Taylor, 1915/1970), and the economic geography of the interwar era was marked by an intensification of this trend. Industrial employers now accelerated their intraurban deconcentration as more efficient horizontal fabrication methods were replacing older techniques requiring multistoried plants—thereby generating

TABLE 3.1. Intrametropolitan Population Growth Trends, 1910–1960

Decade	Central city growth rate	Suburban growth rate	Percent total SMSA* growth in suburbs	Suburban growth per 100 increase in central city population
1910–1920	27.7	20.0	28.4	39.6
1920–1930	24.3	32.3	40.7	68.5
1930–1940	5.6	14.6	59.0	144.0
1940–1950	14.7	35.9	59.3	145.9
1950–1960	10.7	48.5	76.2	320.3

*SMSA, Standard Metropolitan Statistical Area, constituted by the central city and county-level political units of the surrounding suburban ring.

Source: U.S. Census of Population.

greater space needs that were too expensive to satisfy in the high-density inner central city. Newly suburbanizing manufacturers, however, continued their spatial affiliation with intercity rail corridors, because motor trucks were not yet able to operate with their present-day efficiencies and the highway network of the outer ring remained inadequate until the 1950s.

The other major nonresidential activity of interwar suburbia was retailing. Clusters of automobile-oriented stores had first appeared in the urban fringes before World War I. By the early 1920s, the roadside commercial strip had become a common sight in many Southern California suburbs. Retail activities were also featured in dozens of planned automobile suburbs that sprang up in the 1920s, most notably in outer Kansas City's Country Club District where builder Jesse Clyde Nichols opened the nation's first complete shopping center in 1922. But these diversified retail centers spread rather slowly before the 1950s; nonetheless, such chains as Sears & Roebuck and Montgomery Ward quickly discovered that stores situated alongside main suburban highways could be very successful, a harbinger of things to come in post-World War II metropolitan America.

The central city's growth reached its apex in the interwar era and began to level off (Table 3.1) as metropolitan development after 1925 increasingly concentrated in the urban fringe zone that now widely resisted political unification with the city. Whereas the transit infrastructure of the streetcar era remained dominant in the industrial city (see Figure 3.10), the late-arriving automobile was adapted to this high-density urban environment as much as possible, but not without greatly aggravating existing traffic congestion.

FIGURE 3.10. The automobile did not become a major force in the central city until the post–World War II era, but its presence can already be detected by the 1930s. This photograph was taken in St. Paul in 1932. Source: Minnesota Historical Society.

The structure of the American city during the second quarter of the 20th century was best summarized in the well-known *concentric-ring, sector,* and *multiple nuclei* models (reviewed in Harris & Ullman, 1945), which in combination described the generalized spatial organization of urban land usage. The social geography of the core city was also beginning to undergo significant change at this time as the suburban exodus of the middle class was accompanied by the growing influx of African Americans from the South. These parallel migration streams would achieve massive proportions after World War II. The southern newcomers were attracted to the northern city by declining agricultural opportunities in the rural South and by offers of employment in the factories as industrial entrepreneurs sought a new source of cheap labor to replace European immigrants, whose numbers were sharply curtailed by restrictive new legislation after the mid-1920s. But urban whites refused to share residential space with in-migrating blacks, and racial segregation of the metropolitan population swiftly intensified as citywide dual housing markets dictated the formation of ghettoes of nonwhites inside their own distinctive social territories.

Freeway Era (1945–Present)

Unlike the two preceding eras, the post–World War II Freeway Era was not sparked by a revolution in urban transportation. Rather, it represented the coming of age of the automobile culture, which coincided with a historic watershed as a reborn nation emerged from 15 years of economic depression and war. Suddenly the automobile was no longer a luxury or a recreational diversion: it quickly became a necessity for commuting, shopping, and socializing—essential to the successful exploitation of personal opportunities for a rapidly expanding majority of the metropolitan population. People snapped up cars as fast as the reviving peacetime automobile industry could roll them off the assembly lines, and a prodigious highway-building effort was launched, spearheaded by high-speed, limited-access expressways.

Given impetus by the 1956 Interstate Highway Act, these new freeways would soon reshape every corner of urban America as the new suburbs they spawned represented nothing less than the turning inside out of the historic metropolitan city. In retrospect, this massive acceleration of the deconcentration process "cannot be considered a break in long-standing trends, but rather the later, perhaps more dynamic, evolutionary stages of a transformation which was based on a pyramiding of small-scale innovations and underlying social desires" (Sternlieb & Hughes, 1975, p. 12).

The snowballing effect of these changes is expressed spatially in the much-expanded metropolis of postwar America (Figure 3.2, Era IV), whose expressway-dominated infrastructure again produced a star-shaped development pattern reminiscent of the Electric Streetcar Era (Figure 3.2, Era II). A more detailed representation of contemporary intraurban morphology is seen in Figure 3.11, showing the culmination of eight decades of automobile suburbanization. Most striking is the enormous band of growth that has been added since 1950, with freeway sectors pushing the metropolitan frontier deeply into the surrounding zone of exurbia. The huge curvilinear *outer city* that arose within this greatly enlarged suburban ring was most heavily shaped by the circumferential freeway segments that girdled the central city—a near-universal feature of the metropolitan expressway system, originally designed to allow long-distance interstate highways to bypass the congested urban core. Today, well

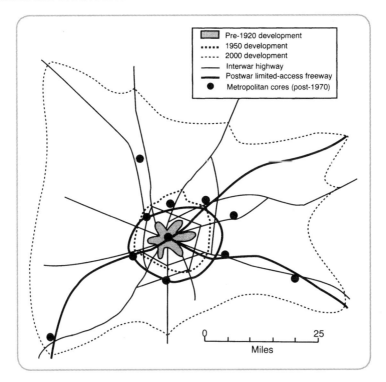

Pre-1920 development
1950 development
2000 development
Interwar highway
Postwar limited-access freeway
Metropolitan cores (post-1970)

FIGURE 3.11. The spatial pattern of growth in automobile suburbia since 1920. Source: From "The role of the suburbs in contemporary metropolitan systems" by P. O. Muller, 1982, in C. M. Christian and R. A. Harper (Eds.), *Modern metropolitan systems* (p. 257), Columbus, OH: Charles E. Merrill. Copyright 1982 by Charles E. Merrill. Reprinted by permission.

over 100 of these expressways form complete *beltways* that are the most heavily traveled roadways in their regions. The prototype high-speed, limited-access circumferential was suburban Boston's Route 128, completed in the early 1950s; by the 1980s, such freeways as Houston's Loop, Atlanta's Perimeter, Chicago's Tri-State Tollway, New York–New Jersey's Garden State Parkway, Miami's Palmetto Expressway, and the Beltways encircling Washington and Baltimore had become some of the best-known urban arteries in the nation.

The maturing freeway system was the primary force that turned the metropolis inside out after 1970 because it eradicated the regionwide centrality advantage of the central city's CBD. Now *any* location on that expressway network could easily be reached by motor vehicle, and intraurban accessibility swiftly became an all-but-ubiquitous spatial good. Ironically, large central cities had encouraged the construction of radial expressways in the 1950s and 1960s because they appeared to enable downtown to remain accessible to the swiftly dispersing suburban population. As one economic activity after another discovered its new locational footlooseness within the freeway metropolis, however, nonresidential deconcentration greatly accelerated. Much of this suburban growth has gravitated toward beltway corridors, and Figure 3.12 displays the typical sequence of land use development along a segment of circumferential I-494 just south of Minneapolis (stretching westward from the site of today's gigantic Mall of America, which opened in 1992).

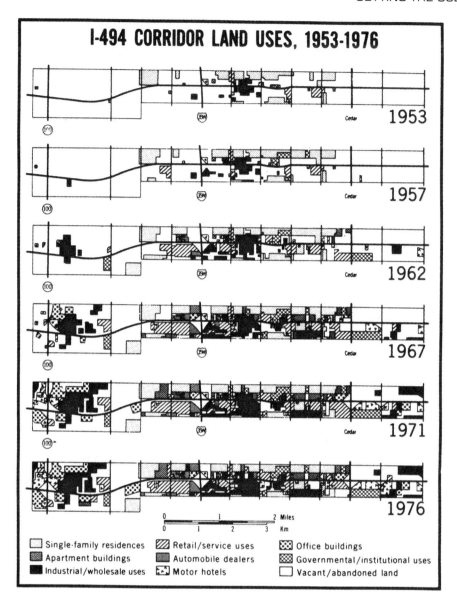

FIGURE 3.12. Land use change in the Interstate 494 corridor south of Minneapolis, 1953–1976. Source: From "The emergence of a new 'downtown'" by T. J. Baerwald, 1978, *Geographical Review, 68*, p. 312. Copyright 1978 by the American Geographical Society. Reprinted by permission.

As high-speed expressways expanded the radius of commuting to encompass the entire dispersed metropolis, residential locational constraints were relaxed as well. No longer were most urbanites required to live within a short distance of their job. Instead, the workplace had now become a locus of opportunity offering access to the best possible residence that a household could afford anywhere within the urbanized area. Thus the heterogeneous patterning of sociospatial clusters that had arisen in prewar automobile suburbia was writ ever larger in the Freeway Era—giving rise to a *mosaic culture* whose component tiles were stratified not only along class lines but also according to age, occupational status, and a host of minor lifestyle differences (Berry, 1981, pp. 64–66).

These developments fostered a great deal of local separatism (for a contemporary assessment, see Bishop, 2008), thereby intensifying the balkanization of metropolitan society as a whole:

> With massive auto transportation, people have found a way to isolate themselves; . . . a way to privacy among their peer group. . . . they have stratified the urban landscape like a checker board, here a piece for the young married, there one for health care, here one for shopping, there one for the swinging jet set, here one for industry, there one for the aged. . . . When people move from square to square, they move purposefully, determinedly. . . . They see nothing except what they are determined to see. Everything else is shut out from their experience. (Schaeffer & Sclar, 1975, p. 119)

After more than seven decades of Freeway-Era change, certain structural transformations have emerged from what was, in retrospect, one of the most tumultuous upheavals in American urban history. Figure 3.11 reveals the existence of several new outlying metropolitan-level cores. Today, such downtown-like concentrations of retailing, business, and light industry have become common landscape features near the major highway intersections of the outer city that now surrounds every large central city. A representative *suburban downtown* of this genre is mapped in Figure 3.13,

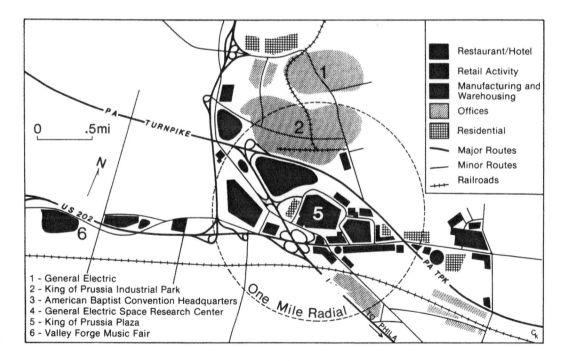

FIGURE 3.13. The internal structure of nonresidential activities in a typical suburban downtown: King of Prussia, Pennsylvania. Source: From *The outer city: Geographical consequences of the urbanization of the suburbs* (p. 41) by P. O. Muller, 1976, Washington, DC: Association of American Geographers, Resource Publication No. 75-2. Copyright 1976 by the Association of American Geographers. Adapted by permission.

revealing the array of high-order activities that have agglomerated around the King of Prussia Plaza shopping center at the most important expressway junction in Philadelphia's northern and western suburbs.[3] Dozens of such diversified activity cores have matured over the past half-century, and suburban downtowns such as Washington's Tysons Corner (Figure 3.14), Houston's Post Oak Galleria, Los Angeles's South Coast Metro, and Chicago's Schaumburg have now achieved national reputations.

In his book-length survey of suburban downtowns—which he calls *edge cities*—Garreau (1991, pp. 6–7) set forth some minimum requirements that such an activity center must meet in order to be classified as an edge city:

1. At least 24,000 jobs
2. 5,000,000-plus square feet of leasable office space
3. 600,000-plus square feet of leasable retail space
4. More jobs than bedrooms
5. An identity as a single place
6. No significant structure more than 30 years old.

FIGURE 3.14. An aerial view of Tyson's Corner in Fairfax County, Virginia, outside Washington, DC, one of the nation's largest suburban downtowns. Source: County of Fairfax (Virginia), Office of Comprehensive Planning.

By the turn of the millennium, more than 200 of these new urban agglomerations had been identified nationwide, "most at least the size of downtown Orlando (in contrast, fewer than 40 [central-city] downtowns are Orlando's size)" (Garreau, 1994, p. 26).

As the suburban downtowns of the outer city achieve economic–geographic parity with each other (as well as with the CBD of the nearby central city), they provide the totality of urban goods and services to their surrounding populations and thereby make each sector of the metropolis an increasingly self-sufficient functional entity. This now-completed transition to a polycentric metropolis of *realms*—the term coined by Vance (1964) to describe the ever-more-independent areas served by new downtown-like activity cores—requires the use of more up-to-date generalizations than are provided by the "classical" concentric-zone, sector, and multiple-nuclei models of urban form. Such an alternative to these obsolete core–periphery models of the interwar metropolis is seen in Figure 3.15.

Another useful model (that builds on the work of Baerwald [1978], Erickson [1983], and others) was developed by Hartshorn and Muller (1989) to interpret the evolution of the suburban spatial economy. Accordingly, the developments of the Freeway Era are subdivided into five growth stages. First was the *bedroom community* stage (1945–1955), dominated by a massive postwar residential building boom but accompanied by only a modest expansion of suburban commercial activity. This was followed by the *independence* stage (1955–1965) during which suburban economic

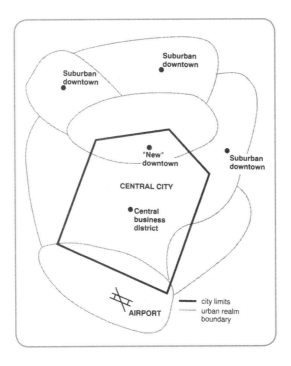

FIGURE 3.15. The generalized layout of urban realms in relation to the central city and suburban downtowns of the polycentric metropolis. Source: From "Suburban downtowns and the transformation of metropolitan Atlanta's business landscape," by T. A. Hartshorn and P. O. Muller, 1989, *Urban Geography, 10,* p. 378. Copyright 1989 by V. H. Winston & Son. Reprinted by permission.

growth accelerated, led by the first wave of industrial and office parks and, after 1960, by the rapid diffusion of regional shopping centers. These malls came into their own during the third stage, *catalytic growth* (1965–1980), attracting myriad office, hotel, and restaurant facilities to cluster around them, and triggering the swift maturation of the suburban economic landscape in these expanding cores and many of the freeway corridors that connected them. The fourth stage, *high-rise/high-technology*, spanned the 1980s and saw the flowering of scores of suburban downtowns increasingly dominated by high-rise office buildings; simultaneously, these burgeoning activity cores attracted high-technology research-and-development facilities, and by 1990 the outer suburban city had become the leading geographic setting for the nation's new postindustrial service economy. The post-1990 period constitutes the fifth stage in which the trends of the 1980s have persisted, and suburban downtowns continue to evolve into *mature urban centers* as their land use complexes steadily diversify and perform ever more important economic, social, and cultural functions that are increasingly international in scope.

URBAN TRANSPORTATION IN THE CHANGING POSTINDUSTRIAL METROPOLIS

This chapter has traced the evolution of the spatial form of the American metropolis in conjunction with the development of its internal transportation system. Four growth eras were identified, each exhibiting a distinctive spatial structure forged in response to a breakthrough in intrametropolitan movement technology. The horsecar, the electric streetcar, the first-generation automobile, and the far-flung automobility unleashed by the freeway network each left its geographic imprints, which cumulatively shaped the metropolitan city we see today.

The last of these four stages has been underway for over 70 years, unlike its two predecessors that spanned 30 and 25 years, respectively. Understandably, one may ask if we have finally entered a fifth, "Post-Freeway" Era, especially in light of today's widespread discussion concerning new urbanization forces at work. Although urban America has undergone a degree of spatial reorganization at the central-city and regional scales since the late 20th century, according to the framework used here the structural layout of the metropolis has changed little over the past three decades. In other words, the maturing polycentric metropolis of realms has now become the cornerstone of the American urban system and is very likely to retain that status well into the future. Thus the transformed metropolitan city of the Freeway Era represents the end product of two centuries of U.S. urban evolution, and should therefore be regarded as an enduring structural entity rather than another temporary phase in the unfolding of a multistage historical process. From a planning standpoint, there are certainly more efficient, more esthetic, and more sustainable urban forms. But this is the reality that transportation policymakers and planners confront in the first quarter of our new century, and their challenge is to make the best of the situation they inherited. As many of the other chapters in this book underscore, creative opportunities abound if the economic resources and political will can be marshaled to use them.

   The stabilizing form of the American metropolis, and the crucial role of transportation in its shaping and spatial extent, are hotly-debated topics among scholars and commentators in the 2010s. It has been widely touted that a cultural and economic "urban renaissance" has been underway since the turn of the millennium. Moreover, many claim that this burgeoning trend favors the central city at the expense of suburbia—whose long arc of population and employment growth supposedly peaked in the mid-2000s before entering a new era of decline triggered by the Great Recession of recent years and its still-abating mortgage-default crisis. Reality, however, presents a rather different picture. As the newest census and other post-2010 data become available, it is increasingly clear that during the past decade an overwhelming majority of population and employment growth in urban America occurred in the suburban ring. Specifically, between 2000 and 2010, 91% of all metropolitan population growth took place in the suburbs (vs. 85% in the previous decade); over the same period, within the 100 largest U.S. metros, the number of jobs within 3 miles of the core-city center fell by 10.4% while the total in the outer ring between 10 and 35 miles away rose by 1.2% (Kneebone, 2013, p. 3). To be sure, certain downtown-area districts of the central city did indeed reinvent themselves as world-class centers of economic and artistic creativity. But "the new global culture [of these relatively few CBD-area components] of the city and its suburbs revealed itself to be the culture of the prosperous educated class"—thereby producing the greatest intrametropolitan inequality in almost a century (Warner & Whittemore, 2012, p. 137).

   Much of that inequality is the result of the nation's recently completed transition to a postindustrial economy and society, in which quaternary (information-related) economic activities increasingly dominate the American labor force. These employers have decidedly expressed a locational preference for the outer metropolitan ring and have been at the forefront of the suburbanization of jobs, a trickle that began to turn into a steady flow during the 1960s. It rapidly became a tidal wave, and as early as 1973 the nation's suburban rings had surpassed the central cities in total employment. Following the surge of the 1970s, this centrifugal trend has persisted over the past four decades, and the newest data for the 35 largest metros in 2013 show the suburban share has reached fully 70%, and cumulatively accounts for 88% of all intrametropolitan jobs created since 1960. A more detailed sectoral breakdown of this historic, post-1970 decentralization is provided in Table 3.2 (using roughly 20-year intervals) for Philadelphia, a metropolis that closely mirrors the national metropolitan economy. The finance/insurance/real estate sector is a particularly good barometer of the magnitude of economic spatial change over the past half-century, because of its traditionally strong ties to the CBD and detachment from intraurban population shifts (at least through the 1980s).

   Above all, quaternary producers, with the prized jobs and advancement opportunities they provide, are attracted to the most prestigious metropolitan locations. Revitalized CBDs have captured their share in recent years, but the evidence to date underscores that the suburban ring retains a commanding lead. This process is best exemplified by the pacesetting electronics/computer industry, which has notably concentrated in *technopoles*—planned technoindustrial agglomerations that create the hardware and software products of the new information-based economy (Castells & Hall, 1994). Technopoles decidedly gravitate toward highly accessible suburban

TABLE 3.2. Suburban Percentage of Major Employment Sectors, Metropolitan Philadelphia, 1970–1998

Employment sector	1970	1978	1988	1998
Manufacturing	**54.5**	**66.0**	<u>**75.0**</u>	<u>**81.5**</u>
Wholesale trade	39.7	**60.5**	<u>**73.6**</u>	<u>**82.0**</u>
Retail trade	**55.9**	**68.0**	<u>**73.0**</u>	**80.3**
Finance/insurance/real estate*	31.0	45.9	**59.2**	**66.5**
Health services**	46.1	**51.8**	**57.4**	**62.5**
Total employment	48.8	**60.2**	<u>**68.3**</u>	<u>**72.2**</u>

Note. Percentages shown in **boldface** exceed the *critical mass* (50%) level; percentages shown in <u>**underlined boldface**</u> exceed the *suburban dominance* (66.7%) level.

*Pre-1998 data also include real estate employment in finance/insurance sector.

**Pre-1998 data do not include social assistance employment in the health services sector.

Source: U.S. Bureau of the Census, *County Business Patterns* (annual).

locations, from Seattle's Redmond to Raleigh-Durham's Research Triangle Park and from Washington's Silicon Alley (Dulles Toll Road Corridor) to Dallas's Telecom Corridor. The prototype, of course, is California's Silicon Valley south of San Francisco, and there is every indication that such R&D/manufacturing complexes are becoming leading components of suburban economic development worldwide (e.g., Tokyo's Tsukuba, Delhi's Gurgaon, Moscow's Skolkovo, and São Paulo's Campinas).

Spearheaded by technopoles, the development of high-order suburban activity centers is also driven by changes in the U.S. economy. A key set of forces involves the dynamic economic and social networks of *globalization,* whose expanding international linkages have triggered new investment flows and entrepreneurial opportunities that have revitalized a number of central-city CBDs (Wilson, 1997). Importantly, these globalization forces simultaneously enable high-order suburban centers to participate strongly as their metropolis evolves into a "world city" (Muller, 1997). They establish their own direct international ties as the regional Internet and telecommunications networks create a grid of local nodes that link suburban downtowns as well as the CBD to the rest of the world; they also become major control points in the global economic system when leading transnational corporations (such as Apple, Exxon Mobil, Toyota USA, and Hewlett Packard) are headquartered in suburban locations. And it should be noted as well that the foreign presence in suburban America is constantly growing, ranging from the ownership of businesses large and small to the emergence of thriving new ethnic communities dominated by affluent professionals.

With the dispersed polycentric metropolis of realms here to stay, today's greatest urban transportation challenge is to improve the daily movement experience of its residents. This entails not only the upgrading and maintenance of transport technologies and infrastructure, but doing so in an environmentally harmonious, sustainable manner. Most planners would undoubtedly agree that the mitigation of congestion and other problems caused by overreliance on the automobile is at the forefront of the quest to achieve greater travel and traffic-flow efficiencies. Clearly, the saturation level of motor vehicles is now being approached throughout metropolitan America, and the lesson has finally been learned that building more freeways is not the solution. Yet the automobile is going to remain the foundation of the widely dispersed urban settlement fabric for decades to come, so the answer may well lie in how cars are used in coordination with other modes of transportation that must be expanded and/or innovated. To an as-yet-unknown extent, that accommodation will involve the reinvention of the automobile; driverless-car technology, in particular, is already becoming reality, but its impact could just as easily lengthen intrametropolitan trips and lead to densification of certain urban land uses (Bilton, 2013).

Let us conclude by spotlighting an ongoing planning experiment that encapsulates all of the issues raised in this section: the retrofitting of Tysons Corner, Virginia, to convert it into a "dense, walkable green city" (Davis, 2009). The reconfiguration of this massive suburban downtown (see Figure 3.14) was triggered by the construction of the Washington Metro's new (Dulles Corridor) Silver Line, whose four stations at Tysons opened in 2014. The immediate aim of this new mode of regional access is to divert about 17,000 commuters (one-seventh of the 120,000+ who daily converge on this activity complex) from their cars onto this transit line, whose stations will be served by a new fleet of local shuttles. The longer-term goal is to increasingly shift Tysons's orientation from cars to people by razing small strip malls and seas of parking lots in order to more tightly cluster office, retail, and other economic activities. As space opens up, new apartment towers will eventually triple the number of housing units, allowing the residential population of this edge city to expand from about 20,000 to 100,000. The remaining open space, to be dominated by greenery and laced with pedestrian and bicycle paths, will house schools as well as other municipal facilities, and emphasize recreation. These ambitious plans, strongly supported by the private sector, may take more than a quarter-century to fully implement. If successful, they will contribute to overcoming a number of current urban transportation challenges, and Tysons could well become the prototype for the suburban downtown of the future that will continue to function as a principal structural anchor of the American metropolis.

ACKNOWLEDGMENTS

In the preparation of the four editions of this chapter, the comments, suggestions, and assistance of the editors and the following are gratefully acknowledged: Raymond A. Mohl of Florida Atlantic University (now of the University of Alabama at Birmingham), Thomas J. Baerwald of the Science Museum of Minnesota (now of the National Science Foundation), Ira M. Sheskin of the University of Miami, Roman A. Cybriwsky of Temple University, John B. Fieser of the Economic Development Administration of the U. S. Department of Commerce (now retired), Sterling R. Wheeler of the Fairfax County (Virginia) Office of Comprehensive Planning, and Sharyn Caprice Goodpaster (now LaCombe) of the City of Houston's Department of Planning and Development (now of the U.S. Department of Transportation's Federal Transit Administration). I am particularly indebted to Truman A. Hartshorn of Georgia State University, my (now retired) research colleague of the past three decades, for allowing me to use materials that were jointly developed in our studies of suburban business centers.

NOTES

1. This chapter is dedicated to the memory of two outstanding transportation geographers, James O. Wheeler and Thomas R. Leinbach, whose finely crafted chapters graced earlier editions of this book. Throughout their distinguished careers, both were research leaders as well as superb colleagues and cherished friends; Jim first steered this urban geographer into the fascinating domain of transportation geography, and my introduction to this hybrid field was greatly enhanced by Tom during an early collaboration on a review article for *Progress in Geography*.

2. This approach, while emphasizing the key role of transportation, does not mean to suggest that movement processes are the only forces shaping intraurban growth and spatial organization. As demonstrated throughout this chapter, urban geographical patterns are also very much the product of social values, land resources, investment capital availability, the actions of private markets, and other infrastructural technologies.

3. Since this map was compiled, this enormous suburban complex has continued its growth. In the early 2000s came a doubling in size of the superregional mall at its heart (labeled 5 in Figure 3.13). In August 2016, the completion of another massive expansion of this retail magnet—known as King of Prussia Plaza—made it the second-largest shopping mall in the United States (only Bloomington, Minnesota's Mall of America contains more selling space).

REFERENCES

Adams, J. S. (1970). Residential structure of Midwestern cities. *Annals of the Association of American Geographers, 60,* 37–62.

Baerwald, T. J. (1978). The emergence of a new "downtown." *Geographical Review, 68,* 308–318.

Banham, R. (1971). *Los Angeles: The architecture of four ecologies.* New York: Penguin Books.

Berry, B. J. L. (1975). The decline of the aging metropolis: Cultural bases and social process. In G. Sternlieb & J. W. Hughes (Eds.), *Postindustrial America: Metropolitan decline and interregional job shifts* (pp. 175–185). New Brunswick, NJ: Rutgers University, Center for Urban Policy Research.

Berry, B. J. L. (1976). *Chicago: Transformations of an urban system.* Cambridge, MA: Ballinger.

Berry, B. J. L. (1981). *Comparative urbanization: Divergent paths in the twentieth century* (2nd ed.). New York: St. Martin's Press.

Bilton, N. (2013, July 8). Disruptions: How driverless cars could reshape cities. *New York Times,* p. B-5.

Bishop, B. (2008). *The big sort: Why the clustering of like-minded America is tearing us apart.* Boston: Houghton Mifflin.

Bruce-Briggs, B. (1977). *The war against the automobile.* New York: Dutton.

Cambridge Historical Commission. (1977). *Northwest Cambridge: Report five: Survey of architectural history in Cambridge.* Cambridge, MA: MIT Press.

Castells, M., & Hall, P. (1994). *Technopoles of the world: The making of twenty-first century industrial complexes.* London and New York: Routledge.

Davis, L. S. (2009, June 22). A (radical) way to fix suburban sprawl. *Time,* pp. 55–57.

Erickson, R. A. (1983). The evolution of the suburban space-economy. *Urban Geography, 4,* 95–121.

Flink, J. J. (1975). *The car culture.* Cambridge, MA: MIT Press.

Garreau, J. (1991). *Edge city: Life on the new frontier.* Garden City, NY: Doubleday.

Garreau, J. (1994, February). Edge cities in profile. *American Demographics,* pp. 24–33.

Harris, C. D., & Ullman, E. L. (1945, November). The nature of cities. *Annals of the American Academy of Political and Social Science, 242,* 7–17.

Hartshorn, T. A., & Muller, P. O. (1989). Suburban downtowns and the transformation of metropolitan Atlanta's business landscape. *Urban Geography, 10,* 375–395.

Kneebone, E. (2013). *Job sprawl stalls: The Great Recession and metropolitan employment location* (Metropolitan Opportunity Series, Report No. 30). Washington, DC: Brookings Institution.

Miller, R. (1982). Household activity patterns in nineteenth-century suburbs: A time–geographic exploration. *Annals of the Association of American Geographers, 72,* 355–371.

Muller, P. O. (1997). The suburban transformation of the globalizing American city. *Annals of the American Academy of Political and Social Science, 551,* 44–58.

Schaeffer, K. H., & Sclar, E. (1975). *Access for all: Transportation and urban growth.* Baltimore: Penguin Books.

Schwartz, J. (1976). The evolution of the suburbs. In P. Dolce (Ed.), *Suburbia: The American dream and dilemma* (pp. 1–36). Garden City, NY: Anchor Press/Doubleday.

Sommers, L. M. (1983). Cities of Western Europe. In S. D. Brunn & J. F. Williams (Eds.), *Cities of the world: World regional urban development* (pp. 84–121). New York: Harper & Row.

Steiner, R. (1981). *Los Angeles: The centrifugal city.* Dubuque, IA: Kendall/Hunt.

Tarr, J. A. (1984). The evolution of the urban infrastructure in the nineteenth and twentieth centuries. In R. Hanson (Ed.), *Perspectives on urban infrastructure* (pp. 4–66). Washington, DC: National Academy Press.

Taylor, G. R. (1970). *Satellite cities: A study of industrial suburbs.* New York: Arno Press. (Original work published 1915)

Vance, Jr., J. E. (1964). *Geography and urban evolution in the San Francisco Bay Area.* Berkeley: University of California, Berkeley, Institute of Governmental Studies.

Warner, Jr., S. B. (1962). *Streetcar suburbs: The process of growth in Boston, 1870–1900.* Cambridge, MA: Harvard University Press/MIT Press.

Warner, Jr., S. B., & Whittemore, A. H. (2012). *American urban form: A representative history.* Cambridge, MA: MIT Press.

Wilson, D. (Special Ed.). (1997). Globalization and the changing U.S. city. *Annals of the American Academy of Political and Social Science, 551,* 8–247.

CHAPTER 4

Impacts of Information and Communication Technology

GIOVANNI CIRCELLA
PATRICIA L. MOKHTARIAN

Information and communications technology (ICT) applications have rapidly changed work organization and social habits in most developed countries as well as in the developing world. The term ICT can embrace numerous and diverse applications: in this chapter we refer primarily to goods and services involved in the production, collection, storage, analysis, and/or transmission of information (whether audio, visual, textual, haptic, or otherwise) in electronic form, ranging from end-user devices such as laptop computers, smartphones, and other Internet-enabled tools, to the network services that move information among users (and/or businesses).

Numerous studies have analyzed the impact of specific technologies in different geographic contexts and in relation to various aspects of urban and regional planning, work organization, and individuals' activity scheduling and travel behavior. The diverse impacts of ICT can act in different directions and produce counteracting effects: for instance, at first a common hope was that information and communication technologies would significantly reduce travel (and its concomitant effects such as energy consumption and emissions) by substituting ICT for physical trips. However, the reality is often more complex. As just one example, the commuting trips that are eliminated by telecommuting may free up time for making other non-commuting trips (which might be made by car instead of transit and therefore use more energy than the commuting trips they replace). For telecommuters who work at home only part of the day, the commute trip remains, and energy consumption may increase because of the additional use of home heating/cooling systems (Kitou & Horvath, 2008). Overall, and as discussed later in greater detail, ICT can have mixed effects on transportation and even lead, under some circumstances, to an *increase* in total travel (Mokhtarian, 2009a). Plate 4.1 visually summarizes the potential interactions among economic activities, ICT supply and demand, and transportation.

In this chapter, we start from the *aggregate* level of the metropolitan region, and discuss how ICTs may alter the urban form of cities through relaxing some of the time/ space constraints that regulate the location of economic activities and residences. ICTs also contribute to changing the organization of the work environment, including office and retail spaces, and more generally affect the organization of cities and regions and the logistics of goods distribution. The second part of the chapter is organized around the *individual* actor, and examines some ways in which ICTs affect people's travel-related decisions. We first discuss the role of technology in affecting *long-term* decisions, for example, where to live, and then address *medium-term* decisions, for example, whether to buy a car. Finally, we explore how ICTs are revolutionizing many aspects of *short-term* travel behavior, such as activity participation and other aspects of trip making.

Other related topics that are not the focus of this chapter are worth highlighting in this introduction. One of these is the important role of online social networks in modern society. Networks such as Facebook, Twitter, and LinkedIn have contributed to restructuring time use and interpersonal relationships, and they affect countless choices that people make every day. The expectation of being constantly "connected" is revolutionizing many familiar aspects of social and economic life. However, the impacts of these networks on the mobility of people and goods are still unclear.

ICTs are also reshaping many aspects of political and social life, such as facilitating greater political and civic engagement among younger generations. Telecommunications have increased the speed at which news reaches voters and activists, changed the political agenda of political parties and leaders, and affected the way leaders communicate. Online social networks (as well as the continuous streaming of news and commentary over the Internet) have established their place in the political arena. Similar changes have happened in the economy, where those businesses that have more easily adjusted to an Internet-based reality have reinforced their presence in the market, while other players have gone out of business or are slowly disappearing.

ICTs have forever changed our society in ways that were barely imaginable only a few years ago, and quantifying the impacts of these changes on urban transportation is not easy. For one thing, the process is still ongoing, and its future results are only partially foreseeable. For another, as we have already seen, impacts arise from numerous sources and can act in opposing directions. Finally, it is difficult to identify the effects specifically attributable to ICT in the complex relationships among urban systems, transportation, and individuals' behavior, and to separate them from other ongoing modifications, such as the impact of economic development, or the concurrent changes in individual preferences and lifestyles observed in many U.S. cities with the resurgence of downtown areas and the revitalization of denser urban neighborhoods (Wachs, 2013). Thus, although we outline some key trends in this chapter, inevitably the continued evolution of technological products and services will lead to additional impacts that are not yet discernable.

ICT AND THE SPATIAL FORM OF CITIES

The current form of urban areas is the result of complex interactions among the numerous needs that influence the location of economic activities and residences.

Living in cities allows individuals to save on transportation costs and provides diversity in the offering of goods and services (e.g., access to theaters, sport centers, restaurants, libraries, stores), facilitating communication and social relations. Similarly, it allows firms to benefit from agglomeration economies, minimize the costs of transporting people and goods, and ensure proper access to information and connection with commercial partners, consumers, and suppliers (O'Sullivan, 2011). The organization of cities is not always efficient in all its dimensions (e.g., as attested by the congestion levels in many urban areas), but nonetheless cities continue to maintain a strong attractive power because of the benefits they provide.

Technology has gradually modified the distance constraints that limit the mobility of goods and people in two ways. First, it has allowed for an increase in travel speed through the development of new means of transportation and/or the improvement of road and rail infrastructures (see Chapter 3). This has led to a contraction of the average travel times needed to reach specific destinations (a process known as *time–space convergence*) (Janelle & Gillespie, 2004) or alternatively to an expansion of the set of destinations that can be reached in a specific unit of time. Second, new technologies have helped make travel cheaper, reducing the costs of moving goods and people over longer distances and making it practical to ship goods farther (or to travel longer distances, for passenger trips) than in the past. Over years, these factors have allowed cities to expand beyond previous limits. The relationship between technological development and urban form has been studied in terms of *technology–land substitution* (Kim, Claus, Rank, & Xiao, 2009). Technological innovations can facilitate greater efficiencies in the use of land, as happened in the process of urbanization in the first half of the 20th century, or they can improve accessibility, reducing the friction of distance and favoring low-density urban expansion, as happened in the second half of the last century.

Despite a consensus that ICTs are capable of further lifting space–time constraints, disagreement exists about the extent of that relaxation (Janelle, 2012; Nijkamp & Salomon, 1989). The most extreme position holds that ICTs portend the "end of geography" (O'Brien, 1992) and the "death of distance" (Cairncross, 2001). *Physical proximity* is a less binding constraint in the decision of where to locate economic activities and/or residences at a time "when dominant forces such as globalization and telecommunications seem to signal that place and the details of the local no longer matter" (Sassen, 2000, p. 144). Relaxation of these constraints would favor a further expansion of urban boundaries, with a reduction of densities encouraged by the non-importance of proximity. Firms and residents would move to places where land (and labor) is cheaper or amenities are more attractive. Because lower densities are associated with greater distances traveled (Ewing & Cervero, 2010), an overall decrease in urban density would further increase the dependence on private vehicles for personal mobility, thus at least partially counteracting the positive effects of ICT as a potential substitute for transportation.

Audirac (2005) summarizes many studies of the dominant impact of ICTs on urban form. She notes that the impact of ICT on urban form could involve several counteracting aspects, which might eventually lead, depending on local conditions, to (1) *decentralization,* for example, through the reduction of density in the city center of an urban area; and/or to (2) a *redistribution of activities* such that population and employment densities increase in lower-density and less-populated areas, and specialized financial and information technology (IT) districts emerge in specific regions (Audirac, 2005; Karlsson et al., 2010).

The effect of ICT on cities could vary significantly with the *scale* of the effects that are observed (as noted earlier by Nijkamp & Salomon, 1989). For instance, ICT reduces the need for material flows or physical travel to destinations, diminishing the need for proximity of back-offices to the headquarters of large corporations. Back offices can then move farther away from central business districts to locations where land is cheaper, or even to remote locations outside of metropolitan areas or to other countries. If this process has significant effects in large metropolitan areas, where rents are high and the pressure to reduce space is strong, the same effects are not necessarily expected in smaller urban areas, where the inertia associated with relocation changes may prevail and the adoption of telecommunications may lead to slower relocation processes or to no restructuring at all. In addition, even if the possibility to substitute in-person meetings and exchange of information with a technological interaction "at a distance" is a benefit of ICT, the need to meet in person and to exchange goods physically still exists. Agglomeration economies and urban amenities retain their attractiveness to businesses and residents. Furthermore, ICT has considerable potential to generate *additional* needs for face-to-face interaction (making concentration more attractive), as increased communication and virtual interactions can increase awareness of opportunities that require travel to fulfill (Mokhtarian, 2009a). Finally, even if ICT relaxes some of the physical constraints on the location of firms and residences, this mainly affects the location of new firms or residences. The relocation of existing businesses or residences will also depend on the prior geographical distribution of activities and resources, real estate prices, and land availability, all of which often cause additional costs and inertia for this relocation process, at least in the short run.

According to some authors, at the metropolitan scale one expected outcome of ICT is to weaken (but by no means eradicate) the localized agglomeration economies that once kept cities more centralized, thereby leading to a polycentric network (a *regional constellation*) of agglomerations that are interconnected via high-speed transmission and digital networks (Audirac, 2005; Tranos, Reggiani, & Nijkamp, 2013). The few places left outside the network and communication grid (e.g., because of topography and/or the high investments required for connectivity, as in remote or mountainous regions) would, however, be increasingly disadvantaged by the lack of access to transportation or communication solutions.

Thus, the electronic revolution is not erasing the value of place (Horan, 2000). Certainly one of the effects of ICT on urban form is the additional *virtual* accessibility afforded to places in addition to the *physical* accessibility afforded by their transportation systems. Telecommunication networks are neither homogeneous nor ubiquitous, and access to them becomes another important factor to be considered in location decisions. Individuals and organizations need to have easy access to electricity, wired or wireless networks, and fast connections for data transfer, as well as to hardware (e.g., cell phone devices, computers, tablets), software, and services (subscription to the needed services, data plans, etc.). Accordingly, the need to access a reliable network—whether ICT or transportation—disadvantages areas that are not adequately served by such networks.

Decentralization has many causes, and the post–World War II acceleration of preexisting trends in that direction certainly predates the advent of modern ICTs. Decentralization has even slowed down in many American cities that have been experiencing a resurgence of downtowns and central areas (Wachs, 2013). Nevertheless, it seems reasonable that some portion of the responsibility for current decentralization does belong with ICT. The case of telecommuting and residential location helps illustrate the causal complexities. If telecommuting allows workers to live farther from work than would be desirable otherwise (Mokhtarian, Collantes, & Gertz, 2004), would an increased ability to telecommute motivate them to do so? Or, on the contrary, might telecommuting provide a solution to longer commutes from a residential location they chose for other reasons?

With the rapid expansion of the availability of telecommunications and ICT-based alternatives to physical activities, ICT has a huge potential to continue reshaping cities in both the short and the long terms. The dominant impacts of ICT on urban form are likely to be a combination of centrifugal or centripetal: ICTs reinforce the importance of cities as the centers of production and consumption while reducing the need for physical proximity.

Impact of ICT on Work Organization

ICT has contributed to a general reorganization of business activities and of the work environment almost everywhere in the increasingly connected world. ICT has permitted the reduction of work space devoted to material objects and people (thanks to the miniaturization or even *dematerialization* of hardware, files, and pieces of information), substituting for many functions that in the past occupied large amounts of space, for example, libraries, server/computer rooms, and storage areas (Kellerman, 2009).

However, rather than necessarily reducing total office space per se, ICT contributes to the reorganization of work activities through a more efficient use of space. It also enables the physical separation of high-end functions, such as central headquarters and management offices, from more routine functions, such as support/technical offices, thus optimizing location choices. Back offices, production, and logistical centers no longer need to be located near headquarters and front offices but can move to cheaper locations outside the central city, where access to the relevant transportation networks (mainly freeways and airports) is also better (Graham & Marvin, 2001). By loosening the effects of distance, ICTs can support the rise of spatially-distributed teams and enable spatial separation between the manufacture and the end use of goods. This process, often criticized for its negative impact on local employment, allows many businesses to reduce their direct workforce through thinner (and "smarter") organizations, while outsourcing some jobs to a network of suppliers and remote providers. Many businesses have relocated to developing countries, where labor is cheaper and regulations are weaker, retaining only higher-salary and more strategic white-collar jobs (such as headquarters staff, legal offices, and research and development) in their home country (Wachs, 2013). Cheaper transportation, real-time communications, and advanced logistic and distribution chains allow efficient communications and on-time delivery of these products with higher economic profitability.

A number of firms permit widespread *telecommuting* (e.g., working remotely from supervision, usually from home or a location close to home) even if this practice is not universally accepted, mainly owing to concerns about individual and team productivity. In addition, telecommunications increase connections among colleagues, and between businesses and clients, supporting the organization of team work even among physically separated team members in geographically dispersed *multiunit organizations* (Denstadli, Julsrud, & Hjorthol, 2012).

What are the implications of these changes for urban transportation? First, despite the role of ICT in promoting virtual accessibility over physical proximity, it seems that a central location, physical interactions, and travel are still important for many businesses because agglomeration economies and face-to-face interactions remain important. Many work activities are still dominated by personal interactions (in-person meetings, business linkages, conference calls, exchange of information), particularly in business units involved in negotiation-type interactions, which are less amenable to telecommunications (Nijkamp & Salomon, 1989). Furthermore, for some firms, central locations allow them to access specific pools of workers and proximity to client firms (Giuliano, 1998). The new agglomerations of ICT-oriented districts (e.g., Silicon Valley), which allow firms to interact and share an integrated pool of workers and local suppliers, bear witness to the benefits of agglomeration economies even in the ICT era (Karlsson et al., 2010). For firms that are less dependent on face-to-face interactions, more flexibility in office location and production sites is possible, and the focus shifts to relevant transportation hubs and subcenters. The location of production activities in less central areas and near suburban transportation hubs brings development to peripheral areas (Giuliano & Small, 1999; Wachs, 2013), which are better served by airports and freeways (and by railway, for freight transportation), but less easily served by public transportation. The increased distances between central business districts (CBDs) and peripheral production centers puts additional pressure on transportation networks, as it increases the reliance on private mobility.

Impact of ICT on Retail Organization

Modern ICTs are changing the urban geography of commercial and retail activity. The Internet has given birth to new ways of buying and selling (including consumer-to-consumer, business-to-consumer, and business-to-business) that were simply not possible before. Furthermore, new technologies can support the recent trend, at least among some segments of society, toward greater *sharing* of durable goods (such as *cars*) rather than private ownership, toward purchasing used items rather than new, toward bartering rather than paying money, and in general toward a reduced level of materialism (a phenomenon referred to as the "sharing economy" or "collaborative consumption"; Belk, 2014).

On the consumer side, ICT helps relax time and space constraints, which affects the organization of individuals' schedules and consequently the spatial distribution of economic activities and residences. But does it also help overcome physical marginality and reduced accessibility? Traditionally, cities have facilitated agglomerations of activities that help overcome time constraints by minimizing distance constraints:

the large concentration of activities in the most central areas increases their *cumulative attraction*. It allows retailers to increase their volumes of sales, and it allows customers to satisfy more needs while minimizing travel distances. Already in 1996, Mitchell (p. 89) reported that "salesperson, customer, and product supplier no longer have to be brought together in the same spot; they just need to establish electronic contact. . . . A geographically distributed, electronically supported consumer transaction system completely replaces the traditional retail store." Accordingly, the advent of virtual retailers such as Amazon.com, which trade solely online, has become a serious threat to the survival of traditional retailers. Online retailers can sell products at lower prices than traditional store retailers because they have lower operating costs, are open 24 hours a day, and are easily accessible to everybody who is familiar with Internet-enabled devices in any location and any weather, overcoming physical disabilities or other limitations on physical travel to stores (Burt & Sparks, 2003; Weltevreden, 2007).

Online shopping has grown steadily over recent years, and the dollar amount it comprises is non-negligible and has continued to grow even during an economic recession, increasing from $138 billion in 2007 to $305 billion in 2014 in the United States. The monetary amount of online shopping sales and the share of total sales are likely to continue to increase relative to the economy as a whole, increasing pressure on traditional retailers. Hybrid forms of retail blending both "bricks" and "clicks" have also emerged, such as physical shops in which wares can be handled and tested, with ICT enabling the rapid customization and delivery of the final product.

Many studies have investigated the impact of ICT, and e-shopping in particular, on traditional retail activities and individual travel behavior. Overall, the impact of ICT on retail activities might involve a number of partial impacts, including effects of *substitution* (replacing trips to a store with e-shopping), *complementarity* (generating additional trips to stores, e.g., to touch, try, and/or buy items seen online), *modification* (adjusting the patterns of preexisting trips), and *neutrality* (no significant impact) (Mokhtarian, 2004; Weltevreden, 2007). Most studies have addressed the impact of ICT on the number of trips to stores (cf. Cao, Xu, & Douma, 2012), with the aim of verifying whether a *substitution* effect exists between online and in-store purchases. The empirical evidence is mixed, finding substitution as well as generation effects, and highlighting the need for further research to overcome a number of methodological and data limitations. For example, most studies do not distinguish clearly enough between impacts resulting from *browsing* the Internet versus those arising from *purchasing* online; similarly, they often fail to identify specific combinations of bricks and clicks, such as the "showroomers" or "free riders" (Couclelis, 2004; Rapp, Baker, Bachrach, Ogilvie, & Beitelspacher, 2015), who use stores for testing products and obtaining advice but then purchase online from different retailers to obtain the cheapest possible price.

There is no doubt that ICT-related shopping channels are better suited to sell some types of *products* or services than others: for example, the sales of retail categories such as books, music, videos, software, travel, and event tickets have quickly declined in traditional stores, causing many physical establishments that specialized in those categories to go out of business. Not only has the standardized nature of these products made it easy to purchase them without having to try them first;

ICT has also (unlike the case for other standardized products such as canned food) changed the very nature of the product itself. It has led to the *dematerialization* of many goods and products, which in some cases (e.g., recorded music) have virtually disappeared altogether.

For some categories such as large appliances and other relatively standardized higher-cost goods, online price comparisons will be inevitable, but physical stores have advantages such as same-day delivery and more personal pre- and post-sale relationships, while for other categories of goods (e.g., clothing), it is possible that ICT will have a neutral or even positive impact on city centers and traditional retail stores (Weltevreden, 2007).

Significant differences might exist in the overall impact of ICT on retail shopping, as a result of personal and geographical factors (Weltevreden & van Rietbergen, 2009). Spatial factors, such as differences in settlement patterns, access to the network and quality of available ICT services, and remoteness of a rural area, certainly affect the opportunities to connect to the Internet as well as the accessibility to retail shops. Consequently, online shopping might play a dual role with respect to geographic location. According to the *innovation–diffusion hypothesis,* people living in urban areas, and therefore with greater store accessibility, may be more likely to e-shop because they are in a better position to know about and adopt innovations. For example, Weltevreden (2007) found that the *penetration* of e-shopping is higher in urban areas of the Netherlands, where people are more technologically savvy and have better access to ICT. However, the *frequency* with which users buy online was found in the same study to be higher in more remote areas where accessibility to local stores is low, in keeping with the *efficiency hypothesis* (Weltevreden, 2007). In other words, people with lower store accessibility might be more likely to e-shop in order to save time: for example, Sinai and Waldfogel (2004) found a more important role of e-shopping among residents located at greater distances from the closest stores. Not surprisingly, the impact of e-shopping is also lower in cities that are served by a better transportation system and in which physical accessibility to stores is higher. Similarly, thanks to ICT, small businesses can increasingly access a larger number of remotely located (or dispersed) consumers that could not be reached by traditional retail stores. Overall, though, other factors like sociodemographic characteristics or attitudinal factors affecting an individual's propensity to engage in e-shopping are believed to dominate over geographical location (Krizek, Li, & Handy, 2005).

Certainly, e-shopping has already caused a complex transformation of retail activities. Some retailers (e.g., bookstores) have been heavily affected by the expansion of ICT-based retail services, virtual stores, and e-commerce, with a stronger decline in activities and retail space. More generally, traditional stores have reacted in various ways. First, retail organizations have integrated online services into their traditional retail business: nowadays, in the United States, no large retailers and few small ones are without an Internet presence, as retailers have adapted their business model toward a combination of "bricks and clicks." Second, an important *adaptation* strategy of retail stores involves their transformation into *entertainment centers* to fully exploit the social elements that are considered a significant component of traditional in-store shopping. The mix of retail, cultural, and other leisure facilities can make shopping areas in city centers, denser areas, and some modern malls more attractive than their virtual alternatives because of the additional amenities offered, including restaurants, cafes, clusters of ethnic- or other-themed shops and services, and other options for social aggregation (the Mall of America, near Minneapolis, Minnesota, is one example, and the largest in size, of such bundling of activities in a modern, multipurpose, commercial mall). Finally, ICTs offer opportunities to boost sales for local businesses, for example, through location-based marketing, that is, information and purchase incentives sent to consumers based on their real-time location. All told, the impacts of ICT on retail activity are diverse and constantly shifting, and it is not possible at this time to neatly summarize a net outcome.

Impact of ICT on Goods Distribution

ICT applications increase the efficiency of goods manufacturing and distribution, and they alter supply chains in many ways. As they lower production and transportation costs, they tend to increase the demand for the associated goods, which in turn generates increased demand for transporting the raw ingredients and finished products, encouraging the distribution of even rather inexpensive goods over long distances. ICTs are also introducing important modifications into the transportation system itself, increasing the advantages associated with the economies of scale and the distribution of goods through large *hub-and-spoke* systems, which allow greater efficiency thanks to their reduced handling costs and an increased level of automation. For example, the massive systems adopted by large shipping companies, express couriers like UPS or FedEx, and traditional postal services all rely on ICT for the automation of transshipment centers.

ICT, and the adoption of e-shopping in particular, has also facilitated a gradual modification in the end-users' level-of-service expectations and generated important modifications in the delivery of goods in urban areas (see Chapter 2). The increased desire for speed, often associated with the use of online shopping, contributes to shifting consumers' preferences toward faster express delivery systems. Similarly, on the supply side, the just-in-time (JIT) supply chain principle made possible by ICT can lead to more frequent deliveries at less-than-truckload (LTL) capacities, and/or the increased reliance on energy-intensive and superior-service modes.

IMPACTS OF ICT ON DISAGGREGATE TRAVEL-RELATED DECISIONS

The first half of this chapter focused on the large-scale spatial impacts of information and communication technology. We turn now to the perspective of the individual decision maker. You need look no farther than yourself to observe the profound influence of ICT on nearly every aspect of life, including your travel-related decisions (e.g., Figure 4.1 shows the rapid growth between 1984 and 2014 in the number of U.S. households that use various types of technologies at home). In this section, we discuss the impacts on *longer-term decisions,* such as migration and residential location, *medium-term decisions* on the ownership and use of various means of transportation, and everyday *shorter-term* decisions on activity engagement and scheduling.

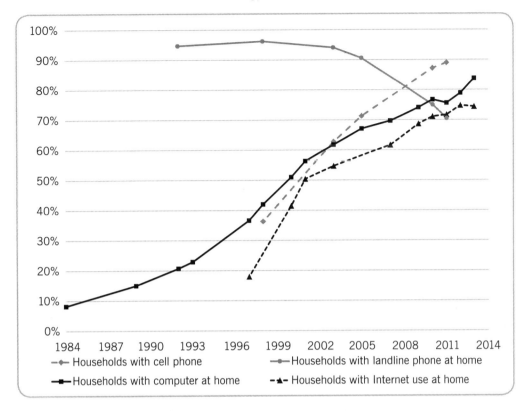

FIGURE 4.1. Percentage of U.S. households using various types of technologies at home: 1984 to 2014. Source: From *Transportation Research Part A, 41*, by S. Choo and P. L. Mokhtarian, "Telecommunications and travel demand and supply: Aggregate structural equation models for the US," pp. 4–18, Copyright 2007, with permission from Elsevier.

Impact of ICT on Long-Term Decisions

Migrations are among the most important long-term decisions in an individual's lifetime and have significant implications for other transportation-related decisions. Usually, they are the result of a strategy to obtain higher income and/or better employment opportunities and career prospects (apart from the very different reasons motivating migrations in cases of war or persecution, which we do not discuss here[1]). Migration decisions are usually well-informed rational choices among economic opportunities at various locations (Yankow, 2003), which are influenced by interpersonal networks, environmental considerations, and location-specific contexts.

ICTs play an important role in all major phases of the migration process (Vilhelmson & Thulin, 2013). They can reinforce a migration propensity, playing the role of an *enabler* or *catalyst*, with different impacts depending on the level of familiarity with the communication technology, as well as the geographic context, cultural and educational background, and sociodemographics. They can also change the spatial preferences and migration patterns of individuals. The internet and related ICT applications provide *augmented awareness* and *increased knowledge* of new opportunities and conditions during the preparation for a potential migration, providing access to information on job or educational opportunities, locations (e.g., distant cities and regions) and processes (e.g., immigration laws). Thus, they reduce the *friction of distance*. They may even stimulate migration in individuals who would not otherwise have considered this option, affecting personal tastes and decisions through multiple composite impacts (Stevenson, 2009; Vilhelmson & Thulin, 2013).

The widespread adoption of mobile phones, e-mail messages, instant messaging (IM), live chat, and online social networks enables a socially and spatially extended network of friends, family, acquaintances, and colleagues. ICTs thereby reinforce organizational and interpersonal ties that connect local labor markets and channel immigrants to particular destinations and into particular occupations. These tools, in addition to the increased availability of low-cost air and intercity bus services, help nourish old ties between immigrants and their places of origin and develop new ties in their new places. They also support temporary migration decisions (e.g., for education or temporary employment), in particular among the young, the well educated, and the most dynamic segments of the population, and increase the level of satisfaction among these connected migrants (Komito, 2011).

Similarly, ICT increases the flexibility of choices and reduces physical constraints associated with both the residential location *processes*, by which individuals make their choices related to residential location, and their *outcom*es, that is, the residential location that is chosen. With respect to the process, the increased social networking enabled by ICT influences all stages, from the collection of preliminary information about a new region and city and the search for information about the real estate market and available units, to the prescreening of possible alternatives, comparison of prices and amenities, exchange of information with a real estate agent and the seller/landlord, and completion of the final transaction. ICT enlarges the range of available options and allows more members of a family, group, or social class to share information and provide feedback, perhaps eventually prompting a relocation close to members of their network, or conversely avoiding the mistakes committed by others. Furthermore, it reduces the number of trips needed to personally visit neighborhoods of potential interest and increases the negotiating power of individuals through providing knowledge of the true prices for the entire population of housing units on the market and those recently sold/rented.

With respect to the outcome, ICT allows more flexibility in the choice about *where* to relocate, for instance, through enabling telecommuting (Shen, 2000). The overall impact of telecommuting on residential location is not easy to investigate: most studies focus on travel patterns observed in the short term, while the impact on residential location generally becomes evident only after several years. It is reasonable to expect that telecommuting (and ICT in general) usually acts as a catalyst, not as a motivating factor per se, making possible a move that is prompted by other reasons, such as accommodating other household members' needs; preferences for larger, cheaper, or simply different lots or housing units; or the desire to be located near vibrant parts of a city/region or other amenities (Mokhtarian et al., 2004; Nijkamp & Salomon, 1989). Still, "to the extent that telecommuting is responsible for releasing a constraint that was preventing further decentralization . . . telecommuting is arguably as culpable as if it were a driver in its own right" (Mokhtarian et al., 2004, p. 1879). In these terms, telecommuting can be considered a potential "cause" of residential relocation.

Impact of ICT on the Ownership and Use of Automobiles and Other Means of Mobility

ICTs directly affect auto-ownership decisions in a number of ways. Even if internet-based e-shopping has only a limited (but increasing) role as a channel for direct sales of vehicles, ICTs are already a central source of information for (and contribute to modifying individual tastes and preferences of) many buyers, presenting experts' opinions and reviews, offers from manufacturers and dealers, and users' experiences and reviews. Similarly, information exchanged with peers through social networks can provide inspiration and information about the adoption of specific automotive technologies (Axsen & Kurani, 2011).

More directly, ICT is central to the growth of carsharing and dynamic ridesharing programs. A common presence in many U.S. and European cities, carsharing offers the availability of a car when needed, without the higher costs associated with the ownership of a private vehicle. Carsharing companies use ICT to manage fleet distribution, usage, and maintenance while customers use an online interface (traditional computers, or smartphone and tablet apps) to check for availability and reserve the desired vehicle at a specific time. Similarly, on-demand ride services like those introduced by Uber or Lyft use ICT to match riders with available drivers in real time. Carsharing and on-demand ride services can affect mode choice as well as car ownership decisions although their overall impacts are not clear yet and may largely depend on local synergistic effects with additional policies and land use variables, such as the availability of public transportation alternatives and land use features in each urban area. For example, carsharing may reduce auto ownership, car trip rates, and travel distance, owing to the increased direct costs and disutility of accessing a vehicle each time there is a need for a trip in areas where other transportation alternatives are feasible. But it might increase the number of vehicles on the road to the extent it provides a lower-cost alternative to individuals and households who could not have afforded a (second) car. Early adopters of carsharing services tended to be higher-income individuals, who often reported car disposal or postponement or complete avoidance of a car purchase to fulfill their mobility needs (Shaheen, Mallory, & Kingsley, 2012). However, such early adopters may not be typical of later entrants to the carsharing market. Meanwhile, the rapid growth of on-demand ride services has disrupted the use of classic taxi services as well as the activity and travel scheduling of many users by providing extended options for short-distance trips (Rayle, Shaheen, Chan, Dai, & Cervero, 2014).

ICTs are central to a variety of other mobility means-sharing approaches that can be adopted throughout the travel-distance spectrum. Bikesharing allows reducing automobile use and it extends the coverage and catchment area of public transit in major cities, making those modes viable for more users. At the longer-distance end, jetsharing services such as Wheels Up have started to proliferate.

Finally, ICT also has an important and growing role in assisting drivers with vehicle operation. Modern vehicles possess a large number of ICT-based features, for entertainment, navigation, telephony, computing, diagnosis, and safety/security. These features are increasingly considered "required" components of the vehicles, rather than optional accessories. ICT-based safety/security systems, in particular, can simplify the operation of vehicles for users with some forms of impairment, thereby extending the independence of elderly and other drivers (Guo, Brake, Edwards, Blythe, & Fairchild, 2010).

At the time of this writing, completely automated vehicles (e.g., Google cars) have already been deployed in on-road testing and are expected to become a commercial reality in the medium-term future. The impacts of such a disruptive technology will be far-reaching and difficult to predict. They will depend heavily on technological, economic, regulatory, legal, and social considerations. On the positive side, autonomous (driverless) vehicles could reduce accidents, smooth traffic flow, decrease fuel consumption rates, increase the effective capacity of the system, and extend the benefits of personal transportation to those who are unable to drive. On the other hand, through lowering the time cost of travel by allowing other activities to be conducted while traveling, as well as through relaxing licensing requirements, they may increase the demand for auto travel, perhaps taking share from public transportation or active modes.

Impact of ICT on Day-to-Day Activity and Travel Decisions

ICTs as a trip replacement strategy have been seen as a solution to many societal problems, including urban congestion, dependence on non-renewable energy sources, air pollution, and greenhouse gas emissions, as well as rural underdevelopment, reduced economic opportunity for the mobility-limited, and the struggle to balance job and family responsibilities (Mokhtarian, Salomon, & Choo, 2005; Salomon, 1998).

ICTs certainly do replace "a lot" of travel, but at the same time, they can generate additional travel as well. ICTs can influence an individual's space–time constraints (for a discussion of Hägerstrand's space–time constraints, see Chapter 1) and the resulting activity participation and travel behavior (cf. Dijst, 2009; Schwanen & Kwan, 2008) in a number of ways (see Table 4.1), including imposing new constraints as well as relaxing some old ones. As with e-shopping, the interaction between ICTs and travel behavior can include several possibilities (see Figure 4.2): ICTs can have no relevant effect on travel (*neutrality*), generate new travel (*complementarity*, or *stimulation*), alter travel that would have occurred anyway (*modification*), or reduce travel (*substitution*) (Salomon, 1986; Salomon & Mokhtarian, 2008).

TABLE 4.1. The Impact of ICTs on Hägerstrand's Constraint Categories

Constraint	Definition (Dijst, 2009)	Example of Relaxation by ICT	Example of Imposition by ICT
Capability	Biological, mental, and instrumental restrictions	ICT-facilitated multitasking "creates time"	Lack of computer/Internet literacy can increase social exclusion
Coupling	Synchronization and synchorization of individuals, instruments, and materials	Don't have to be at a fixed location to make or receive a phone call	Must have physical products and services in place (phones, electricity, transmission equipment, etc.)
Authority	Regulation of access to space	Don't have to shop when the store is physically open	Prohibition on mobile phone use under certain circumstances (in theater, while driving, in flight, etc.)

Source: Mokhtarian (2009b).

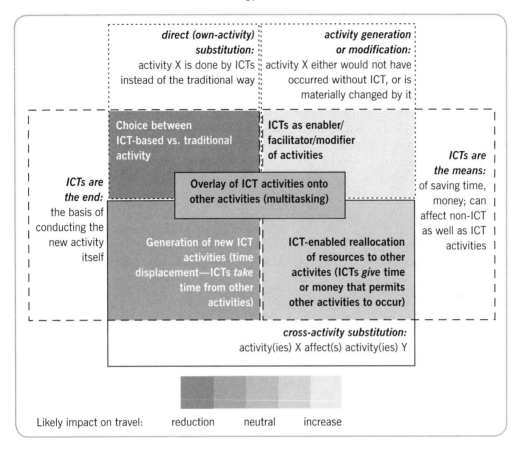

FIGURE 4.2. Interactions among ICT and activities, and likely impacts on amount of travel. Source: Adapted and expanded from *Transportation*, "The impacts of ICT on leisure activities and travel: A conceptual exploration," Vol. 33, 2007, pp. 263–289, by P. L. Mokhtarian, I. Salomon, and S. L. Handy. With permission from Springer. Also see Salomon and Mokhtarian (2008).

Mokhtarian (2009a) discusses a number of reasons that favor complementarity as the dominant impact. She first notes that *not all ICT-based activities reduce travel* because (1) not all activities have an ICT counterpart; (2) even when an ICT alternative exists in theory, it may not be practically feasible; (3) even when feasible, ICT is not always a desirable substitute; (4) in particular, travel carries a positive utility in its own right, not just as a means of accessing specific locations; and (5) not all uses of ICT constitute a replacement of travel. She then presents several reasons why *ICT actively increases travel.* In the short run, (6) ICT saves time and/or money for other activities, some of which may involve travel. Specifically, (7) it permits travel to be sold more cheaply (e.g., through last-minute airline deals, or airfare and hotel bundles); (8) through Intelligent Transportation Systems (ITS) and other technologies it increases the efficiency (and thus the effective capacity) of the transportation system, making travel less costly and therefore more attractive; and (9) personal ICT use can increase the productivity and/or enjoyment of travel time, thereby also increasing the attractiveness and/or decreasing the disutility of travel. Importantly, (10) ICT directly stimulates additional travel, through its ability to inspire and to facilitate transactions (among other ways).

In the long run, (11) ICT is an engine driving the increasing globalization of commerce; and (12) it facilitates shifts to more decentralized and lower-density land use patterns.

On the other hand, *ICT may also reduce travel* in the following ways: (1) it may directly substitute for making a trip; (2) it consumes time (and/or money) that might otherwise be spent traveling; (3) when travel becomes more costly, difficult, or dangerous, ICT substitution increases; (4) it can be deployed to make shared means of transportation more attractive (reducing drive-alone trips); and (5) it can reduce unnecessary travel (such as when "letting your fingers do the walking" prevents driving to several stores to look for an item) (Mokhtarian, 2009a; Mokhtarian & Tal, 2013).

In the following paragraphs, we sketch some likely impacts of ICT on selected trip purposes.

Business Travel

The expectation that ICT would substitute for business travel dates back at least to the invention of the telephone and has been reinforced by the potential attributed to videoconferencing for eliminating trips, thanks to its time- and cost-saving advantages. ICT can also allow contact to be maintained when travel becomes dangerous, difficult, or impossible. Still, business travel has continued to grow steadily during the years of widespread ICT. As Mokhtarian (2009a, pp. 8–9) notes, business people "travel ever farther and more frequently to develop new clients and serve existing ones, . . . inter-firm collaborations and geographically dispersed project teams have increased over time, cheaper labor and raw materials make it cost-effective to transport them from farther away, and the worldwide customer base created through internet-based marketing as well as more conventional channels generates greater travel in the transport of finished products to the consumer." A number of scholars have commented on the continued need for face-to-face interaction despite the increasing effectiveness of ICT as a substitute. Research also suggests that videoconferencing and ICT in general substitute for some types of face-to-face interactions (Denstadli et al., 2012), e.g., intracompany business trips, while they do not substantially reduce other types of business trips.

Commuting

Many studies have examined the impact of telecommuting on travel. For people who must live somewhere because of a spouse's job, for example, telecommuting may allow the opportunity to accept (or keep) a job he or she may not have considered feasible otherwise or to reduce the amount of commuting he or she would otherwise have had to do. Similarly, the location-independence offered by ICT has given rise to new types of jobs (e.g., full-time telecommuting IT programmers) that were not possible before. The adoption of telecommuting has increased in recent years in the United States. According to the U.S. Census, the share of working at home as the usual "means of transportation to work" has grown from 3.3% in 2000 to a 4.4% average for 2010–2014 (*www.census.gov/population/www/cen2000/briefs/phc-t35/ tables/tab01-1.pdf* and *http://factfinder.census.gov/faces/tableservices/jsf/pages/ productview.xhtml?pid=ACS_13_5YR_S0801&prodType=table*, both accessed

October 13, 2016). These numbers include home-based businesses and exclude less-frequent telecommuters. A measure of the intensity (not just adoption) of telecommuting is obtained from the 2015 American Time Use Survey, which shows that among the 96.6 million wage and salary workers (i.e., not including the self-employed in unincorporated businesses) who worked on a given day (including weekends), 15.0% (14.5 million) worked only at home (computed from *www.bls.gov/news.release/atus.t07.htm,* accessed October 13, 2016), while an additional 5.1% (4.9 million) worked at home for part of the day in addition to physically traveling to their workplace. The average number of hours worked at home across both groups, however, was only 3.1. Overall, teleworkers can include a number of distinct groups, such as *substituters* (the classic "telecommuters" who substitute some of their commuting trips with telecommuting), *supplementers* (who work at home in addition to the regular hours worked in the office or other work location), *telecenter-based workers* (a less-common category after the spread of modern technological devices), *remote back-office workers, home-based business workers, field workers, mobile workers, long-distance telecommuters,* and *distributed team members,* among others (Mokhtarian & Tal, 2013). Most empirical studies of the impact of telework on transportation have focused on the classic type of telecommuter, the substituter, for whom ICT does generally replace travel. But the travel impacts for the remaining categories can differ dramatically, from no impact at all for supplementers, to ambiguous impacts for home-based business workers (as it depends on whether their alternative is working full- or part-time at a job requiring commuting, or not working at all), to increases in travel for many mobile workers and others.

The literature debates whether telecommuting might indirectly generate more travel, through prompting employees to move farther away from work since they do not have to commute as often. Telecommuters do live farther from work than otherwise similar non-telecommuters, on average, but that could be the *cause* of their telecommuting rather than the *effect* of it. Whatever the dominant direction of causality, telecommuting can compensate for those longer one-way commutes (Mokhtarian et al., 2004; Muhammad, Ottens, Ettema, & de Jong, 2007). To properly assess the aggregate impacts as a whole, more comprehensive studies should investigate the impacts of all forms of teleworking. So far, the evidence indicates that telecommuting reduces travel, even if only to a small degree (Choo, Mokhtarian, & Salomon,, 2005; Helminen & Ristimäki, 2007). The impact specifically on *peak-period* travel is probably larger but is difficult to document since traffic reductions of any noticeable degree are likely to be met by the "triple convergence" of shifts in mode, route, and departure time to partly or completely counteract any freed capacity (Downs, 2005).

Leisure

Early studies that examined the impact of ICT, and in particular of social networks, on the participation in leisure activities found that ICT-savvy and frequent Internet users report "greater sociability and interconnectivity," but "primarily because they are more educated, wealthier and younger" (Nie, 2001, p. 428). However, there was concern that ICTs in general and Internet use in particular may reduce interpersonal interaction and communication by substituting solitary computer pursuits for physical contact. The majority of recent research supports the opposite conclusion: ICT use tends to complement social activity and a greater adoption of Internet-based activities is usually associated with more time spent with friends and acquaintances (Robinson & Martin, 2010). Furthermore, online social networks and improved mobile telecommunications affect *how* leisure activities and trips are organized, allowing micro-coordination of time and location for social gatherings and last-minute schedule adjustments. Mokhtarian, Salomon, and Handy (2006) suggest that the dominant effect is to expand an individual's choice set, making more leisure options available to an individual.

ICT has the potential to significantly influence trip objectives and purposes, mode choice, origins and destinations, and the specific route that is chosen for a trip (Kellerman, 2009). As Mokhtarian and Tal (2013, p. 250) point out, "the same types of roles of ICT generally apply across destination, mode, and route choices": ICT can serve as inspiration, information provider, explanatory variable, and/or as one of the alternatives. ICT also influences decisions related to the time and duration of activities, through the related mechanisms of fragmentation and multitasking. In the following paragraphs we explore these effects further.

Destination Choice

ICT provides a powerful channel for disseminating information about potential destinations, even in the context of daily travel. The intensity of this phenomenon is increased by the availability of Internet-based marketing, social networks, instant sharing of photos and videos, GPS technology, and location-aware mobile devices, all of which can generate considerable new travel. The internet allows easy collection of information on a variety of candidate destinations for a given activity (e.g., eating out) and it provides information about how to reach these destinations, contact the vendor/service provider, and make instant reservations. Destinations that are less technologically friendly, or for which less information is available, might see a reduction in their popularity owing to their reduced attractiveness or travelers' lower awareness of them. Furthermore, ICT provides access to travel information, both before and during the trip, allowing for easier en-route modifications or overall substitution of a destination. Finally, sometimes ICT is per se a desirable attribute in the selection of a destination, in particular among some groups of users, as in the case of the increased attractiveness of cafes that offer free WiFi.

Mode Choice

ICT can offer inspiration to travel by a specific travel mode, for instance, through the promotion of transit and bicycling for commuting trips. ICT can provide pre-trip information about multiple travel modes available to reach a destination, allowing the user to make a more informed decision about his or her means of transportation and also including modes that he or she would have not considered otherwise. Internet-based services such as Google Maps provide travel times and directions to reach a specific destination by driving, transit, walking, cycling, and even by air. Advanced traveler information systems (ATIS) provide real-time information to drivers regarding travel times on different routes, availability of parking spaces, current and planned disruptions on the road network, etc. Similarly, public transportation operators provide accurate real-time information on expected waiting time and in-vehicle travel time at public transportation stops or before the beginning of a trip through the transit agency websites and other web-based or smartphone applications (Watkins, Ferris, Borning, Rutherford, & Layton, 2011). Most travelers find uncertainty about travel time uncomfortable; by reducing uncertainty about travel attributes, ICT can increase the popularity of a particular mode. ICT offers convenient platforms for rideshare matching, whether ad hoc or longer-term, and increases the efficiency of some modes and their potential attractiveness and patronage among users. For example, taxi operators use ICT to optimize fleet dispatching, and valet parking facilities offer text messaging services or mobile apps to recall a car at the desired time. All these solutions reduce waiting time and significantly increase the attractiveness of these modes or services.

For longer-distance trips, new entrants to the intercity bus coach market such as Megabus and Bolt Bus rely on the Internet for marketing to prospective users and for nearly "just in time" matching of demand to supply, resulting in the ability to sell previously unused capacity at low prices (a model that airlines are increasingly using). Services such as these, as well as the carsharing and on-demand ride services discussed earlier, simply could not exist without modern ICT and point to an important way in which ICT may generate travel—by lowering its monetary and transaction costs.

Route Choice

ICTs, and particularly Advanced Traveler Information Systems (ATIS), have a strong role in affecting route choice before and during a trip, for example, prompting a route change mid-trip in case of congestion and advising travelers to take an alternate route. More broadly, Nagurney, Dong, and Mokhtarian (2002) have discussed the role of ICT-based solutions in affecting travel choices with regard to the complete path optimization problem, where the available network comprises both the physical transportation network and the virtual ICT-based ways of accessing a destination. ICT can also have an inspirational role, affecting the selection of the travel route, as in the cases of scenic, historical, or cultural interests, and ICT can affect the trip experience of passengers, as, for example, through entertainment systems that deflect attention from the unpleasantness of an unappealing route.

Spatiotemporal Fragmentation of Activities

ICT has promoted an increased fragmentation of activities in time and space (Couclelis, 2004; Hubers, Schwanen, & Dijst, 2008). By facilitating the conduct of some activities "anywhere, anytime," ICT promotes the blurring of space–time boundaries between and among activities associated with home, work, shopping, and entertainment. Increased fragmentation usually implies a shorter duration of the fragments of time dedicated to activities (but a higher number of activities, which may generate a larger number of trips, for those activities that cannot be carried out "virtually"). Certainly, ICTs (e.g., mobile communications) increase time fragmentation, by providing virtual presence at a distance and more frequent contacts with a larger network of peers, for example, through instant messaging on smartphones or tablets. But how does this virtual presence affect travel? ICTs facilitate travel by reducing the "pain of absence" (Mokhtarian & Tal, 2013). Several technological devices, from "nanny-cams" to medical-alert systems to daily blogs and frequent tweets, help reassure a traveler that all is well at home and conversely (White & White, 2007). At the margin, this process supports travel that might otherwise have been suppressed in favor of staying home (e.g., Mascheroni, 2007 highlights "the consistent use of new media while on the move" as an "incitement to travel"). Travel-based multitasking can affect how people evaluate the utility of travel time (Ettema & Vershuren, 2007), such that travel time may be considered to be anything from completely unproductive to partially or fully productive or even "ultraproductive" (Lyons & Urry, 2005), depending on a combination of factors and on the specific travel mode.

The possibility to increase the *utility* (or reduce the *disutility*) of travel time might significantly enhance travelers' experiences and become a factor in the mode choice decision. For example, travelers might be willing to travel by public transportation even at the expense of longer total travel time if they can make positive use of their time on board or while waiting at a terminal. Similarly, the ability to make productive use of ICT devices during a trip, for example, by accessing the Internet through a wireless connection on trains, increases the perceived utility of travel time, reduces the valuation of travel time savings, and contributes to increasing public transit patronage (Dong, Mokhtarian, Circella, & Allison, 2015). The increased availability of ICT-enabled devices also, however, creates additional risk of distraction while driving, thus highlighting some of the potentially negative externalities of ICT adoption in relation to traveling.

CONCLUSIONS

The impacts of information and communication technologies (ICTs) permeate society, spanning from the reorganization of cities and large businesses to modifications in an individual's daily activities. ICTs are contributing to reshaping the way we work, live, and interact with each other; the way we participate in social and leisure activities; and the way we travel. Definitively identifying the impacts of ICT in each of these areas may be impossible because of the many overlapping relationships between (1) the development and adoption of ICT and (2) the headlong rate of development of modern technologies, which continuously modifies the available applications and services.

In many cases, a clear dominant effect of ICT on either reducing or stimulating travel cannot be confirmed. The specific impacts vary depending on local contexts and other concurrent causes. ICTs have contributed to reshaping work organization through increased opportunities for distributed team work and the devolution of important functions to remote locations, and have contributed to reshaping the demand for space in urban areas. While many residents are moving back into livelier and more central parts of American cities, firms require reduced space for front offices in these areas and are reorganizing their production activities closer to important transportation hubs, in particular airports and freeway corridors. Similar changes are affecting the organization of retail stores, which are increasingly integrated in mixed forms with their virtual counterparts ("bricks and clicks"), while the physical organization of traditional stores has often evolved toward the model of entertainment centers, which are less subject to competition with e-shopping.

ICT innovations have profoundly modified individuals' personal decisions and travel behavior. For example, ICTs allow teleworkers to accept jobs far from home without residential relocation or to have more freedom in choosing a residential location by reducing constraints on commuting trips. Modern ICTs are also modifying users' relationships with the adoption and use of private vehicles. ICT-enabled car-sharing, on-demand ride services, and bikesharing services are a familiar presence in many urban areas: they substitute for the ownership and use of a private vehicle, allow users to enjoy automobility when needed while avoiding the fixed costs of owning a vehicle, and extend the area of coverage of public transportation.

For a long time, policymakers have hoped ICT would be a valid substitute for physical trips as a way to reduce traffic congestion and increase transportation sustainability in urban areas. We have shown, however, that ICT can increase travel demand through both its direct and indirect effects on individuals' behaviors. For instance, ICT can (1) generate additional business travel via the increased number of ICT-based businesses; (2) reduce transportation costs and allow reinvesting some of the money and time savings in additional travel; and (3), in the longer term, promote economic growth and stimulate new activities that generate additional travel.

ICT also provides alternatives to travel, and it increases people's freedom to eventually choose not to travel. It increases the efficiency of transportation so that more travel can be accommodated within the existing infrastructure. ICT has a central role in all these strategies: The way public policies will be implemented in future years will greatly affect its capability to reduce environmental externalities of transportation and ensure safe and reliable transportation options that satisfy travelers' mobility needs. To date, only limited aspects of the relationships between ICT and transportation have been uncovered: research has mainly focused on specific impacts of ICT on urban form and the organization of some activities, and on specific components of individuals' decisions and travel behavior—primarily commuting trips and to a lesser extent business, leisure, and shopping trips. Many other impacts of ICT remain to be explored, including effects on congestion, fuel consumption, and greenhouse gas emissions.

Future technologies have the potential to further induce dramatic changes in lifestyles and travel behavior, for example, through technologies for driverless vehicles that will automate the use of roads, or technologies that will make point-to-point air travel feasible and affordable to a mass market. Important shifts are also associated with the type of technologies that will be used, with a larger prevalence of two-way communication solutions: not only will ICT provide services to users (e.g., information about travel time), but it will also be used increasingly for interactions between users and providers of services, and among users, for example, through increased automatic collection of personal data and the introduction of additional location-aware services.

However, potential threats associated with the adoption of ICT relate to privacy issues, as in the case of the automatic collection of personal data through mobile communication devices, and equity, due to the increased technological gap suffered by non-ICT users. The latter topic, in particular, deserves renewed attention: New technologies are contributing to reshaping the world in unprecedented ways, but this has come at the expense of those individuals who, owing to economic conditions, physical disabilities, or other reasons, do not have access (permanently or even temporarily) to many of these technological solutions. In an increasingly globalized and connected world, policymakers should pay increased attention to the way technological innovations affect these populations.[2] Technological development will continue to evolve and will further reshape transportation in the 21st century. In this continuous transition, it will be the responsibility of future generations of transportation planners and policymakers to make sure that nobody is left behind in the modern digital society.

NOTES

1. Increased connectivity and mobile telecommunications are increasingly reported to have a central role in modern migration flows, including in the case of refugees escaping from persecutions and civil wars. The recent immigration flows originating from the Middle East—Syria and Iraq in particular—probably represent the first case of a cell phone-driven migration flow, with migrants using cell phones, instant messaging apps, and mapping services to navigate themselves, and help others, on their path toward *refugee* status in Western democracies.

2. For example, Mokhtarian and Tal (2013) discuss the impact of technological solutions and Internet-based provision of travel time information for public transportation, which have almost entirely replaced the use of printed timetables for technologically savvy users. However, non-ICT users might be affected negatively if all resources and policies are solely invested in supporting ICT-based improvement plans.

REFERENCES

Audirac, I. (2005). Information technology and urban form: Challenges to smart growth. *International Regional Science Review, 28*(2), 119–145.

Axsen, J., & Kurani, K. S. (2011). Interpersonal influence in the early plug-in hybrid market: Observing social interactions with an exploratory multi-method approach. *Transportation Research Part D: Transport and Environment, 16*(2), 150–159.

Belk, R. (2014). You are what you can access: Sharing and collaborative consumption online. *Journal of Business Research, 67,* 1595–1600.

Burt, S., & Sparks, L. (2003). E-commerce and the retail process: A review. *Journal of Retailing and Consumer Services, 10*(5), 275–286.

Cairncross, F. (2001). *The death of distance: How the communications revolution is changing our lives.* Boston: Harvard Business School Press.

Cao, X. J., Xu, Z., & Douma, F. (2012). The interactions between e-shopping and traditional in-store shopping: An application of structural equations model. *Transportation, 39*(5), 957–974.

Choo, S., & Mokhtarian, P. L. (2007). Telecommunications and travel demand and supply: Aggregate structural equation models for the US. *Transportation Research Part A, 41,* 4–18.

Choo, S., Mokhtarian, P. L., & Salomon, I. (2005). Does telecommuting reduce vehicle-miles traveled?: An aggregate time series analysis for the US. *Transportation, 32*(1), 37–64.

Couclelis, H. (2004). Pizza over the Internet: E-commerce, the fragmentation of activity and the tyranny of the region. *Entrepreneurship and Regional Development, 16*(1), 41–54.

Denstadli, J. M., Julsrud, T. E., & Hjorthol, R. J. (2012). Videoconferencing as a mode of communication: A comparative study of the use of videoconferencing and face-to-face meetings. *Journal of Business and Technical Communication, 26*(1), 65–91.

Dijst, M. (2009). ICT and social networks: Towards a situational perspective on the interaction between corporeal and connected presence. The expanding sphere of travel behaviour research. In R. Kitamura, T. Yoshii, & T. Yamamoto (Eds.), *Selected papers from the 11th International Conference on Travel Behaviour Research* (pp. 45–75). Bingley, UK: Emerald Group.

Dong, Z., Mokhtarian, P. L., Circella, G., & Allison, J. (2015). The estimation of changes in rail ridership through an onboard survey: Did free Wi-Fi make a difference to Amtrak's Capitol Corridor service? *Transportation, 42*(1), 123–142.

Downs, A. (2005). *Still stuck in traffic: Coping with peak-hour traffic congestion.* Washington, DC: Brookings Institution.

Ettema, D., & Verschuren, L. (2007). Multitasking and value of travel time savings. *Transportation Research Record: Journal of the Transportation Research Board, 2010*(1), 19–25.

Ewing, R., & Cervero, R. (2010). Travel and the built environment: A meta-analysis. *Journal of the American Planning Association, 76*(3), 265–294.

Giuliano, G. (1998). Information technology, work patterns and intra-metropolitan location: A case study. *Urban Studies, 35*(7), 1077–1095.

Giuliano, G., & Small, K. A., (1999). The determinants of growth of employment subcenters. *Journal of Transport Geography, 7*(3), 189–201.

Graham, S., & Marvin, S. (2001). *Splintering urbanism: Networked infrastructures, technological mobilities and the urban condition.* London: Routledge.

Guo, A. W., Brake, J. F., Edwards, S. J., Blythe, P. T., & Fairchild, R. G. (2010). The application of in-vehicle systems for elderly drivers. *European Transport Research Review, 2*(3), 165–174.

Helminen, V., & Ristimäki, M. (2007). Relationships between commuting distance, frequency and telework in Finland. *Journal of Transport Geography, 15*(5), 331–342.

Horan, T. A. (2000). *Digital places: Building our city of bits.* Washington, DC: ULI–The Urban Land Institute.

Hubers, C., Schwanen, T., & Dijst, M. (2008). ICT and temporal fragmentation of activities: An analytical framework and initial empirical findings. *Tijdschrift voor Economische en Sociale Geografie, 99*(5), 528–546.

Janelle, D. G. (2012). Space-adjusting technologies and the social ecologies of place: Review and research agenda. *International Journal of Geographical Information Science, 26*(12), 2239–2251.

Janelle, D. G., & Gillespie, A. (2004). Space–time constructs for linking information and communication technologies with issues in sustainable transportation. *Transport Reviews, 24*(6), 665–677.

Karlsson, C., Maier, G., Trippl, M., Siedschlag, I., Owen, R., & Murphy, G. (2010). *ICT and regional economic dynamics: A literature review.* Luxembourg: JRC Scientific and Technical Reports, Publications Office of the European Union.

Kellerman, A. (2009). The end of spatial reorganization?: Urban landscapes of personal mobilities in the information age. *Journal of Urban Technology, 16*(1), 47–61.

Kim, T. J., Claus, M., Rank, J. S., & Xiao, Y. (2009). Technology and cities: Processes of technology–land substitution in the twentieth century. *Journal of Urban Technology, 16*(1), 63–89.

Kitou, E., & Horvath, A. (2008). External air pollution costs of telework. *International Journal of Life Cycle Assessment, 13*(2), 155–165.

Komito, L. (2011). Social media and migration: Virtual community 2.0. *Journal of the American Society for Information Science and Technology, 62*(6), 1075–1086.

Krizek, K. J., Li, Y., & Handy, S. L. (2005). Spatial attributes and patterns of use in household-related information and communications technology activity. *Transportation Research Record: Journal of the Transportation Research Board, 1926*(1), 252–259.

Lyons, G. & Urry, J. (2005). Travel time use in the information age. *Transportation Research Part A: Policy and Practice, 39*(2), 257–276.

Mascheroni, G. (2007). Global nomads' network and mobile sociality: Exploring new media uses on the move. *Information, Community and Society, 10*(4), 527–546.

Mitchell, W. J. (1996). *City of bits*. Cambridge, MA: MIT Press.

Mokhtarian, P. L. (2004). A conceptual analysis of the transportation impacts of B2C E-commerce. *Transportation, 31*(3), 257–284.

Mokhtarian, P. L. (2009a). If telecommunication is such a good substitute for travel, why does congestion continue to get worse? *Transportation Letters, 1*(1), 1–17.

Mokhtarian, P. L. (2009b). Social networks and telecommunications: The expanding sphere of travel behaviour research. In R. Kitamura, T. Yoshii, & T. Yamamoto (Eds.), *Selected papers from the 11th International Conference on Travel Behaviour Research* (pp. 429–438). Bingley, UK: Emerald Group.

Mokhtarian, P. L., Collantes, G. O., & Gertz, C. (2004). Telecommuting, residential location, and commute-distance traveled: Evidence from State of California employees. *Environment and Planning A, 36,* 1877–1897.

Mokhtarian, P. L., Salomon, I., & Choo, S. (2005). Measuring the measurable: Why can't we agree on the number of telecommuters in the US? *Quality and Quantity, 39*(4), 423–452.

Mokhtarian, P. L., Salomon, I., & Handy, S. L. (2006). The impacts of ICT on leisure activities and travel: A conceptual exploration. *Transportation, 33*(3), 263–289.

Mokhtarian, P. L., & Tal, G. (2013). Impacts of ICT on travel behavior: A tapestry of relationships. In J.-P. Rodrigue, T. Notteboom, & J. Shaw (Eds.), *The Sage handbook of transport studies* (pp. 241–260). London: Sage.

Muhammad, S., Ottens, H. F., Ettema, D., & de Jong, T. (2007). Telecommuting and residential locational preferences: A case study of the Netherlands. *Journal of Housing and the Built Environment, 22*(4), 339–358.

Nagurney, A., Dong, J., & Mokhtarian, P. L. (2002). Multicriteria network equilibrium modeling with variable weights for decision-making in the information age with applications to telecommuting and teleshopping. *Journal of Economic Dynamics and Control, 26*(9), 1629–1650.

Nie, N. H. (2001). Sociability, interpersonal relations, and the Internet: Reconciling conflicting findings. *American Behavioral Scientist, 45*(3), 420–435.

Nijkamp, P., & Salomon, I. (1989). Future spatial impacts of telecommunications. *Transportation Planning and Technology, 13*(4), 275–287.

O'Brien, R. (1992). *Global financial integration: The end of geography*. London: Council on Foreign Relations Press.

O'Sullivan, A. (2011). *Urban economics* (8th ed.). New York: McGraw-Hill Education.

Rapp, A., Baker, T. L., Bachrach, D. G., Ogilvie, J., & Beitelspacher, L. S. (2015). Perceived customer showrooming behavior and the effect on retail salesperson self-efficacy and performance. *Journal of Retailing, 91*(2), 358–369.

Rayle, L., Shaheen, S., Chan, N., Dai, D., & Cervero, R. (2014). App-based, on-demand ride services: Comparing taxi and ridesourcing trips and user characteristics in San Francisco (Working paper). Berkeley: University of California Transportation Center (UCTC). Retrieved October 13, 2016, from *http://76.12.4.249/artman2/uploads/1/RidesourcingWhitePaper_Nov2014Update.pdf*.

Robinson, J. P., & Martin, S. (2010). IT use and declining social capital?: More cold water from the General Social Survey (GSS) and the American Time-Use Survey (ATUS). *Social Science Computer Review, 28*(1), 45–63.

Salomon, I. (1986). Telecommunications and travel relationships: A review. *Transportation Research Part A: General, 20*(3), 223–238.

Salomon, I. (1998). Technological change and social forecasting: The case of telecommuting as a travel substitute. *Transportation Research Part C: Emerging Technologies, 6*(1), 17–45.

Salomon, I., & Mokhtarian, P. L. (2008). Can telecommunications help solve transportation problems?: A decade later: Are the prospects any better? In D. A. Hensher & K. J. Button (Eds.), *The handbook of transport modelling* (pp. 519–540). Oxford, UK: Elsevier.

Sassen, S. (2000). New frontiers facing urban sociology at the millennium. *British Journal of Sociology, 51*(1), 143–159.

Schwanen, T., & Kwan, M.-P. (2008). The Internet, mobile phone and space–time constraints. *Geoforum, 39*(3), 1362–1377.

Shaheen, S. A., Mallery, M. A., & Kingsley, K. J. (2012). Personal vehicle sharing services in North America. *Research in Transportation Business and Management, 3*, 71–81.

Shen, Q. (2000). New telecommunications and residential location flexibility. *Environment and Planning A, 32*(8), 1445–1464.

Sinai, T., & Waldfogel, J. (2004). Geography and the Internet: Is the Internet a substitute or a complement for cities? *Journal of Urban Economics, 56*(1), 1–24.

Stevenson, B. (2009). The Internet and job search. In D. H. Autor (Ed.), *Studies of labor market intermediation* (pp. 67–86). Chicago: University of Chicago Press.

Tranos, E., Reggiani, A., & Nijkamp, P. (2013). Accessibility of cities in the digital economy. *Cities, 30*, 59–67.

Vilhelmson, B., & Thulin, E. (2013). Does the Internet encourage people to move?: Investigating Swedish young adults' internal migration experiences and plans. *Geoforum, 47*, 209–216.

Wachs, M. (2013). Turning cities inside out: Transportation and the resurgence of downtowns in North America. *Transportation, 40*(6), 1159–1172.

Watkins, K. E., Ferris, B., Borning, A., Rutherford, G. S., & Layton, D. (2011). Where is my bus?: Impact of mobile real-time information on the perceived and actual wait time of transit riders. *Transportation Research Part A: Policy and Practice, 45*(8), 839–848.

Weltevreden, J. W. (2007). Substitution or complementarity?: How the Internet changes city centre shopping. *Journal of Retailing and Consumer Services, 14*(3), 192–207.

Weltevreden, J. W., & van Rietbergen, T. (2009). The implications of e-shopping for in-store shopping at various shopping locations in the Netherlands. *Environment and Planning B, Planning and Design, 36*(2), 279.

White, N. R., & White, P. B. (2007). Home and away: Tourists in a connected world. *Annals of Tourism Research, 34*(1), 88–104.

Yankow, J. J. (2003). Migration, job change, and wage growth: A new perspective on the pecuniary return to geographic mobility. *Journal of Regional Science, 43*(3), 483–516.

PART II

PLANNING FOR MOVEMENT WITHIN CITIES

CHAPTER 5

Theories and Models in Transportation Planning

HARVEY J. MILLER

Transportation planning is the process of designing, building, and managing transportation systems to maximize good outcomes and minimize bad outcomes. Good outcomes comprise an efficient, equitable, and sustainable system that enhances economic productivity and community livability. Bad outcomes include congestion, poor air quality, environmental degradation, public health problems such as traffic fatalities, reduced economic productivity, social exclusion, and loss of community.

In order to plan for better outcomes, planners and policy analysts need to understand the current state of a transportation system, formulate effective policies and plans, and predict the impacts of these plans. Suppose a metropolitan region is considering how to improve its transportation system over the next 20–30 years. This requires answering both short-term and long-term questions such as:

- How well is the current transportation system performing?
- What changes to the system could improve performance in the short term?
- What will be the likely impacts of these changes?
- What kinds of alternative futures do citizens and planners envision for their communities and regions?
- How do transportation systems relate to each of these alternative futures?
- What changes to the transportation system will be needed to meet the demand for transportation 20–30 years hence?

Transportation models help decision makers to answer these questions. The questions above involve describing the current state of the transportation system, identifying factors that affect the system, considering alternative futures, and predicting future states of the system based on different planning scenarios. *Models* are procedures for summarizing, explaining, and predicting the behavior of transportation systems. Modeling techniques typically include mathematics and statistics, but increasingly also include simulation and data exploration. These methods allow summaries, explanations, and predictions to be obtained in a reasonably objective and transparent manner. Although the ultimate decisions are often subject to other factors such as fiscal constraints, politics, preferences and values, models provide a common basis for discussing these complex issues in light of scientific evidence.

Models are selective representations of reality, and as such they require careful thinking about which aspects of the real world are important enough to include given the policy and planning questions at hand. Theory provides this foundation. A *theory* is a way of thinking about a phenomenon: what components are critical and how these components interact; often a theory aims to explain something. In turn, models are focused, pragmatic representations of theories that can be manipulated using procedures such as mathematics, statistics, or algorithms. Therefore, understanding the results from models requires understanding not only the modeling technique but also the underlying theory and what it says about the system being investigated.

For example, as a planner you might wish to encourage bicycling as a travel mode in your local community. You would like to construct a model to help identify the factors that would lead to more bicycling and gauge their relative impacts. What should comprise this model? One way to think about this question is through the economic theory of markets. *Market theory* focuses on aggregate outcomes—especially costs—that result from the balance of supply and demand in a system. A mode-share model based on market theory would emphasize the relative costs (in time and money) of available modes and how the supply of infrastructure and services (e.g., bike lanes, dedicated signals, bikeshare systems, bike racks on buses) determine these costs.

An alternative perspective is *activity theory*. Based in consumer theory, activity theory emphasizes the underlying activities such as work, shopping, recreation, and so on, that require mobility for access and participation. An activity-based model would highlight individual differences in home and work locations, scheduling constraints, household organization and life-cycle factors that make bicycling a better mobility option for some individuals than others (e.g., younger vs. older, parents vs. nonparents, males vs. females, poor vs. affluent). Which perspective—and model—is better? As we will see below, there are several dimensions to consider when asking this question, but a short answer is—it depends on the questions being asked. In many cases, both perspectives are valuable.

This chapter provides an overview of transportation theory and modeling as they relate to transportation planning. The chapter begins with a description of market theory and activity theory; as mentioned above, these are two major theoretical perspectives on transportation systems. The third section of the chapter provides a brief overview of modeling and how it fits into the transportation planning process. The fourth section of this chapter discusses two major modeling techniques that follow from these theories. *Trip-based models* are based on market theory and model transportation systems at an aggregate level as flows between locations (zones or areas) through a network. *Activity-based analysis* is a disaggregate approach that follows from activity theory: it treats transportation systems as comprising individuals using transportation and communication technologies to participate in activities that are distributed sparsely in time and space. The fifth section of this chapter considers the advantages and disadvantages of these two modeling approaches. Section six concludes the chapter with some brief summary comments.

THEORY

Because transportation has physical, economic, social, and psychological dimensions, it is difficult to identify a single theory that can encompass all aspects of transportation. We will examine two theories that highlight key aspects of transportation systems at different levels of granularity (detail). The *theory of markets* from economics illustrates an aggregate approach that treats transportation systems as balancing transportation supply with travel demand. *Activity theory* is a disaggregate approach that focuses on individual activities in space and time as the driver of travel demand.

Market Theory

A *market* is a venue where the sellers of a good or service can meet with buyers to conduct transactions involving the exchange of the good or service at a specified price. The interaction between supply and demand determines the market price and quantity sold.

Market Equilibrium

Figure 5.1 illustrates the theoretical relationship between quantity supplied, demanded, and price in a market. Market equilibrium occurs where the supply and demand curves intersect: the price that emerges is also referred to as the *equilibrium* or *market-clearing price*. In Figure 5.1, P and Q indicate the equilibrium price and quantity supplied/demanded, respectively. Economic theory dictates that the market equilibrium price provides the most efficient allocation of resources: it is where the marginal (incremental) revenue gained from producing the good equals the marginal cost that consumers are willing to pay. This process describes the famous "Invisible Hand" identified by 18th-century economist Adam Smith: the natural adjustment of supply and demand leading to a market that allocates resources efficiently.

Market theory suggests that a transportation system should achieve this type of equilibrium between travel demand and transportation supply. Travel demand exists at different levels and varies spatially; it includes total amount of travel in an area, the demand for travel between specific origin–destination pairs, the demand among available modes, and demands for particular routes within each mode. The supply of transportation in an area encompasses infrastructure and services. The market prices that people pay for transportation can include time, money, or a generalized cost that includes both money and time.

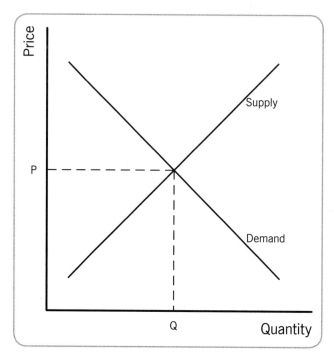

FIGURE 5.1. Market theory and equilibrium price.

Externalities and Market Failures

The tidy story of the Invisible Hand can be confounded by *market failures*. A market failure occurs when the allocation of resources is not efficient, that is, when the supply and demand curves intersect where marginal cost and marginal revenue are not equal. One source of market failure is *externalities* or "spillover" effects. As noted in Chapter 1, an externality is a market cost or benefit not transmitted through prices and incurred by someone not involved in the transaction. For example, a new toll road provides benefits to users at a price (the toll). Nearby residents and businesses may enjoy positive externalities—such as increased economic activity and property values resulting from increases in traffic and accessibility—but proximate residents can also suffer negative externalities such as increased air pollution, noise, and congestion.

Market theory suggests that negative externalities lead to overconsumption of a good or service because the price paid by consumers is lower than its true cost. The private automobile is a good example. Because the prices and taxes paid by automobile owners do not cover the full societal costs of driving, which include congestion, accidents, air pollution, and climate change, people drive more than they would if these externalities were included in the prices and taxes associated with driving (Levinson, 2005; Litman & Burwell, 2006).

Goods and services can also have positive externalities, which lead to underprovision of the market good: the equilibrium price and quantity are both lower than the equilibrium price and quantity that would obtain if the social benefits were captured in prices. When positive spillover effects remain unpriced, suppliers have little incentive to meet demand since some people enjoy the benefits without paying the cost.

Economists refer to some types of goods and services with positive externalities as *public goods*. Public goods are *nonexclusive* (it is difficult to prevent outsiders from enjoying their benefits) and *nonrivalrous* (one person's consumption of the good does not exclude another's consumption). Transportation is at best a quasi-public good; for example, the benefits of transportation extend beyond its direct users, but transportation facilities have limited capacity, which invites rivalry, particularly in cities during peak hours, for example, for a seat on a bus or space on a roadway. These aspects—especially the nonexclusive property—mean that the market will underprovide the good, requiring the public sector to become involved by providing the good directly or by offering incentives to producers. An example is public transportation. Although the benefits—including lower congestion, better air quality, and access to work for employees—extend beyond its users, public transportation often must be subsidized by public funds because otherwise it would be under-provided by the market. However, these subsidies may require higher taxes or mean that existing funds cannot be spent on other public goods and services; therefore, whether or not public transit is a good use of public funds is a matter of values and preferences and requires public discussion.

Market Theory and Transportation Systems

Market theory is useful for thinking about transportation systems, their costs and benefits, and the effects of policy, planning, and pricing on outcomes. However, transportation systems have peculiar characteristics that can make balancing demand and supply complex in reality (Ortúzar & Willumsen, 2011).

Travel demand is an *epiphenomenon,* in the sense that travel is a derived demand, as explained in Chapter 1. People typically do not demand transportation for itself; rather, they travel in order to satisfy required or desired activities such as work, recreation, shopping, and socializing.

Transportation demands are highly differentiated with respect to space and time. Some locations generate and attract a large number of trips, while others are associated with much lower levels of demand. Similarly, a great deal of the demand for transportation is concentrated during a few hours of the day such as the morning and afternoon peak or "rush" hours. The variable nature of transportation demands with respect to space and time can make it difficult to balance transportation demand with supply. Since transportation is a service and not a good, it is not possible to stockpile it for use in times and places with higher demand.

A transportation system is also a complex system of diverse components and actors, not the simple interaction of buyers and sellers as described in market theory. Components include fixed assets (infrastructure), mobile units (vehicles), and rules for operations. Typically, the infrastructure and vehicles are not owned or operated by the same entity. This situation can create a complex set of interactions among the wide range of stakeholders involved in transportation systems, including travelers, transport operators, government, and construction companies.

The nature of human activity and travel is also changing dramatically. Enabled by low-cost and convenient mobility, people in many regions of the world are doing more activities at more times and places than did their ancestors. Encouraging these high mobility levels are information and communication technologies (ICTs; see Chapter 4). Rather than a simple substitute for travel, the information and connectivity afforded by ICTs increase the demand for activities and social interaction, and therefore travel. ICTs also allow easy, fluid reallocation of activities. These capabilities, combined with social changes such as two-earner households, work spreading beyond its traditional 9:00 A.M.–5:00 P.M. Monday through Friday confines, shorter-term part-time employment, and the rise of nontraditional households, means that travel patterns are also becoming more complex. Human mobility patterns are no longer dominated by single-purpose, home-based commuting and shopping trips. These factors mean that we must focus on individuals and their different activities in order to answer some key policy and planning questions (McNally & Rindt, 2007).

Activity Theory

Activity theory focuses on human activities in geographic space and time as the foundation for understanding, planning, and managing transportation and related systems. Activity theory has its roots in the analysis of human time use. This includes the microeconomic theory of time allocation originally due to Becker (1965), the empirical time-use studies of Chapin (1974), and the time geographic framework of Hägerstrand (1970). An influential paper by Jones (1979) articulated the activity-based approach to understanding mobility behavior based on these conceptual foundations. The central idea is that time is a scarce resource that must be allocated among activities and mobility and communication are needed to participate in these activities (Axhausen & Garling, 1992; Ellegård & Svedin, 2012; Golob, Backmann, & Zahavi, 1981; McNally & Rindt, 2007; Pinjari & Bhat, 2011).

Two core concepts from time geography that are at the heart of activity theory are the *space–time path* and the *space–time prism*. These represent actual and potential mobility, respectively. Figure 5.2 illustrates a space–time path among activity *stations*: these are places where activities can occur, characterized by locations in space and durations in time. In Figure 5.2, the location and length of the tubes corresponding to each station reflects its location in space and availability in time. The space–time path highlights the sparse availability of activities in space and time and the individual's need to allocate scarce time to access and participate in activities. For example, in Figure 5.2 both the locations and time availability of a medical appointment and retail shopping hours, plus the need to be at home and work at specific times, dictate where and when the person can participate in health care and shopping activities. The space–time prism, introduced in Chapter 1 of this text, is a measure of accessibility; it encompasses all locations that can be reached during some time interval given constraints on the object's speed. In other words, the space–time prism is the envelope of all possible space–time paths between two locations in space and time.

In addition to continuous space, it is also possible to define space–time prisms within transportation networks (Kuijpers & Othman, 2009; Miller, 1991). Also, while space–time paths and prisms traditionally focus on physical mobility and accessibility, their representation of these behaviors at the individual level allows more natural integration with behavior and interactions in *cyberspace*—the virtual space

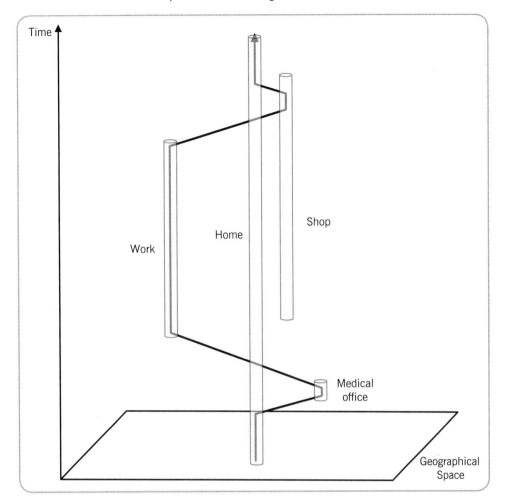

FIGURE 5.2. A space–time path among activity stations.

implied by networked ICTs. Interaction and accessibility in cyberspace can be treated as relationships among space–time paths and prisms, with their relative locations in time and space facilitating different types of telecommunication (Couclelis, 2009; Miller, 2005). For example, telephone conversations can occur between two paths or prisms only at the same time, while email communication can connect one path/ prism to any point in the future of the second path/prism (Yu & Shaw, 2008).

MODELING

Effective and responsible planning requires the ability to understand the current situation, identify good strategies for improving the situation, and predict the impacts of these strategies on transportation outcomes using criteria such as efficiency, equity, sustainability, and resilience. Conceptual frameworks and theories such as market theory and activity theory are useful for thinking about transportation, but such frameworks cannot by themselves provide the estimates and predictions needed before expensive and far-reaching infrastructure, policies, and plans are implemented. Models provide the scientific tools for understanding transportation systems and for predicting their future behavior under different scenarios.

What Is a Model?

A *model* is a simplification of a portion of the real world. Models are useful for understanding how that portion of the real world behaves and for predicting its future (Ford, 2009). In studying social and economic systems such as transportation, the modeler develops a simplified representation that retains key relationships within the system while stripping away less critical aspects. The modeler can then manipulate the core components and processes to understand how the real-world system behaves and to generate predictions about its future.

For example, if a planner wants to determine how to increase ridership on public transport, he or she can develop a model that represents the main factors that influence people's decisions to use one mode or another (such as car ownership and the trade-offs between money, time, and level of service—an approach rooted in market theory). He or she can then collect data in his or her city on people's use of different modes and on the factors he or she believes are important in affecting mode use. He or she can then use some of the data to calibrate (statistically estimate) the model parameters. These parameters, or weights associated with each variable, help explain ridership by identifying the relative importance of each factor in people's observed mode choices. He or she should also verify that the model fits reality well by comparing the model predictions against data not used in the calibration process (since using the same data for both calibration and verification biases the outcome toward accepting the model). If he or she is satisfied with the model, he or she can manipulate policy-relevant variables—such as fares, schedules, and routes—within the model and generate predictions of ridership under these scenarios. If he or she is not satisfied, he or she needs to rethink the data collected, the variables chosen, the modeling technique, or perhaps even the underlying theory about what determines mode choice.

Model Design

Determining the core components of the system that should be included in a model depends on several considerations (Ortúzar & Willumsen, 2011; Wilson, 1974). As mentioned above, theory is important because it can help us to identify the important variables and how they are related to each other.

A primary consideration is the *purpose* of the model. Purposes for modeling include *explanation* and *prediction*. These are related but different. An explanatory model attempts to represent real-world processes as faithfully as possible, while predictive models simply try to maximize predictive accuracy, regardless of whether the model faithfully reproduces the real world.

For example, an analyst developing a model for explaining why people choose to ride public transit would select a model structure and variables based on theories and hypotheses about the preferences and constraints that underlie people's mode choice. In contrast, an analyst who is interested only in forecasting ridership might include as many variables as might conceivably be correlated with ridership, regardless of theory, in order to maximize model fit. However, this approach can be dangerous because correlation does not mean causality: causality can be inferred only if the correlation has an explanation and is not rooted in a spurious relationship.

To illustrate: an analyst might have noticed that public transit ridership in a city seems to vary with the price of milk: higher milk prices are associated with higher public transit ridership levels. But the relationship between transit ridership and milk prices has no basis in travel theory; that they vary together is a coincidence or perhaps reflects some underlying factor that affects both (such as higher consumer prices overall, incentivizing people to drive less). Unless there is an explanation for the relationship between the price of milk and transit ridership, using the price of milk as an explanatory variable would be misleading as it suggests that policymakers can influence public transit ridership by manipulating the price of milk. Predictive models that are not based in explanation are of limited utility; in particular, they are not likely to be transferrable to other places or time periods.

In developing a model and selecting variables to include in a given model, another consideration is the *policy relevance* of variables; that is, can a variable be manipulated in practice? For example, travelers' moods and emotions may be important to mode choice, and the analyst may want to include variables measuring moods and emotions for that reason. But planners cannot easily manipulate moods. However, they are able to change some aspects of transit service, such as the scheduled time between buses. In developing a model, the analyst often must balance theoretical consistency with policy relevance. But it is always important to include the variables that matter (i.e., are consistent with theory)—whether they are policy-relevant or not—in the relationship being modeled in order to generate reliable predictions.

Model purpose also dictates the appropriate level of *aggregation* (e.g., Does the model use data for zones or households?) and representation of *time* (e.g., Does the model operate at the level of seconds, days, weeks, or years?). In theory, models should be as disaggregate as possible: for example, ideally we should represent every individual and vehicle moving within a city on a second-by-second basis when modeling the urban transportation system. While this is becoming increasingly possible (as discussed below), it is not always appropriate. Sometimes we need broader explanations and predictions for policy questions at regional or greater geographic scales and over longer time frames. Simpler models are easier to understand and can be useful as initial steps in the analysis or as inputs into more detailed models. Other factors affecting the level of spatial and temporal aggregation are the availability of data and the time frame for the analysis: sometimes a quicker but cruder answer is more useful than a lengthier but more detailed answer.

Data and available techniques are practical but crucial concerns. A sophisticated model may be useful for basic science, but it is not as useful in applications such as transportation planning if it requires data that are expensive or difficult to collect, maintain, and update over time. Similarly, a powerful and detailed simulation of travel-demand patterns in a city is not practical if it requires high-performance computers and infrastructure and expertise that are not readily available. These practical concerns must be balanced with theory and policy requirements to determine the appropriate model for the questions at hand.

Types of Models

As simplified representations of reality, models can be expressed in a variety of forms. For example, one could construct a physical model of a vehicle to test how it performs under different physical conditions, including crashes. However, the most convenient and powerful type of modeling is *mathematical modeling:* models expressed in the form of equations. Mathematics provides systematic and reliable procedures for generating explanations and predictions through the rules of logical inference. Mathematics is not the only way to do this, but the combination of easy manipulation, rigor, and power make this type of modeling "unreasonably effective" (paraphrasing the famous phrase by physicist Eugene Wigner; see Casti, 1993).

In the public transit example above, the hypothetical planner developed a mode choice model expressed in mathematical form and used statistical inference to estimate its parameters and test how closely it represents the situation in his or her city. Mathematics also allows him or her to predict future outcomes under new values for the variables, including levels of confidence in these predictions. For example, he or she could change the fares, level of service, and schedules to predict new ridership levels under proposed future scenarios. Without mathematics, it is difficult to see how to generate these types of explanations and predictions other than experimenting with fares and service in the real world—an approach that can be costly, not to mention unethical.

A more recent trend is the rise of *computational modeling.* Rather than estimate the relationships among variables by constructing equations and estimating their parameters from data, computational modeling uses numerical methods to simulate the dynamics and evolution of a system over time. Simulation combines theory and experimentation: the analyst programs a simulation model reflecting a theory about the real-world system and then experiments with simulation to see how it behaves under different settings and conditions. Computational models have become increasingly prevalent in science because they require fewer assumptions than do mathematical models. Also, as computers have become more powerful they can model complex, real-world systems at high levels of detail—often for individual travelers (Openshaw, 2000)

Another trend is the rise of *data-driven science* due to the dramatic collapse in costs of collecting, storing, and processing digital data (Hey, Tansley, & Tolle, 2009). A data-driven approach starts with examining data (e.g., individual vehicles in large metro areas) for patterns that suggest hypotheses to explore (e.g., How are workday versus weekend travel patterns different? Do these patterns change over the course of the year?). These hypotheses can then be examined more closely using traditional mathematical models and statistical techniques. Thus, data-driven modeling is most closely associated with visualization, data mining, and other exploratory techniques (Miller, 2010). But one cannot work directly with raw data; effective data exploration requires analytical summaries and measures that can be easily understood.

Model Evaluation

What determines a good model? There are several considerations; the appropriate mix and weight of each dimension depends on the application. *Theory* is important: a model is more believable if it is based on a strong theoretical foundation. *Accuracy* is of course important: a model is more believable if its predictions match reality well. *Parsimony* can also help distinguish between good and bad models. Simpler models are better: a good model explains the most variation with the fewest variables. More parsimonious models also tend to be easier to implement and interpret. Finally, we must also consider the model's *functionality*; how useful will it be in solving a problem and pointing to a relevant policy? Can the model give us answers to the questions we need to answer in a given policy or planning context? Ultimately, models are decision-support tools, whether the decisions involve basic or applied scientific questions. Good models help scientists and planners make better decisions.

TRANSPORTATION MODELS

Like many human-related phenomena, transportation is complex, with a diverse range of stakeholders and motivations and a huge number of intricate, interacting parts. Transportation systems reflect the dynamic and complex nature of economies, demographics, lifestyles, weather, climate, and both the built and physical environments. No single theory or model can neatly encompass all aspects of transportation. The choice of a model is driven by the considerations outlined above, and any model will do some, but not all, things well.

Two major types of transportation models are often used in planning. *Trip-based models* are aggregate models of travel demand based on market theory. *Activity-based models* include disaggregate models and data analytic techniques based on activity theory. Being rooted in different theories, these models ask different types of questions about transportation systems and are therefore relevant for different types of analysis. As will be discussed in more detail below, trip-based models have a longer history and are better suited for traditional policy and planning questions surrounding infrastructure provision, such as where to build new roads or rail lines. Activity-based models are better suited for policy and planning questions surrounding *travel demand management* or strategies to reduce and redistribute travel demand for better collective outcomes such as lower congestion, less resource consumption, and lower pollution.

Trip-Based Models

A *trip* is the movement of a person, good, and/or vehicle from an origin to a destination, where origins and destinations are areas such as traffic analysis zones, postal units, or census tracts. Trips are influenced by positive factors at the origins and destination (e.g., the number of households at origins and the number of jobs at destinations in the case of commuting trips) and the negative effects of travel costs such as distance, time, and monetary expenditures such as tolls or fares.

Although trip-based models aggregate individual movements by areas or zones, they nevertheless assume that trips are the outcome of a rational decision-making process, in which travelers attempt to maximize *utility,* where utility is a function of expected travel benefits minus travel costs (see Chapter 1). Trip-based models also tend to be analytical rather than computational in that these models can be solved manually using the tools of algebra and calculus. The rational-choice paradigm described above facilitates analytical tractability by allowing relatively simple model forms and methods for aggregating from individual to collective choices. Later in this chapter, we will see examples of activity-based models that are computational in that they involve numeric simulation.

The Four Components of Travel Demand

A trip-based approach to travel-demand modeling typically divides the travel demand in a study area into four components: (1) *trip generation,* or the total number of trips in the study area; (2) *trip distribution,* or the allocation of trips among destination zones; (3) *modal split,* or the proportion of trips made by each available transportation mode between origin–destination pairs; and (4) *network assignment,* or the route choice within each mode (Ortúzar & Willumsen 2011; Southworth & Garrow 2010).

Trip generation (TG) models estimate the total level of travel demand for the study region. This includes the total trip outflows from origin zones and total trip inflows to destination zones. A typical approach is a regression model that estimates the number of trips leaving from an origin zone and arriving at a destination zone as a function of the overall travel cost and zone-specific factors. For example, if a planner is modeling commuting flows, the zone-specific factors might include the number of workers at origin zones and the number of jobs at destination zones.

Trip distribution (TD) models allocate the total travel demand in the study area to the travel demands between specific origin–destination pairs. A common TD model is the spatial interaction or "gravity" model. These models trade off the travel cost, measured in time, distance, or money, against the attractiveness of traveling from an origin zone to a destination zone. These models can be constrained by the total outflows and inflows estimated in a TG model (see Fotheringham & O'Kelly, 1989; Haynes & Fotheringham, 1984; Wilson, 1971).

Mode split (MS) models allocate the travel demand between an origin–destination pair among the available travel modes between each pair. These models typically depend on the travel cost associated with each mode and mode-specific factors such as level of-service, safety, and comfort. Common methods for modeling modal split include discrete choice models: these estimate mode choice probabilities assuming utility-maximizing behavior (Ben-Akiva & Lerman, 1985).

Network allocation (NA) models allocate the modal demands between an origin–destination pair among the possible routes within each mode. This allows the planner to predict flows and travel times in the network. Since transportation networks are typically congested, individuals' route choices affect others' choices by imposing higher travel costs in the form of slower travel. Network assignment methods solve for the combined network flow by finding a stable pattern or equilibrium.

The most common equilibrium is the *user-optimal* (UO) principle, originally due to Wardrop (1952). This requires that, at network equilibrium, no traveler can reduce his or her travel costs by unilaterally changing routes (Smith, 1979). There are also other network flow principles include *stochastic user optimal, system optimal,* and *dynamically optimal* (see Fernandez & Friesz, 1983; Friesz, Luque, Tobin, & Wie,, 1989; Peeta & Ziliaskopoulos, 2001; Ran & Boyce, 1994; Sheffi, 1985).

Solving for the Combined Travel Demand Pattern

Estimating the travel demand pattern in a study area requires solving for the four travel demand components. Traditionally, the most common strategy for solving the combined pattern is the *sequential* or *four-step approach*. Sequential travel demand models estimate the four travel demand components sequentially and feed the results from one model to the next in the sequence. A common sequence is TG-TD-MS-NA. Figure 5.3 illustrates the sequential approach along with principal data inputs in each stage.

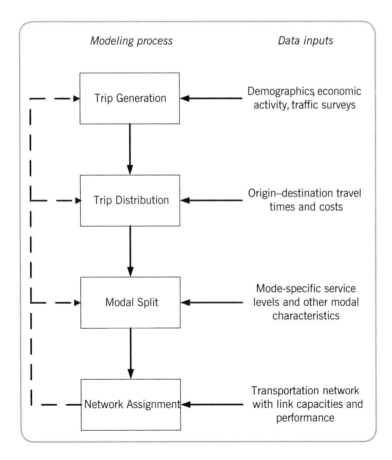

FIGURE 5.3. The sequential or "four-step" travel demand modeling strategy (after South-worth & Garrow, 2010).

Sequential travel demand models have several issues, including the following. A major problem is inconsistency among the four models as they are solved in sequence. The NA model changes network travel costs that result from congestion; these costs are no longer consistent with the travel costs used for the preceding TG, TD, and MS phases. A common response is to use *feedback loops* (the dotted arrows in Figure 5.3) and feed answers from the final steps back to the previous steps for additional rounds of estimation. However, market theory dictates that all four of the travel demand components should converge to a combined equilibrium pattern. Feedback loops do not guarantee convergence to an equilibrium solution; in fact, they do not guarantee convergence to any solution at all (Florian, Nguyen, & Ferland, 1975). Another problem is that prediction errors from components are compounded in each subsequent stage, potentially leading to substantial errors in later stages (Boyce, Zhang, & Lupa, 1994; COMSIS, 1996; Fernandez & Friesz, 1983; Sheppard, 1995).

Equilibrium travel demand models attempt to reconcile the flaws of the sequential approach by embedding the four travel demand components explicitly into a market equilibrium framework (Fernandez & Friesz, 1983; Sheffi, 1985). Examples of equilibrium travel demand models include integrated MS/NA models (Dafermos, 1980; Evans, 1976), TD/MS/NA models (Florian & Nguyen, 1978) and TG/TD/MS/ NA models (Dafermos, 1982; Oppenheim, 1995; Safwat & Magnanti, 1988; Sheffi & Daganzo, 1980). Extending this, *integrated land-use/transportation models* often use the equilibrium approach to explain and predict the coevolution of transportation and land use patterns (see, e.g., de la Barra, 1989; Echenique et al., 1990; Putnam, 1983).

Activity-Based Analysis

Activity-based analysis (ABA) methods focus on individual activities in space and time as the foundation for understanding human mobility as well as broader human–environmental systems such as cities. ABA seeks to understand how individuals' needs or desires to participate in activities available at different locations and times leads to observed travel demand. ABA also seeks to understand individual and social differences in the constraints on mobility and activity participation, for example, where people live and work; scheduling constraints such as the need to be at work or conduct childcare and other household maintenance activities; the locations and timings of activities; the availability and appropriateness of different transportation modes; and the availability and ability to use information and communication technologies to conduct and coordinate activities. Because of this shift in focus from travel demand to the underlying activities and from the aggregate to the individual, ABA is better suited for determining methods to reduce or shift travel demand patterns to better outcomes as well as determining the social impact of transportation and related policies.

ABA methods have a long tradition in transportation planning, but they were not practical for policy and planning until recently. The rise of *location-aware technologies* (LATs) such as cell phones and GPS receivers that can report geographic location densely with respect to time has generated an explosion of data on individual activities, mobility, and communication in space and time. More powerful computing environments combined with advances in *geographic information systems* (GIS), *spatial database management systems* (SDBMS), and *mobile objects database management systems* (MODBMS) have created new capabilities for handling individual-level data, making ABA more practical for policy and planning.

We will first discuss mobility data collection. Although LATs have made this task easier, there are still issues to be resolved, including constructing space–time paths from sampled locations, dealing with error, and inferring the activities that correspond to the paths (since the activities associated with the mobility data are sometimes not available—an example is cell phone data).

We will next discuss methods for summarizing and analyzing mobility data to support data-driven science. Exploratory analysis and visualization of mobility data can help the policy analyst or planner get a sense of the current system and identify variables and trends that are worthy of further analysis. Such data can also help planners communicate information about the current state of the system to stakeholders and the general public. However, exploratory analysis and visualization of mobility data requires methods for summarizing these data; we will discuss common measures and analytics available for mobility data.

Mobility data are useful as inputs to activity-based travel demand models; we will discuss these last in this section. The ultimate goal of activity-based models is the same as trip-based models: to explain travel demands and to predict future demands under different planning scenarios. Because mobility data are rooted in human activities, they are better able than are aggregate zonal data to capture complex interactions between transportation and ICTs. Because they are for individuals, mobility data can capture social differences in transportation (such as the impacts of changes in public transit schedules on young vs. elderly people, poor vs. affluent people, or men vs. women).

Mobility Data Collection

Traditionally, data on people's daily activities in time and space were collected by asking a sample of people to record manually details about their trips made over the course of a day or occasionally longer. Problems with collecting travel data via these travel-activity diaries include nonparticipation biases (certain types of people are less willing to participate in a study than are others), recall biases (people selectively forget certain types of activities such as mailing a letter), and accidental or willful inaccuracies (people may underreport trips to the liquor store or the lottery kiosk). GPS receivers, mobile phones, and other LATs allow more accurate and higher-volume data to support space–time path analysis although they also raise privacy concerns.

A space–time path is not directly available from LAT data: LATs generate a temporal sequence of point locations in one of several ways (Andrienko, Andienko, Pelekis, & Spaccapietra, 2008; Ratti, Pulselli, Williams, & Frenchman, 2006). *Event-based recording* captures the time and location when a specified event occurs, for example, a person texting or calling using a mobile phone. *Time-based recording* captures mobile object positions at regular time intervals; this is typical of GPS receivers. In *change-based recording* a record is made when the position of the object is sufficiently different from the previous location. *Location-based recording* occurs when a mobile object comes close to locations where sensors are located; examples include stationary radiofrequency identification (RFID) and Bluetooth sensors (see Versichele, Neutens, DelaFontaine, & Van de Weghe, 2012).

The ease of collecting space–time path data often comes at the expense of *path semantics* such as what activity the person pursued at the trip destination. Semantics can be recovered by overlaying paths with georeferenced land use and infrastructure data. This requires decomposing the time–space path into a sequence of moves and stops and annotating these sequences based on map matching with background geographic information (see Liu, Janssens, Wets, & Cools, 2013; Spaccapietra et al., 2008), but these methods can produce errors related to data inaccuracies and intrinsic ambiguities. For example, GPS positional error means that it can be hard to tell if a person is inside or outside a coffee house. Also, if inside the coffee house, what was the person doing—drinking coffee, working, socializing, or some combination of the above? Often, there is no unambiguous link between locations and activities, especially in an era with near-ubiquitous access to information and communication where people can work, shop, and socialize anyplace they have Internet access (Couclelis, 2000; Lenz & Nobis, 2007).

Mobility Analytics

As mentioned above, mobility data are rich but complicated and voluminous: space–time paths are complex and can easily number in the thousands for a small sample of people. In order to make sense of these data, it may be necessary to summarize, group, or aggregate the paths of individuals, but—in contrast to the aggregate approaches described in the previous section—the basis of the summary, grouping, or aggregation is usually some aspect of the traveler or of his or her travel pattern other than a shared origin or destination zone. Strategies include (Andrienko et al., 2008; Long & Nelson, 2013; Miller, 2014):

• Summarizing basic mobility properties such as the distribution of speeds and directions of travelers in different parts of the city at different times, or whether mobility patterns seem to be directed (tend to follow straight lines) or exploratory (tend to be sinuous) (see Liu, Andris, & Ratti, 2010);

• Identifying and grouping together mobility patterns that are similar to each other based on geometry (see Alt & Godau, 1995; Maheshwari, Sack, Shahbaz, & Zarrabi-Zadeh, 2011; Vlachos, Kollios, & Gunopulos, 2002);

• Locating mobility "hot spots" or areas with high mobility activity at different times of the day, days of the week, seasons of the years, and so forth (see Demšar & Virrantaus 2010; Downs, 2010).

Path summaries can be integrated and mapped with other georeferenced data such as land use, activity locations, transportation infrastructure and services, and sociodemographic data for visualization and further analysis. For example, a planner could summarize the space–time paths by individuals' age, gender, and socioeconomic status and integrate these data with transportation and land use data to understand how these different social groups utilize different parts of a city and whether the current transportation system is meeting their needs.

Besides providing summaries of mobility data, mobility analytical techniques can provide inputs into *activity-based models* of travel demand. Activity-based travel demand models attempt to explain and/or predict travel demand using human activities in space and time as a basis. Because they are based in human activities, activity-based models are "data-hungry" and require individual-level activity data.

Activity-Based Models

Figure 5.4 illustrates a general framework for activity-based models (Ben-Akiva & Bowman, 1998). *Urban dynamics* influence the opportunities for activity participation, including the locations of housing and businesses, and the infrastructure and services that influence these activities and their locations. Individual activity and travel decisions occur within the context of households (broadly defined) as a basis for joint decision making, including shared activities and the allocation of maintenance activities such as shopping and childcare. *Mobility and lifestyle* decisions occur irregularly and infrequently over annual to decadal time frames; these include household composition and allocation of maintenance activities, work force participation, automobile ownership, commuting mode decisions, and so on. *Activity and travel scheduling* decisions occur at more regular and frequent intervals; this involves the sequencing of activities, selection of activity locations, and the times and modes for travel. *Implementation and rescheduling* decisions occur in real time in response to conditions in the transportation system. Together, urban dynamics combined with household activity decisions affect *transportation system dynamics*; these in turn affect urban dynamics and individual activity participation and mobility.

As Figure 5.4 implies, activity-based models deal with complex and intricate dynamics operating at different time scales. There are two major approaches to modeling these dynamics. *Econometric models* focus on the linkages between lifestyle and mobility decisions and activity and travel scheduling decisions, typically from a static perspective (i.e., at a particular point in time). *Simulation models* imitate the dynamics and evolution of the system over time using computational techniques; these models tend to be more comprehensive and capture more of the interactions depicted in Figure 5.4. Note that these approaches are not mutually exclusive: they are often combined into hybrid models. For example, it is common to use an econometric model as a component in a broader simulation model (e.g., Bhat, Guo, Srinivasan, & Sivakuma, 2004; Pendyala, Kitamura, Kikuchi, Yamamoto, & Fujii, 2005).

Econometric models are the oldest activity-based modeling strategy; these derive from extending trip-based models to capture more complex activity and travel patterns. Some of the activity-travel patterns addressed by these models include (1) the number of *tours* (multipurpose/multistop trips) per period of time for different purposes such as work, shopping, and recreation; (2) the primary destination, travel modes, and timings of a tour; and (3) the sequence of stops within a tour (Ben-Akiva & Bowman, 1998; Pinjari & Bhat, 2011). Econometric models are mathematical and statistical: the analyst hypothesizes factors that influence the choice of activity-travel patterns (such as socioeconomic status, household size and organization, the

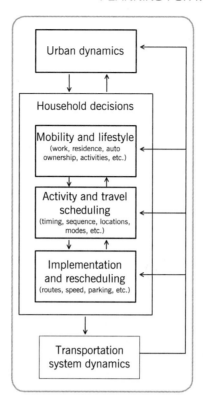

FIGURE 5.4. Framework for activity-based models (after Ben-Akiva & Bowman, 1998).

availability of an automobile, scheduling constraints such as work and household maintenance duties, and monetary and time costs required for different choices) and states this in the form of a utility function. He or she must also make assumptions about unobserved, random effects not included in the model—Is this error independent across choices, or are some choices correlated? (E.g., when choosing among automobile, bus, and train travel modes, choices between bus and train tend to be correlated since they are both forms of public transit. Therefore, unobserved factors influencing these choices will also be correlated.) Given a hypothesized utility function and assumptions about error, he or she can estimate the parameters of these relationships using observed data, assuming that individuals make utility-maximizing choices (see, e.g., Bhat, 1998; Bhat & Misra, 1999; Bowman & Ben-Akiva, 2001; Gliebe & Koppelman, 2002; Kockelman, 2001).

Unlike econometric models, simulation models are computational rather than mathematical. Rather than estimate the statistical relationships between independent variables and activity-travel patterns, these models replicate computationally the dynamics of the activity system over time. There are several simulation strategies depending on the activity system components being modeled. *Computational process models* (CPM) attempt to capture the decision heuristics (simple rules) used by individuals during activity implementation and rescheduling in the face of prevailing travel conditions and unplanned events (e.g., Recker, McNally, & Root, 1986a, 1986b). The core of a CPM is a set of condition–action pairs, often stated in the format of "if–then" rules that specify how individuals react to environmental conditions (such as *if <late> and <going to work> then <do not go to cafe>*). The heuristics can be extracted from empirical data using data-mining techniques (see Arentze, Hofman, van Mourik, & Timmermans, 2000).

Microsimulation and agent-based models (ABMs) are more comprehensive and attempt to replicate the overall system dynamics based on simulating the behaviors and interactions at disaggregate levels, often at the individual level (Arentze et al., 2000; Bonabeau, 2002; Buliung & Kanaroglou, 2007). In particular, ABMs allow the simulation of the spatial behavior of individuals (represented by software agents) and their interactions within an environment (such as a simulated highway, public transit system, or entire city).

An example microsimulation model is the *Transportation Analysis and Simulation System* (TRANSIMS), an open-source software environment for modeling activities and travel at the individual level. TRANSIMS generates a synthetic population whose demographics and locations match a city's empirical population distribution. An activity module generates activity schedules for the population based on activity survey data, and a route planner generates trip plans consisting of modes, routes, and planned departures to satisfy the planned activities. A transportation microsimulator executes plans created by the route planner and simulates traffic dynamics at the individual vehicle level. These simulated real-time travel conditions feed back to the route planning and activity generator, requiring possible adjustments in activity schedules and planned routes (Bonabeau, 2002; Cetin, Nagel, Raney, & Voellmy, 2002; Rilett, 2001).

Econometric and simulation models have contrasting advantages and disadvantages. Advantages of econometric models are a rigorous theoretical foundation and mature methodologies for model specification, calibration, and validation. Weaknesses include the empirically suspect assumption that individuals are perfectly informed, rational utility maximizers, and a lack of solid behavioral theory to support many of these models (Timmermans, Arenze, & Joh, 2002). Simulation methods are more flexible and can capture heuristics that can better represent decision making in the real world. They provide explicit representations of individual-level behaviors that drive systems such as transportation, the ability to develop and test behavioral theory, better understanding of macrolevel processes produced by individual-level behaviors, and the ability to link processes operating at different temporal scales (e.g., travel demand, economic activity, jobs and housing markets, and environmental changes). However, simulation models do not have mature theory and techniques for calibration and testing, making it difficult to distinguish between good-fitting and bad-fitting models (Buliung & Kanaroglou, 2007; McNally & Rindt, 2007).

Simulation models can lend themselves to expert engagement and judgment better than traditional, analytical models since the analyst can experiment with the model and see how the outcomes change in response (Bonabeau, 2002). However, it can be difficult to make sense of these models since they generate a large amount of data that must be analyzed and understood. In other words, the results of a simulation model are implicit and must be extracted using visualization and statistical analysis of the data generated by the model. In contrast, econometric models provide explicit, directly available numeric estimates of the relationships among variables.

TRIP-BASED VERSUS ACTIVITY-BASED MODELS IN TRANSPORTATION PLANNING

Trip-based and activity-based models reflect different perspectives on transportation systems, in terms of theory, parsimony, data, methods of analysis, and policy relevance. These differences make each approach appropriate for certain planning applications. Table 5.1 summarizes the two modeling approaches with respect to theory, parsimony, and policy relevance.

Theory

The trip-based approach treats each trip as an event that is isolated not only from underlying activities, but also from other trips. Therefore, the organization of trips and their resulting interrelationships are ignored in all four components of trip-based models (Bhat & Koppelman, 2003). In contrast, since they are rooted in individual behavior, ABA can better capture complex travel episodes such as multipurpose, multistop, and multimodal travel. For similar reasons, activity-based approaches are better at capturing the social, economic, cultural, geographic, and technological context that influence activities and therefore travel decisions. This includes household organization, social networks, ICTs, and social differences in scheduling constraints. Finally, activity-based approaches allow a wider range of decision heuristics than utility-maximizing trip-based models (McNally & Rindt, 2007).

TABLE 5.1. Trip-Based versus Activity-Based Models

Trip-based models	Activity-based models
Theory	
• Focus on trips as isolated events independent from each other and other human activities	• Focus on human activities as the fundamental driver of mobility and travel demand
Parsimony	
• Easier to implement and furnish with data • Well-established calibration and verification procedures	• More intricate and complex to implement and furnish with data • Simulation models generate implicit information that must be extracted from the model results
Policy relevance	
• More appropriate for evaluating "hard policies": infrastructure construction, management, and renewal	• More appropriate for evaluating "soft policies": managing the demand for travel via incentives and penalties • Integrate well with emissions and other physical models • Better at representing diversity and addressing social justice issues

Parsimony

The trip-based approach has clear advantages in terms of parsimony: these models are easy to implement and furnish with data (Algers, Eliasson, & Mattsson, 2005). The methods and practices for model construction, parameter estimation, and model verification are well established. Also crucial is that trip-based models are analytical rather than computational simulations: they generate explicit results and measurements, with specific error ranges and bands. In contrast, the numerical simulation methods that are popular in the activity-based approach generate large databases of simulated futures with implicit answers that must be extracted through data summary, exploration, and visualization methods.

Except for the econometric models, activity-based approaches are less mature than trip-based approaches: theories and methods for model specification, estimation, and verification are not as developed or standardized as in the trip-based approach. However, activity-based approaches harmonize well with the rise of computational and data-driven science. As noted previously in this chapter, it has become easier to collect, store, and analyze individual-level activity data through dedicated GPS devices, mobile phones, and other location-aware technologies. Although computational and data-driven science does not provide explicit answers, simulating futures and exploring data can provide insights and raise new questions that can be investigated through more focused and mature confirmatory methods.

Policy Relevance

With respect to usefulness for policy and planning, the trip-based approach is more appropriate than are activity-based approaches for determining where to construct infrastructure and to expand capacity to facilitate travel (Algers, Eliasson, & Mattsson, 2005). This fits in well with the traditional "predict and provide" paradigm in transportation planning (Bannister & Button, 1992; Owens, 1995). However, contemporary policy and planning initiatives also focus on *managing* travel demand because of increasing concerns about the negative externalities associated with mobility such as congestion, air quality degradation, depletion of nonrenewable resources, poor public health, and loss of community.

Activity-based approaches are more appropriate than are trip-based approaches for evaluating the potential effectiveness of different travel demand management strategies. Because they address the fundamental behaviors that determine travel, they are better at capturing non-capital-improvement measures such as ridesharing incentives, congestion pricing, employer-based demand management schemes, and other so-called soft policies, as well as nontransportation solutions to accessibility problems. They also integrate well with evolving methods for estimating environmental impacts of transportation such as individual-level emissions models (Bamberg, Fuiji, Friman, & Gärling,. 2011; Bannister, 2011; Bhat & Koppelman, 2003; Malayath & Verma, 2013). Finally, since activity-based approaches are better at accounting for diversity in the population (e.g., groups can be distinguished based on their mobility patterns rather than simply their zone of residence), they are better suited for examining the differential impacts of travel demand management strategies for addressing community livability and social justice concerns (Algers, Eliasson, & Mattsson, 2005).

The differing functionality of trip-based versus activity-based views does not necessarily imply that one is superior. Although planners need to be better at managing travel demand, there will always be a need to manage and renew infrastructure. Both approaches have a role in transportation policy and planning.

CONCLUSION

This chapter provides an overview of transportation theory, modeling, and how these support the transportation planning process. Two major theories of transportation—market theory and activity theory—lead to two different families of models and techniques with complementary perspectives on transportation systems. The trip-based approach based on market theory is easy to implement and has a well-developed set of techniques for model building and verification; these models are better suited than are activity-based approaches for questions surrounding infrastructure and capacity. Activity-based analysis, a set of modeling, simulation, and data exploration techniques, is often more complex to implement and understand, but represent human behavior better and are therefore better suited to answer increasingly complicated questions surrounding policy and planning for travel demand management.

Beyond implementing and testing models, another challenge to the transportation modeler is *communication*. As noted previously in this chapter, models are support techniques to help planners make better decisions. As more diverse groups of stakeholders become involved in transportation planning, knowledge *delivery* becomes as important as modeling in transportation planning. *Collaborative spatial decision support systems* integrate models, geographic visualization, and multicriteria analysis techniques to help diverse stakeholders reach group consensus based on scientific evidence (Jankowski, Nyerges, Smith, Moore, & Horvath, 1997). Combining these methods with ICTs and social media can allow greater participation in collaborative transportation planning (Jankowski & Stasik 1997; Peng, 2001; Rinner, Kebler, & Andrulis, 2008). However, technology is not sufficient to guarantee effective knowledge delivery in the planning process; also crucial are skills for communicating the theory and science underlying the models to nonscientifically trained audiences (see Baron, 2010).

REFERENCES

Algers, S., Eliasson, J., & Mattsson, L.-G. (2005). Is it time to use activity-based urban transport models?: A discussion of planning needs and modelling possibilities. *Annals of Regional Science, 39,* 767–789.

Alt, H., & Godau, M. (1995). Computing the Fréchet distance between two polygonal curves. *International Journal of Computational Geometry and Applications, 5,* 75–91.

Andrienko, N., Andienko, G., Pelekis, N., & Spaccapietra, S. (2008). Basic concepts of movement data. In F. Giannotti & D. Pedreschi (Eds.), *Mobility, data mining and privacy* (pp. 15–38). Heidelberg, Germany: Springer.

Arentze, T., Hofman, F., van Mourik, H., & Timmermans, H. (2000). ALBATROSS: Multiagent, rule-based model of activity pattern decisions. *Transportation Research Record*, No. 1706, 136–144.

Axhausen K. W., & Gärling, T. (1992). Activity-based approaches to travel analysis: Conceptual frameworks, models, and research problems. *Transport Reviews, 12,* 323–341.

Bamberg, S., Fuiji, S., Friman, M., & Gärling, T. (2011). Behaviour theory and soft transport policy measures. *Transport Policy, 18,* 228–235.

Banister, D., & Button, K. (1992). *Transport, the environment and sustainable development,* London: Spon.

Bannister, D. (2007). The sustainable mobility paradigm. *Transport Policy, 15,* 73–80.

Baron, N. (2010). *Escape from the ivory tower: A guide to making your science matter.* Washington, DC: Island Press.

Becker, G. S. (1965). A theory of the allocation of time. *Economic Journal, 75,* 493–517.

Ben-Akiva, M., & Bowman, J. L. (1998). Activity based travel demand model systems. In P. Marcotte & S. Nguyen (Eds.), *Equilibrium and advanced transportation models* (pp. 27–46). Boston: Kluwer Academic.

Ben-Akiva, M. E., & Lerman, S. R. (1985). *Discrete choice analysis: Theory and application to travel demand.* Cambridge, MA: MIT Press.

Bhat, C. R. (1998). A model of post home-arrival activity participation behavior. *Transportation Research Part B: Methodological, 32,* 387–400.

Bhat, C. R., Guo, J. Y., Srinivasan, S., & Sivakuma, A. (2004). Comprehensive econometric microsimulator for daily activity-travel patterns. *Transportation Research Record,* No. 1894, 57–66.

Bhat, C. R., & Koppelman, F. S. (2003). Activity-based modeling of travel demand. In R. Hall (Ed.), *Handbook of transportation science* (2nd ed., pp. 39–66). Norwell, MA: Kluwer Academic.

Bhat, C. R., & Misra, R. (1999). Discretionary activity time allocation of individuals between in-home and out-of-home and between weekdays and weekends. *Transportation, 26,* 193–229.

Bonabeau, E. (2002). Agent-based modeling: Methods and techniques for simulating human systems. *Proceedings of the National Academy of Sciences, 99,* 7280–7287.

Bowman, J. L., & Ben-Akiva, M. E. (2001). Activity-based disaggregate travel demand model system with activity schedules. *Transportation Research A: Policy and Practice, 35,* 1–28.

Boyce, D. E., Zhang, Y.-F., & Lupa, M. R. (1994). Introducing feedback into four-step travel forecasting procedure versus equilibrium solution of combined model. *Transportation Research Record,* No. 1443, 65–74.

Buliung, R. N., & Kanaroglou, P. S. (2007). Activity–travel behaviour research: Conceptual issues, state of the art, and emerging perspectives on behavioural analysis and simulation modeling. *Transport Reviews, 27,* 151–187.

Casti, J. L. (1993). *Searching for certainty: What scientists can know about the future.* New York: Morrow.

Cetin, N., Nagel, K., Raney, B., & Voellmy, A. (2002). Large-scale multi-agent transportation simulations. *Computer Physics Communications, 147,* 559–564.

Chapin, F. S. (1974). *Human activity patterns in the city: Things people do in time and space.* London: Wiley.

COMSIS. (1996). *Incorporating feedback in travel forecasting: Methods, pitfalls and common concerns* (Final Report to Federal Highway Administration, U.S. Department of Transportation, Contract number DTFH61-93-C-00216). Washington, DC: COMSIS Corporation.

Couclelis, H. (2000). From sustainable transportation to sustainable accessibility: Can we avoid a new *Tragedy of the Commons?* In D. G. Janelle & D. C. Hodge (Eds.), *Information, place and cyberspace: Issues in accessibility* (pp. 341–356). Berlin, Germany: Springer.

Couclelis, H. (2009). Rethinking time geography in the information age. *Environment and Planning A, 41,* 1556–1575.

Dafermos, S. (1980). Traffic equilibrium and variational inequalities. *Transportation Science, 14,* 42–54.

Dafermos, S. (1982). The general multimodal network equilibrium problem with elastic demand. *Networks, 12,* 57–72.

de la Barra, T. (1989). *Integrated land use and transport modelling.* Cambridge, UK: Cambridge University Press.

Demšar, U., & Virrantaus, K. (2010). Space–time density of trajectories: Exploring spatio-temporal patterns in movement data. *International Journal of Geographical Information Science, 24,* 1527–1542

Downs, J. A. (2010). Time–geographic density estimation for moving point objects. In S. I. Fabrikant, T. Reichenbacher, M. van Kreveld, & C. Schliederet (Eds.), *Geographic information science* (Lecture Notes in Computer Science 6292, pp, 16–26). Berlin/Heidelberg, Germany: Springer-Verlag.

Echenique, M. H., Flowerdew, A. D., Hunt, J. D., Mayo, T. R., Skidmore, I. J., & Simmonds, D. C. (1990). The MEPLAN models of Bilbao, Leeds and Dortmund. *Transport Reviews, 10,* 309–322.

Ellegård, K., & Svedin, U. (2012). Torsten Hägerstrand's time–geography as the cradle of the activity approach in transport geography. *Journal of Transport Geography, 23,* 17–25.

Evans, S. P. (1976). Derivation and analysis of some models for combining trip distribution and assignment. *Transportation Research, 10,* 37–57.

Fernandez, J. E., & Friesz, T. L. (1983). Equilibrium predictions in transportation markets: The state of the art. *Transportation Research B, 17B,* 155–172.

Florian, M., & Nguyen, S. (1978). A combined trip distribution, modal split and trip assignment model. *Transportation Research B, 12,* 241–246.

Florian, M., Nguyen, S., & Ferland, J. (1975). On the combined distribution-assignment of traffic. *Transportation Science, 9,* 43–53.

Ford, A. (2009). *Modeling the environment* (2nd ed.). Washington, DC: Island Press.

Fotheringham, A. S., & O'Kelly, M. E. (1989). *Spatial interaction models: Formulations and applications.* Dordrecht. The Netherlands: Kuhwer Academic.

Friesz, T. L., Luque, J., Tobin, R. L., & Wie, B.-W. (1989). Dynamic network traffic assignment considered as a continuous time optimal control problem. *Operations Research, 37,* 893–901.

Gliebe, J. L., & Koppelman, F. S. (2002). A model of joint activity participation between household members. *Transportation, 29,* 49–72.

Golob, T. F., Beckmann, M. J., & Zahavi, Y. (1981). A utility-theory travel demand modeling incorporating travel budgets. *Transportation Research B, 15B,* 375–389.

Hägerstrand, T. (1970). What about people in regional science? *Papers of the Regional Science Association, 24,* 1–12.

Haynes, K. E., & Fotheringham, A. S. (1984). *Gravity and spatial interaction models.* Beverly Hills, CA: Sage.

Hey, T., Tansley, S., & Tolle, K. (2009). *The fourth paradigm: Data-intensive scientific discovery.* Redmond, WA: Microsoft Research.

Jankowski, P., Nyerges, T. L., Smith, A., Moore, T. J., & Horvath, E. (1997). Spatial group choice: A SDSS tool for collaborative spatial decision-making. *International Journal of Geographical Information Science, 11,* 577–602.

Jankowski, P., & Stasik, M. (1997). Design considerations for space and time distributed collaborative spatial decision making. *Journal of Geographic Information and Decision Analysis, 1,* 1–8.

Jones, P. M. (1979). New approaches to understanding travel behaviour: The human activity approach. In D. A. Hensher & P. R. Stopher (Eds.), *Behavioural travel modeling* (pp. 55–80). London: Croom-Helm.

Kockelman, K. M. (2001). A model for time- and budget-constrained activity demand analysis. *Transportation Research B: Methodological, 35,* 255–269.

Kuijpers, B., & Othman, W. (2009). Modeling uncertainty of moving objects on road networks via space–time prisms. *International Journal of Geographical Information Science, 23,* 1095–1117.

Lenz, B., & Nobis, C. (2007). The changing allocation of activities in space and time by the use of ICT—"Fragmentation" as a new concept and empirical results. *Transportation Research Part A: Policy and Practice, 41,* 190–204.

Levinson, D. (2005). Micro-foundations of congestion and pricing: A game theory perspective. *Transportation Research Part A, 39,* 691–704.

Litman, T., & Burwell, D. (2006). Issues in sustainable transportation. *Environment and Sustainable Development, 6,* 331–347.

Liu, F., Janssens, D., Wets, G., & Cools, M. (2013). Annotating mobile phone location data with activity purposes using machine learning algorithms. *Expert Systems with Applications, 40,* 3299–3311.

Liu, L., Andris, C., & Ratti, C. (2010). Uncovering cabdrivers' behavior patterns from their digital traces. *Computers, Environment and Urban Systems, 34*(6), 541–548.

Long, J. A., & Nelson, T. A. (2013). A review of quantitative methods for movement data. *International Journal of Geographical Information Science, 27,* 292–318.

Maheshwari, A., Sack, J.-R., Shahbaz, K., & Zarrabi-Zadeh, H. (2011). Fréchet distance with speed limits. *Computational Geometry, 44,* 110–120.

Malayath, M., & Verma, A. (2013). Activity based travel demand models as a tool for evaluating sustainable transportation policies. *Research in Transportation Economics, 38,* 45–66.

McNally, M. G., & Rindt, C. R. (2007). *The activity-based approach* (Working paper UCI-ITS-AS-WP-07-1). Irvine: Institute of Transportation Studies, University of California–Irvine.

Miller, H. J. (1991). Modeling accessibility using space–time prism concepts within geographical information systems. *International Journal of Geographical Information Systems, 5,* 287–301.

Miller, H. J. (2005). Necessary space–time conditions for human interaction. *Environment and Planning B: Planning and Design, 32,* 381–401.

Miller, H. J. (2010). The data avalanche is here: Shouldn't we be digging? *Journal of Regional Science, 50,* 181–201.

Miller, H. J. (2014). Activity-based analysis. In M. Fischer & P. Nijkamp (Eds.), *Handbook of regional science* (pp. 705–724). Berlin, Germany: Springer.

Openshaw, S. (2000). Geocomputation. In S. Openshaw & R. J. Abrahart (Eds.), *GeoComputation* (pp. 293–312). London: Taylor & Francis.

Oppenheim, N. (1995). *Urban travel demand modeling.* New York: Wiley.

Ortúzar, J., & Willumsen, L. G. (2011). *Modelling transport* (4th ed.). New York: Wiley.

Owens, S. (1995). From "predict and provide" to "predict and prevent"?: Pricing and planning in transport policy. *Transport Policy, 2,* 43–49,

Peeta, S., & Ziliaskopoulos, A. K. (2001). Foundations of dynamic traffic assignment: The past, the present and the future. *Networks and Spatial Economics, 1,* 233–265.

Pendyala, R. M., Kitamura, R., Kikuchi, A., Yamamoto, T., & Fujii, S. (2005). Florida activity mobility simulator: Overview and preliminary validation results. *Transportation Research Record, 1921,* 123–130.

Peng, Z.-R. (2001). Internet GIS for public participation. *Environment and Planning B: Planning and Design, 28,* 889–905.

Pinjari, A. R., & Bhat, C. R. (2011). Activity-based travel demand analysis. In A. de Palma, R. Lindsey, E. Quinet, & R. Vickerman (Eds.), *Handbook in transport economics* (pp. 213–248). Cheltenham, UK: Elgar.

Putnam, S. H. (1983). *Integrated urban models.* London: Pion.

Ran, B., & Boyce, D. E. (1994). *Modeling dynamic transportation networks: An intelligent transportation system oriented approach.* Berlin, Germany: Springer

Ratti, C., Pulselli, R. M., Williams, S., & Frenchman, D. (2006). Mobile landscapes: Using location data from cell-phones for urban analysis. *Environment and Planning B: Planning and Design, 33,* 727–748.

Recker, W. W., McNally, M. G., & Root, G. S. (1986a). A model of complex travel behavior: Part I. Theoretical development. *Transportation Research A, 20,* 307–318.

Recker, W. W., McNally, M. G., & Root, G. S. (1986b). A model of complex travel behavior: Part II. An operational model. *Transportation Research A, 20,* 319–330.

Rilett, L. R. (2001). Transportation planning and TRANSIMS microsimulation model: Preparing for the transition. *Transportation Research Record*, No. 1777, 84–92.

Rinner, C., Kebler, C., & Andrulis, S. (2008). The use of Web 2.0 concepts to support deliberation in spatial decision-making. *Computers, Environment and Urban Systems, 32*, 386–395.

Safwat, K. N. A., & Magnanti, T. L. (1988). A combined trip generation, trip distribution, modal split, and trip assignment model. *Transportation Science, 18*, 14–30.

Sheffi, Y. (1985). *Urban transportation networks: Equilibrium analysis with mathematical programming methods.* Englewood Cliffs, NJ: Prentice-Hall.

Sheffi, Y., & Daganzo, C. F. (1980). Computation of equilibrium over transportation networks: The case of disaggregate demand models. *Transportation Science, 14*, 155–173.

Sheppard, E. (1995). Modeling and predicting aggregate flows. In S. Hanson (Ed.), *The geography of urban transportation* (2nd ed., pp. 101–128). New York: Guilford Press.

Smith, M. J. (1979). The existence, uniqueness and stability of traffic equilibria. *Transportation Research B, 13B*, 295–304.

Southworth, F., & Garrow, L. (2010). Travel demand modeling. In J. J. Cochran (Ed.), *Wiley encyclopedia of operations research and management science* (pp. 1–13). New York: Wiley.

Spaccapietra, S., Parent, C., Damiani, M.-L., de Macedo, J. A. F., Porto, F., & Vangenot, C. (2008). A conceptual view on trajectories. *Data and Knowledge Engineering, 65*, 126–146.

Timmermans, H. J. P., Arenze, T., & Joh, C.-H. (2002). Analyzing space–time behavior: New approaches to old problems. *Progress in Human Geography, 26*, 175–190.

Versichele, M., Neutens, T., Delafontaine, M., & Van de Weghe, N. (2012). The use of Bluetooth for analysing spatiotemporal dynamics of human movement at mass events: A case study of the Ghent Festivities. *Applied Geography, 32*, 208–220.

Vlachos, M., Kollios, G., & Gunopulos, D. (2002). Discovering similar multidimensional trajectories. In R. Agrawal, K. Dittrich, & A. H. H. Ngu (Eds.), *Proceedings of the 18th International Conference on Data Engineering, 2002* (pp. 673–684). Washington, DC: IEEE Computer Society.

Wardrop, J. G. (1952). Some theoretical aspects of road traffic research. *Proceedings of the Institute of Civil Engineers, Part II, 1*(36), 325–362.

Wilson, A. G. (1971). A family of spatial interaction models, and associated developments. *Environment and Planning A, 3*, 1–32.

Wilson, A. G. (1974). *Urban and regional models in geography and planning.* New York: Wiley.

Yu, H., & Shaw, S.-L. (2008). Exploring potential human activities in physical and virtual spaces: A spatio-temporal GIS approach. *International Journal of Geographic Information Science, 22*, 409–430.

CHAPTER 6

Regional Transportation Planning

GIAN-CLAUDIA SCIARA
SUSAN HANDY

The history of transportation planning in the United States reflects an ongoing struggle to find the appropriate geographic scale for addressing transportation challenges and build the institutions for doing so. As daily travel—both passenger and freight—increasingly extended beyond the boundaries of a single town or city over the 20th century, the need for government involvement at scales beyond cities and towns became ever more apparent. Over a century ago, state governments, followed soon thereafter by the federal government, took on the responsibility of planning, building, and maintaining a highway network. At the same time, cities grappled with downtown congestion problems that accompanied the automobile's rapid adoption and private transit providers' increasing ailments. Recognition grew that transportation—and other pressing issues—ought to be considered regionally. But it was not until state highway departments extended interstate construction activities into urban areas that the institutional framework for contemporary regional transportation planning emerged.

Metropolitan regions—comprised of cities and counties across which residential and employment opportunities yield a fairly self-contained set of commute flows—are an appropriate scale for planning, building, and improving the systems that handle daily travel needs in today's world. About one-quarter of commute trips start and end within a central city; the rest spread throughout the metropolitan area, with nearly half of commute trips starting and ending in the suburbs (Polzin & Pisarski, 2015). Regional shopping centers and unique retail destinations draw customers from throughout the metropolitan region. Regions and regional economies generate flows of people and goods among neighboring cities, and these flows depend on an efficient regional transportation system. Reflecting these intraregional interdependencies, 43% of lane miles on the interstate highway system are within rather than between urban areas, as are nearly two-thirds of the vehicle miles of travel on this system (U.S. Department of Transportation, 2012).

The metropolitan region is also an appropriate scale for transportation planning because problems associated with the system do not respect local political boundaries. Pollutants emitted in one city spread throughout the region and often beyond. Congestion in one city hampers the movement of goods to and from other cities and limits residents' ability to travel to and from jobs and services in other cities. The negative impacts on health resulting from air pollutants and restricted access impose costs throughout the region. None of these problems can be solved city-by-city; instead, their solutions require cooperation across cities at a regional scale.

In the United States, regional transportation planning is carried out by metropolitan planning organizations (MPOs). Federal law requires the establishment of MPOs in all urbanized areas of over 50,000 population and vests them with responsibility for regional transportation planning, a process for making plans and decisions about investments in the region's transportation system, at present and in the future. This chapter reviews the history of regional transportation planning in the United States, then describes how MPOs meet their responsibilities, and, finally, assesses the degree to which they provide the regional leadership that was envisioned.

HISTORY

As cities expanded in population and geographic size in the late 19th and early 20th centuries, planners increasingly called for regional approaches to metropolitan-scale problems. These included not just traffic congestion and goods movement, but also water management, overcrowding, and sanitation. Scotsman Patrick Geddes and his U.S. followers—Lewis Mumford, Clarence Stein, Benton MacKaye, Stuart Chase, and others—saw the antidote to urban ills not in city-centered solutions but in deliberate regional planning. They advocated for regional surveys, garden-city style regional plans weaving together city and hinterland, and economic and social interventions to remedy congestion, inefficiency, and deterioration in industrial cities. These early planners and their successors also recognized that the scale and complexity of housing, transportation, sanitation, and other challenges of America's rapidly expanding cities exceeded the limits of a single city's decision making.

However, visions for regional-scale planning in transportation and other sectors were largely unrealized given the absence of institutional mechanisms for their pursuit in the U.S. federalist system of government. Grounded in a fundamental wariness of unitary government, the U.S. federalist system has evolved to share decision making and policy responsibilities among federal, state, and local governments. The absence of a clear framework for delegating government authority for responding to regional problems is a long-standing challenge that remains unresolved today. Powers not delegated to the federal government by the Constitution are reserved to the states, and state governments in turn delegate to local governments authority in certain domains, such as providing basic municipal services and infrastructure and engaging in land use planning. Thus, the structural limits of U.S. federalism leave regions in a twilight zone (Advisory Commission on Intergovernmental Relations, 1972; Banfield & Grodzins, 1958).

In spite of various 20th-century efforts to establish regional institutions and contemporary acknowledgment of regional problems, few examples exist of regional planning agencies with legislative or taxing authority. State and local governments are instead vested with these authorities and understandably are reluctant to cede them. Absent formal regional governments, regional interests in the United States have been pursued through various ad hoc mechanisms, including administrative; procedural; functional, or sectoral; and political, or structural, approaches (Table 6.1). The federally defined framework for regional transportation planning in the United States that emerged in the 1960s has followed a largely procedural approach.

The Emergence of Federally Required Regional Transportation Planning

In the decades prior to the 1960s, federal and state transportation planning and construction had focused on highways providing rural connectivity. State investment, powered by widespread adoption of state gas taxes beginning in the 1920s, focused on building and maintaining state highways (Bloom & Bennett, 1998). Early federal transportation aid to states also sought to ameliorate rural isolation, and state highway departments spent their federal dollars in rural areas. Meanwhile, local governments were responsible for circulation in urban areas; they struggled with severe traffic congestion and with unstable ridership and ailing finances among urban transit systems, typically still privately owned (Jones, 1985). Thus, state and local decision making for transportation occurred on largely separate tracks. Even when federal aid became available in the 1940s for *urban* extensions to the early federal system, existing state-oriented decision-making channels did not include urban officials.

At midcentury, however, this situation began to change. Some cities, including Chicago, Detroit, and San Diego, undertook pioneering urban transportation studies to understand broader metropolitan transportation needs across cities and suburbs

TABLE 6.1. Governance Mechanisms for Regionalism

	Approach	Application
Administrative Regionalism	Administrative regions are a tool of decentralization for federal administrations.	Federal agencies with multistate regional offices (e.g., FTA).
Procedural Regionalism	Federal requirements for intergovernmental cooperation.	The 3-C process in metropolitan transportation planning.
Functional/Sectoral Regionalism	An entity for formal government action is created and organized around a single purpose or service.	Special districts or special authorities: e.g., Delaware Basin (1960s); Appalachian Regional Development Commission (1965).
Political/Structural Regionalism ("Metropolitanism")	Formal government consolidation to achieve metropolitan-/regional-scaled government jurisdiction with purview over multiple functions and services.	Territorial annexations by cities; city–county consolidations; and elected regional governments.
Informal/Collaborative Regionalism; "New Regionalism"	Regional interests are pursued not through top-down, unitary, or formal government, but via informal, collaborative efforts of government agencies, the private sector, citizens' groups, and NGOs.	Water resources collaboratives; other collaborative regional initiatives.

Sources: Advisory Commission on Intergovernmental Relations (1972); Banfield & Grodzins (1958); Fishman (2000); Weir (2000).

and to predict future travel demand and highway expansion needs for the entire region. The federal government encouraged the development of regional councils and more comprehensive planning in metropolitan areas with the 1954 Housing Act and its Section 701 planning grants (Gage, 1988; Mladinov, 1963). Additionally, urban officials and advocates increasingly sought to recast federal transportation spending patterns that had emphasized rural highways (National Committee on Urban Transportation [NCUT], 1958). From these beginnings, a series of forces combined as the second half of the century progressed to underscore the urgent need for regional transportation activities, to create explicit support for them, and to formally establish regional transportation planning in the United States.

The dramatic expansion of federal highway aid with the 1956 Highway Act and ensuing construction of the interstate system brought into high relief the lack of a forum for engagement between state highway planners and urban interests. Cities from San Francisco to New Orleans to New York witnessed the devastating impacts of federally financed, state-led urban freeway projects. In what became known as the "freeway revolts," community groups, academics, and elected officials protested the power imbalance and technical assumptions that had enabled state highway agencies to pursue projects at odds with urban interests (Lupo, Colcord, & Fowler, 1971).

Together, the freeway revolts (Altshuler & Luberoff, 2003; Leavitt, 1970; Lewis, 1999), urban lobbying (NCUT, 1958), and state–urban battles over transit funding (Banfield, 1961) all lent momentum to calls for local government involvement in federal and state investment decision making. Congress and federal officials took such calls seriously (Holmes, 1973; Mohl, 2008), and the early 1960s marked a turning point.

Through the 1962 Federal Aid Highway Act, Congress established a foothold for metropolitan transportation planning and decision making in federal law. Under the act, the federal government would fund transportation projects in urbanized areas only when projects were based on "a continuing comprehensive transportation planning process carried on cooperatively by States and local communities." This so-called 3-C process makes intuitive sense and is still required today: regional transportation decisions based on planning that continues over time, that engages both local officials and states in a cooperative process, and that is comprehensive in scope.

Under the 3-C framework, regional transportation planning is embedded in, not independent of, other levels of government, as evidenced by MPOs' structure and the decidedly intergovernmental environment in which they operate. Federal directives aim to ensure some uniformity across the country with respect to transportation system planning and design. Federal funding for transportation projects in metropolitan regions passes through the state first, with varying implications, and the state must approve an MPO's plans and spending programs. Regions spanning state borders may be served by separate MPOs, as in New York's tri-state metro area, or a single, multistate MPO with participation from respective state departments of transportation (DOTs).

The composition of an MPO's decision-making board is determined by agreement for each region and is usually comprised largely of member local governments, represented by their elected officials. However, following the long U.S. tradition of "local control," cities and counties continue to make local land use and development decisions for their individual jurisdictions, often with only secondary consideration of regional transportation impacts. Local governments also supply a growing share of funding for major transportation projects (Goldman & Wachs, 2003), thereby increasing their influence in regional planning matters. The responsibility for crafting regional transportation plans lies with MPOs, yet realization of those plans is circumscribed by independent actions of the federal, state, and local governments.

The MPO and Its Evolving Role

While short on 3-C implementation instructions, the 1962 law sought to broaden decision making, shifting some power from state highway departments to urban officials. Successive federal transportation legislation and regulations in the early 1970s operationalized the vision, requiring that states and local governments cooperatively establish MPOs as largely advisory bodies responsible for regional transportation planning. (See Tables 6.2 and 6.3.) The environmental movement added weight to calls for broadened decision making beyond state highway departments and their concerns. The 1969 National Environmental Policy Act (NEPA) required that environmental impact assessments be completed for proposed highways (and other federally funded projects), and that alternatives to a proposed highway or other project be identified and their environmental impacts evaluated. The NEPA process that has evolved fosters transportation decision making that balances consideration of environmental factors with engineering and transportation needs. It also provides a formal way for citizens to participate in decision making through its opportunities for public review and comment. Citizens and environmental groups have used lawsuits over NEPA compliance as a powerful tool to challenge proposed highway and other transportation infrastructure projects and to gain concessions for environmental mitigation. Chapter 11 discusses NEPA and the environmental review process in greater detail.

The Clean Air Act (CAA) of 1970 and subsequent amendments also led to greater emphasis on regional-scale transportation planning and enhanced the role of MPOs. The 1970 CAA led to the establishment of national air quality standards and requirements that states formulate plans to achieve these standards. The 1990 Clean Air Act Amendments (CAAA) strengthened these requirements; for regions failing to demonstrate air quality conformity, the CAAA increased sanctions including withholding of federal transportation funding. The requirements called on the use of travel-demand forecasting models to assess the impact of the MPOs plans on air quality. The models used in transportation planning today draw on early forecasting models from mid-century transportation studies that defined the widely adopted "four-step" approach (see Chapter 5).

The 1991 Intermodal Surface Transportation Efficiency Act (ISTEA) was a major turning point in the evolution of U.S. regional transportation planning. This legislation shifted transportation decision making more visibly from the state to the regional scale, not only locating more decision making authority with MPOs but also giving them flexibility to meet local needs. In particular, several interrelated ISTEA directives altered the substance of regional planning and the allocation of funds to enhance the MPO role. These included requirements that both long-term

TABLE 6.2. Key Characteristics of Metropolitan Planning Organizations

Formation

- Required in urbanized areas with population of 50,000 or more.
- Designated by joint agreement between the governor and local governments in region.
- Local government signatories to the agreement must represent at least 75% of the region's population.

MPO Structure

- May be an independent organization, or part of a regional council or governments, a city or county government office, or state transportation department.

Composition

- Governing or policy board.
- Technical and citizen advisory committees to advise board.
- Executive director, board-appointed.
- Professional transportation planning and technical staff.

Governing Board

- Consists of local elected officials, transportation operating agencies, state (DOT) officials, and other appropriate stakeholders.
- Board membership, voting rights, and voting rules determined by agreement between the governor and local governments in region.

Funding for MPO Planning* (Median Amount)

- Federal funding: $448,000
- State support: $37,000
- Local support (e.g., sales, gas, & property tax): $52,000

*Reported in U.S. Government Accountability Office (2009a, 2009b).

plans and near-term investment programs realistically reflect anticipated funding levels (fiscal constraint); that proposed transportation system improvements support federal air quality goals (conformity); and that MPOs actively involve the public in the development of plans and investment choices (public involvement). In addition, ISTEA funding provisions allowed regions more flexibility and control in funding local transportation priorities, furthering multimodal planning. The law established federal funding pots explicitly for bicycling and walking investments and enabled funds formerly available only for highway investments to support transit, nonmotorized modes, and demand management instead. Additionally, MPOs gained authority to more directly control the expenditure of some funds.

Subsequent federal surface transportation authorization bills have generally reinforced the MPO role in regional transportation planning, while also streamlining some provisions and adding others. For instance, the 2012 authorization law, Moving Ahead for Progress in the 21st Century, referred to as MAP-21, has put many ISTEA-era planning requirements in place, but put new emphasis on performance measurement in planning. The 2015 authorization law, the Fixing America's Surface Transportation Act, called the FAST-Act, continues these features.

TABLE 6.3. Metropolitan Planning Organizations in Profile

	Sacramento	Jacksonville	Chicago	St. Louis	Austin
MPO	Sacramento Area Council of Governments (SACOG)	North Florida Transportation Planning Organization (North Florida TPO)	Chicago Metropolitan Agency for Planning (CMAP)	East–West Gateway Council of Governments (EWGCOG);	Capital Area Metropolitan Planning Organization (CAMPO)
Regional population (2010)	2,274,557	1,274,426	8,444,660	2,571,253	1,759,122
Regional size (sq. mi.)	6,189	2,681	4,096	4,586	5,307
No. of counties in region	6	4	7	7	6
Counties in region	El Dorado, Placer, Sacramento, Sutter, Yolo, Yuba	Clay, Duval, Nassau, St. Johns	Cook, DuPage, Kane, Kendall, Lake, McHenry, Will	MO: Franklin, Jefferson, St. Charles, St. Louis IL: Madison, Monroe, St. Clair	Bastrop, Burnet, Caldwell, Hays, Travis, Williamson
No. of cities in region	22	12	283	41	49
TIP Expenditure (all sources; most recent TIP)	$3.3 billion (FY 2010/2011–2013/2014)	$3.8 billion (FY 2013/2014–2017/2018)	$14.2 billion (FY 2014–2019)	$1.8 billion (FY 2015–2018)	$1.1 billion (FY 2015–2018)
MPO structure	MPO within COG	Independent MPO	MPO within regional planning organization	MPO within COG	MPO is hosted by city
Combined MPO/COG board	Yes	No	No	Yes	No
MPO board size (voting members)	31 total; 1 member/county (2 from Sacramento County), 1 member/city (3 from Sacramento)	15 total; 4 from member counties, 4 from Jacksonville, 1 from St. Augustine, 1 beach communities representative, 5 from transportation agencies	18 total; 1 from Chicago, 1 from other cities, 7 from counties, 9 from transportation agencies	24 total; 3 from IL cities, 5 from IL counties, 2 from MO cities, 9 from MO counties, 4 regional citizens	19 total; 1 member/county (4 from Travis County), 1 member/city (4 from Austin), 2 from transportation agencies
Staff size (FTE)	50 (COG)	9	92	56 (COG)	17

One of the ongoing debates since ISTEA has focused on the appropriate role of the federal government in developing the surface transportation system. Some interests have pushed for devolution of the federal program to the states, arguing that completion of the interstate system means federal involvement in transportation is no longer needed. Such debates could have serious implications for regional transportation planning and MPOs, which exist through the federal requirements that devolution could make moot. At the same time, however, the ongoing shift of U.S. transportation funding responsibility to states and local governments is weakening the role of the MPO.

THE REGIONAL PLANNING PROCESS

Practically speaking, the purpose of regional transportation planning in the United States is to inform and to make decisions about how to invest federal funds in regional transportation systems. What facilities or services should be improved, enhanced, or added to best serve regional access and mobility in the future? Consequently, establishing a guiding vision for the region's desired future is central to the transportation planning process. This vision should inform the goals and standards for judging both existing conditions and possible futures. If current trends suggest that a region's likely future will not match its desired future, that region can make investments and adopt policies that will better direct it toward that desired future.

This model of planning is visible in the two key steps at the heart of regional transportation planning: development of the long-range transportation plan (LRTP) and of the short-term transportation improvement program (TIP), akin to a capital investment program. Federal policy requires these steps, establishes key parameters for their execution, and makes the receipt of federal transportation funds contingent upon their approval. The content of both documents is largely up to the MPOs, but federal law dictates critical components of the process, including the use of public involvement, travel demand forecasting, air quality conformity analysis, and performance measures in LRTP and TIP development. Fulfillment of each step involves review and approval at multiple levels, including approval by the MPO board, the state, and the U.S. Department of Transportation, providing for involvement of many stakeholders along the way.

Figure 6.1 depicts the general flow of, and important inputs to, this process. Regional goals and visions crafted considering public input, available budgets, and federal requirements in turn inform the development of alternative strategies, long-range plans, and near-term investments to improve operation of the transportation system.

The Long-Range Transportation Plan (LRTP)

The LRTP articulates the region's vision for its transportation system for 20 or more years in the future and serves to guide transportation investments and other policies within the region. Reflecting the 3-C approach, these plans are comprehensive in scope, addressing a full range of transportation modes for passenger and freight

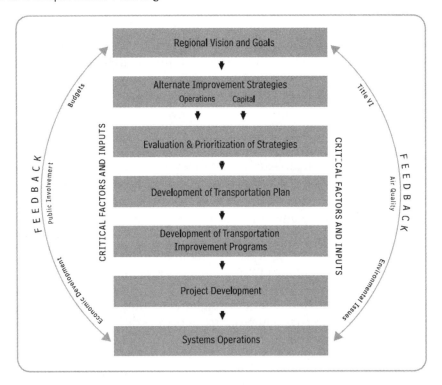

FIGURE 6.1. Regional transportation planning process.

demand. Although focused on the long-term future, they represent continuing planning and are updated at least every 5 years following federal requirements. They are developed cooperatively with state, local, and other regional agencies.

In developing their long-range plans, MPOs are increasingly adopting a "visioning" framework (Ames, 1993) that asks four critical questions: (1) Where are we? (2) Where are we going? (3) Where do we want to be? (4) How do we get there? Spurred by growing awareness of the environmental, economic, and social impacts of unhampered growth and road expansion and by the need to explore the costs and benefits of different future development paths, the approach discourages reactive planning. Rather than planning transportation systems simply to accommodate anticipated population growth and travel demand, vision-based planning weighs alternative future scenarios, using regional plans instead to strategically influence how future growth and mobility unfold and to facilitate participatory planning. Widely praised visioning processes in Portland, Salt Lake City, and Sacramento in the 1990s and 2000s inspired other regions to follow their lead.

Following the visioning framework, the MPO first assesses current conditions and identifies problems and opportunities. Second, it examines trends under current policies and uses travel-demand forecasting models (discussed below and in Chapter 5) to assess system performance, assuming that no new investments will be made (called the "no build" scenario). To answer the third question, the MPO develops its guiding vision as the basis for the LRTP goals. Although MPOs are free to set their own goals, federal requirements to provide for public involvement and

use performance targets influence the goal-setting process. Fourth, the MPO identifies a plan of action for realizing its vision, forming a list of proposed investments
and policies for the plan period. MPOs may examine multiple investment scenarios
when developing their plans; travel-demand forecasting models are used to assess the
contribution of different investment scenarios toward meeting plan goals and realizing the regional vision. The list of proposed investments becomes the core of the
long-range plan.

When identifying proposed investments, MPOs are required by federal law to
show that the long-range plan—and any funded program ensuing from it later—
demonstrate both "conformity" with federal air quality standards and "fiscal
constraint." To be considered "fiscally constrained," a long-range plan must limit
proposed investments to the amount of funding likely to be available over the plan
period. Prior to ISTEA, long-range regional transportation plans typically included a
long "wish list" of projects proposed by local governments, the costs of which would
far exceed the funding anticipated over the life of the plan. ISTEA's requirement for
"fiscal constraint," both within the long-range plan and subsequent investment programs, compelled MPOs to forecast future regional funding and to winnow proposed
projects accordingly. In partial compromise, however, the law also allowed MPOs
to include a second list of "illustrative projects" that would be implemented should
future revenues exceed forecasts.

Additionally, long-range plans are also subject to federal air quality requirements,
articulated in the Clean Air Act and its amendments, for transportation-related pollutants such as ground-level ozone, carbon monoxide, small particulate matter, and
sulfur dioxide. When the levels of these pollutants in a region exceed national air
quality standards, the regional transportation plan and subsequent spending programs, if financed even partly with federal funds, must conform to wider state plans
to lessen air pollution and meet those national standards. Under the CAAA (1990)
and ISTEA (1991), regions unable to show air quality conformity could experience
sanctions, including withholding of federal transportation funds.

The Transportation Improvement Program (TIP)

The LRTP, once it has been approved, is implemented through the development of the
transportation improvement program (TIP). Among the projects listed in the LRTP,
the MPO prioritizes those investments to be made in the near term, over the next 3–5
years. To prioritize projects for TIP inclusion, MPOs apply evaluation criteria that,
ideally, are tied to the LRTP performance targets. Different criteria may be used to
prioritize projects in different funding categories, and MPOs vary as to the formality
and transparency of their scoring procedures.

The MPO lists these "programmed" projects in the TIP along with their funding
sources, ensuring fiscal constraint. The MPO also performs a conformity analysis of
the TIP. Using modeled forecasts of travel patterns and emissions, it tests whether the
programmed transportation projects will help or hinder regional efforts to meet specified emissions standards under the Clean Air Act. Not merely procedural requirements, fiscal constraint and conformity rules enable MPOs to assert regional priorities more firmly when selecting projects than if they had to include all projects

proposed by local governments. For example, some MPOs have leveraged fiscal constraint and conformity requirements to counter congressional earmarks for parochial projects that would disrupt regional transportation priorities (Sciara, 2012).

To be included in the TIP, a project must be sufficiently well developed to ensure that its cost estimate is fairly solid and that its execution or construction can advance within the 3- to-5-year TIP timeframe. Thus, project programming is linked to project development. The agencies responsible for building projects and operating the transportation system—the "implementing agencies"—develop individual projects from the concept stage, through design, environmental review, and engineering, and ultimately to construction and operation. States own and operate the highway system; cities and counties own and operate local roads; and state, multicounty, or local entities may own and operate rail transit and bus systems, airports, and maritime ports. As implementing agencies, they may propose projects for LRTP and TIP inclusion. MPOs are not implementing agencies and cannot include a project in the TIP unless an implementing agency will sponsor it.

In developing the TIP, MPOs program, or budget, funds for specific investments. Yet, with rare exceptions, they do not generate their own revenue for such investments. Instead, MPOs rely on the array of federal, state, and local funds available, and they generally have little independent authority over how most of those funds are expended. Most federal highway funds flow to state departments of transportation, through so-called core programs such as the National Highway Performance Program focused on the interstate and national highway systems. Similarly, most federal transit funds flow directly to transit agencies, making operators most influential in decisions about their expenditure.

While state DOTs and MPOs coordinate the expenditure of highways funds in urban regions, the DOTs generally take the lead in identifying the projects. MPOs do not control the funds but may negotiate with the state over their use (Edner & McDowell, 2002). Some states elect to "suballocate" at least some federal funds, passing their control to substate regions. Yet funds may be suballocated to state DOT subdivisions, not MPOs, and suballocation practices may be driven less by regional need than by the politics of geographical distribution, including state DOT district structure, historical funding trends, and statehouse politics (Hill, Geyer, & Puentes, 2003).

Prior to ISTEA, MPOs could exercise far less flexibility when programming available funds for desired projects. Federal funding was formerly distributed to specific categories of highways; funding from one category could not be spent on highways in another category, and certainly not on nonhighway investments. Highly differentiated funding siloes gave the federal government significant influence over the kinds of projects funded. By overhauling the structure of federal funding, ISTEA enabled DOTs and MPOs to transfer some existing funds to different categories, depending on where they were most needed. Furthermore, ISTEA gave large MPOs direct responsibility for programming the urban share of Surface Transportation Program (STP) funds, and it created the Congestion Mitigation and Air Quality Improvement Program (CMAQ) to support varied investments—including highways, transit, nonmotorized modes, or even demand management programs—provided they lessen air pollution in regions challenged to meet federal air quality standards. While CMAQ funds flow to the state, many states suballocate them to the MPOs in urban regions where they are needed.

Some MPOs are using this discretion over flexible transportation funds to support regional transportation and land use goals (Sciara & Handy, 2013). In the San Francisco Bay Area, the Metropolitan Transportation Commission's Transportation for Livable Communities program was ground-breaking in this regard, and its successor, the One Bay Area Grant program, furthers this innovation. These initiatives use flexible federal dollars as incentive grants to encourage planning and capital investments that support transportation alternatives and development in priority areas. They are also bolstered by MPO-board policies ensuring that other funds are distributed following similar principles.

Key Requirements of the Regional Transportation Planning Process

The regional transportation planning process first emerged in an era when automobiles were seen as symbols of modernity and system efficiency was the overriding concern. Engineers and planners sought to minimize travel time and maximize vehicle throughput, and road building was the primary strategy (Meyer & Miller, 2001; Wachs, 2004). The broader range of goals and strategies now under consideration (see Chapter 1) is both a reflection of and an impetus for revisions to MPO planning processes (e.g., Transportation Research Board [TRB], 2001). In some cases, federal policy has pushed MPOs to evolve, and in some cases MPOs are innovating beyond federal requirements. Federal law prescribes key components of the process; public involvement, performance-based planning, and travel demand forecasting are discussed here.

Public Involvement

While the public had little say in decisions about federal transportation investments prior to the 1960s, the freeway revolt increased conviction that citizens and their local officials should play a role in shaping future investments. The 3-C planning process as envisioned in the 1962 Highway Act would involve local communities and states in transportation planning and decision making, yet subsequent planning regulations failed to specify how that should occur, and many MPOs gave public involvement short shrift.

ISTEA placed renewed emphasis on public involvement by requiring MPOs to develop a public participation plan and instituting an MPO certification process that considered the effectiveness of this plan. Additionally, the impetus for public involvement in planning was further strengthened by President Clinton's 1994 executive order on environmental justice. Based on Article VI of the 1964 Civil Rights Act, the order requires all federal agencies to include environmental justice in their mission, by requiring consideration of a federal policy's positive and negative impacts on minority and low-income communities and by requiring agencies to ensure "full and fair participation by all potentially affected communities in the transportation decision-making process" (Federal Highway Administration [FHWA], 2000). These enhanced public involvement requirements implicitly acknowledged that the transportation planning process had theretofore focused too narrowly on the traditional concerns of transportation agencies, not necessarily on the goals and values of the public they serve (Poorman, 2001). Over time, increased public involvement has gone hand-in-hand with a broadening of transportation planning goals in many regions (Handy, 2008).

Public involvement activities often help to inform the MPO's vision of "where do we want to be" and to set goals, and MPOs employ many different techniques to achieve this aim. At a minimum, MPOs are required to hold public hearings in developing the LRTP and the TIP. Often, these hearings are held late in the process, following the release of draft documents and prior to their final adoption.

Public involvement *should* inform all stages of the planning process, and MPOs have increasingly focused on involving the public earlier in the planning process. Common approaches include forming citizen advisory committees, holding public workshops, and administering opinion polls (Howard/Stein–Hudson Associates, 2011; Howard/Stein–Hudson Associates & Parsons Brinckerhoff Quade and Douglas, 1996; Morris & Fragala, 2010). Federal law encourages the use of visualization techniques to describe the plan and requires MPOs to make the plan accessible electronically, as through a website. Increasingly, MPOs are incorporating web-based and social media tools to increase citizen engagement; these practices are evolving rapidly (Bregman & Watkins, 2013).

Achieving meaningful public involvement is not easy. Even in the face of vigorous public involvement efforts, participants may not be representative of the broader citizenry; voices from vulnerable and less empowered communities may be absent given language, cultural, and socioeconomic barriers (Post, Buckley, Schuh & Jernigan, Inc., 2006); and organized interest groups may dominate the process. Even when public participation is robust, stakeholders may disagree as to whether better investment decisions have resulted and what should count as better. Case study-based guidance on best practices for public involvement is abundant, but systematic evaluation of the impacts of public involvement is rare. One study of state-level citizen participation efforts suggests encouragingly that citizen inclusion in decision making increases agency effectiveness and cost efficiency, while also allowing citizens to increase their understanding of transportation decision making (Neshkova & Guo, 2012).

Performance-Based Planning

Reflecting concern over public sector decision making, performance, and accountability, a movement toward performance-based transportation planning emerged in the late 1990s (Hartgen & Neumann, 2002; TRB, 2001), leading to a new federal requirement in MAP-21 that MPOs adopt a performance-based approach to transportation planning. Under MAP-21, an MPO's performance-based approach must support national transportation goals (Table 6.4) and use performance measures and associated targets as tools for objectively assessing how both the LRTP and TIP will help to achieve the region's goals (Grant et al., 2014). Thus, performance-based planning is a way to ensure that a region is moving toward its vision of "where we want to be" and that stakeholders agree on how to measure its progress. Additionally, whereas

TABLE 6.4. U.S. Department of Transportation Goals under MAP-21
Safety: To achieve a significant reduction in traffic fatalities and serious injuries on all public roads.
Infrastructure Condition: To maintain the highway infrastructure asset system in a state of good repair.
Congestion Reduction: To achieve a significant reduction in congestion on the National Highway System.
System Reliability: To improve the efficiency of the surface transportation system.
Freight Movement and Economic Vitality: To improve the national freight network, strengthen the ability of rural communities to access national and international trade markets, and support regional economic development.
Environmental Sustainability: To enhance the performance of the transportation system while protecting and enhancing the natural environment.
Reduced Project Delivery Delays: To reduce project costs, promote jobs and the economy, and expedite the movement of people and goods by accelerating project completion through eliminating delays in the project development and delivery process, including reducing regulatory burdens and improving agencies' work practices.

MPO performance has largely been defined with respect to procedural requirements, performance-based planning orients the evaluation of MPO accomplishments toward outcomes (U.S. Government Accountability Office [U.S. GAO], 2009a).

Developing appropriate performance measures that are clearly tied to the goals of the MPO is critical to the success of performance-based planning. Planners have traditionally measured transportation system performance with so-called level-of-service (LOS) standards, as defined by the Highway Capacity Manual (TRB, 2000), an appropriate approach if the goal is to reduce traffic congestion. However, the broadening of transportation planning goals that began with ISTEA is spurring the development and use of alternative measures of system service to reflect the travel experience of cyclists, pedestrians, and other nonauto modes users as well (Matute & Pincetl, 2013). In addition, some MPOs are now developing broader measures of investment outcomes, such as changes in accessibility to jobs or reductions in health care costs. Challenges MPOs face in developing and using performance measures include lack of data (Pickrell & Neumann, 2001), little standardization of measures for broader goals, and institutional changes needed to ensure their effective application (Hartgen & Neumann, 2002).

Travel Demand Forecasting Models

Travel demand models are a tool used by MPOs to answer the question "Where are we going?" and to assess whether proposed transportation strategies and projects will meet the region's goals, informing the question "How do we get there?" Travel modeling techniques were developed in the 1950s to forecast whether proposed highways would meet projected demand (Rosenbloom, 1988) and evolved into the well-developed and widely used four-step modeling approach, described in Chapter 5. MPOs have invested substantial resources in developing these models, updating and improving them over time, and applying them in developing their regional transportation plans (Johnston, 2004).

Model outputs, generated by an MPO's technical staff or outside consultants, are used to inform the MPO policy board's decisions. For instance, when the region considers various combinations of roadway and transit improvements for the LRTP, the model can forecast how each combination might affect overall congestion, given assumptions about other investments, future population growth, and economic development. Such models are essential to performance-based planning, in that MPOs can use them to forecast whether the LRTP is likely to meet its future targets.

However, because travel demand models were designed to produce forecasts of LOS, they speak most directly to the goal of roadway congestion reduction and are best suited for evaluating proposed investments that aim toward this goal. The models do not easily represent certain kinds of proposed investments or policies, generating criticisms among advocates for the environment and for economically vulnerable groups, and calls for more state-of-the-art modeling techniques. For example, because most regional models do not currently include bicycling and walking as modes, they cannot be used to assess the impact of proposed bicycling and pedestrian infrastructure investments without "postprocessing" of model outputs. In addition, performance measures tied to goals other than congestion reduction are not standard model outputs. Hence, as they consider broader transportation goals, MPOs are increasingly using model outputs to derive new performance measures that reflect the wider range of goals in their plans. For example, these models have been used to assess the impact of proposed investments on accessibility to jobs for lower-income workers.

The use of travel-demand forecasting models in the regional transportation planning process also presents a challenge from a public involvement standpoint (Handy, 2008). These models are often described as a "black box," the complex inner workings of which are known only to the modelers themselves (Beimborn & Kennedy, 1996). On the one hand, modelers need to acknowledge the judgment that goes into their work to avoid the "illusion of technical objectivity" (Wachs, 1987). On the other hand, members of the public may feel that modelers are inserting their own biases (or those of their clients or employers) into the models, leading to forecasts that support a favored alternative. For instance, if models forecast only a subset of performance measures, the public as well as decision makers may have an incomplete picture upon which to base decisions and be unsure how to weigh the forecasted measures against considerations for which they do not have forecasts. In general, observers note a weak link between modeling results and decision making (Meyer & Miller, 2001), despite the considerable sums invested in these models.

Impending improvements in travel-demand modeling, such as activity-based models and microsimulation models, described in Chapter 5, may help with both of these challenges. Some argue that activity-based models are more intuitive and thus less opaque for participants in the planning process. In addition, they enable new performance measures that may better match the broader range of goals in today's plans. For example, activity-based models can be used to estimate accessibility measures based on all activities and trips throughout the day rather than just peak-hour commute trips (Dong, Ben-Akiva, Bowman, & Walker, 2006). Microsimulation models can be used to estimate benefits and impacts for different subsets of the population, defined by characteristics such as gender, income, auto availability, or household structure (Castiglione, Hiatt, Chang, & Charlton, 2006). It is unlikely that all performance measures can be derived from even these new travel demand models, but they offer potential. Such improvements are essential for maintaining the technical legitmacy of MPOs in the eyes of stakeholders, who increasingly challenge traditional travel-demand models.

CHALLENGES IN PROVIDING REGIONAL LEADERSHIP

Along with requirements for regional transportation planning, the federal government established MPOs as a way to strengthen across metropolitan areas the voices of local governments in what had been a largely state-driven highway planning process. Later, the MPO process was envisaged as a widely participatory forum for creating a unified regional transportation vision and for providing the leadership needed to realize it (McDowell, 1999; Turnbull, 2007). However, many scholarly assessments of U.S. regional transportation planning suggest that reality falls short of this hope (Gage & McDowell, 1995; Wolf & Farquhar, 2005). Although vested with the important responsibilities of producing the LRTP and the TIP, MPOs face significant barriers to asserting leadership at the regional level (Mallett, 2010).

At the core of many of these barriers is the broad reluctance within the U.S. federalist system of government to empower regional-scale government institutions to address public problems. The processes and institutions through which federal policy requires regional plannning in the United States represent a compromise position: the reality of travel flows and of transportation system performance requires regional-scale planning for major infrastructure investments, but federal policy confers little real authority to regional bodies, leaving the choice to empower regions to individual states. State governments have varied track records when it comes to top-down support for regional-scale decision making. For transportation investments, ISTEA and subsequent legislation has transferred only partial decision-making power in metropolitan regions from state DOTs to MPOs, and states have resisted giving it up (Goldman & Deakin, 2000). At the same time, local governments have routinely promoted their own interests in the regional planning process over those of the region, and many have begun to raise and allocate their own locally generated transportation funding. Sandwiched between state government and local governments, MPOs operate in a decidedly intergovernmental environment that puts checks on regional efforts. Nevertheless, many MPOs have successfully used levers available to them to foster a truly regional vision and to advance the investments that would support it.

Constraints on Regional Planning

Federal law empowers MPOs to conduct long-range transportation planning, lead the processes to establish a regional vision, set goals, generate and evaluate plan and project alternatives, and prioritize projects for funding. While this authority is not insignificant, it is also limited in several domains critically important for realizing regional plans. In various ways, the authority to make final decisions or take decisive action resides squarely *outside* the MPO.

First, as noted earlier, it is "implementing agencies," not MPOs, that have authority over construction and operation of the transportation system and that propose projects for LRTP and TIP inclusion. MPOs cannot fund a project without an implementing agency sponsor, whether the state DOT, a public transit property, or a local transportation or public works department. Furthermore, the governor and the state DOT retain veto authority over MPO-selected projects (McDowell, 1999).

Second, despite increasing recognition that land development shapes travel demand and that regional transportation planning should consider alternative development patterns in identifying and prioritizing investments, land use control remains firmly within the purview of local governments, not regional entities such as councils of governments or MPOs. Cities and counties craft the general plans and zoning regulations that guide growth and hence influence travel demand, and they approve subdivision and development applications. These choices are based on local priorities and, increasingly, on the drive to generate local revenue from land use (Burnes, Newmark, & White, 2014; Lewis, 2001; Wassmer, 2006), not the regional plan.

A third check on the authority of regional planning is the absence of regional sources of revenue to support transportation investment. The widening gap between federal transportation funds for metropolitan investment and the need for such investment makes nonfederal transportation funds increasingly important. State and local governments are empowered to raise revenues by imposing taxes and fees, but most MPOs are not (Sciara & Wachs, 2007). Regional authorities that operate transit systems, toll roads, or ports and airports may have revenue-generating capacity, but their proceeds typically support that particular modal system and not wider regional transportation facilities or services. Local and state governments are likely to use the revenue they generate and control to make transportation improvements scaled to provide local or state benefits, leaving a resource vacuum for investments that provide dispersed regional benefits.

Transportation funding from local governments, in the form of local-option tax revenues, play an increasingly important role in supporting transportation projects in metropolitan areas (Goldman & Wachs, 2003), a trend particularly visible in California. Through ballot initiatives, more than 20 California "self-help" counties have dedicated quarter- or half-cent sales tax increments to pay for predefined transportation improvements, raising close to $4 billion for such investments in 2014 alone. While locally based tax measures can significantly increase the scale of transportation investment possible in an urban region and can work in favor of transit, as evident in Denver and Los Angeles, they may also develop alongside, not within, the regional planning process. Lest they fail at the ballot box, tax measures must support projects with voter appeal, making regional benefit a secondary consideration. In addition, state legislation empowers counties and cities to pursue tax revenues for transportation but typically does not require those revenues to pay for projects in the regional plan, creating ambiguity for regional transportation planning that is required but inconsistently supported with available resources.

The level of internal resources available to provide for MPO staff and technical activities can be another check on regional planning. To support MPO operations, federal law devotes a small portion (historically about 1–1.25%) of federal highway funds to metropolitan planning. While about 80% of MPOs rely on federal support for the majority of their planning activities, most also supplement their administrative budgets with funds from the state and the region's member local governments (U.S. GAO, 2009b). Whether an MPO is a stand-alone entity or is housed within another agency may also influence its internal resources, for instance, by shaping how it recruits technical staff or its ability to share resources. Whereas in the early 1970s MPOs were largely housed within state DOTs, this model has waned. By 2010, MPOs were commonly housed within city or county governments or regional councils of governments (Bond & Kramer, 2011). An MPO's internal resources determine whether it is able to move beyond simply fulfilling federal requirements to proactively address broad regional transportation questions.

Finally, realization of a region's long-term transportation vision rests in part on many discrete decisions made not only by the MPO board as a group but also by cities or transit operators acting independently. Whether those decisions support the regional vision or not may depend on whether efforts to create a regional transportation system are viewed as appropriate and desirable. For MPO boards, political legitimacy is especially dependent on the willingness of member jurisdictions to weigh decisions from a regional perspective. Yet the tension between local and regional interests is built into the structure of regional planning. Local elected officials are a dominant force on most MPO boards (Bond & Kramer, 2010), which ensures local communities a voice in regional planning, yet members may put local concerns ahead of regional interests (Goetz, Dempsey, & Larson, 2002). As a result, MPOs have been known to distribute funding across local jurisdictions according to geopolitcal considerations rather than prioritizing projects based on regional needs (see Chapter 10).

The legitimacy of regional planning may also suffer if MPO boards are not considered representative of the regions they serve. One analysis of MPO board composition shows that central cities are underrepresented relative to suburban areas on a representative-to-population basis (Goldman & Deakin, 2000). Furthermore, MPO voting structures have the potential to bias funding decisions toward highway projects and away from investments in transit (Nelson, Sanchez, Wolf, & Farquhar, 2004; Sanchez, 2006). In addition, transit agencies have also had limited representation on MPO boards (Hoover, McDowell, & Sciara, 2004), though MAP-21 established new requirements for transit agency representation on MPO boards may help to correct the imbalance.

At the same time, the fundamental legitimacy of long-range transportation planning (O'Toole, 2008) and regional planning itself—and, indeed, of planning of any sort—has been questioned in the early 21st century United States by certain libertarian political interests, most notably the Tea Party, that seek to diminish government's role across broad sectors of public engagement (Frick, 2013). These challenges may have gained traction in part because of the ambiguous position of regional-scale planning in the United States, transportation or otherwise, coupled with the strong "home rule" tradition empowering local governments.

Making It Work

To carry out their regional planning responsibilities, MPOs thus depend on state and local agencies, but MPOs do retain some leverage over them, stemming from specific federal directives. First, only projects listed in the LRTP are eligible to be programmed in the TIP. This means that state, local, and transit agencies must engage with the regional planning process, at least if they seek federal funding for their projects. Second, the requirement that the list of projects in the LRTP be fiscally constrained means that the MPO must decide what projects to include rather than simply listing every project proposed. Fiscal constraint creates an opportunity to apply criteria that favor regional rather than local interests. Third, air quality conformity requirements mean that those regional interests include reducing the emissions of air pollutants and give MPOs the power to say no to projects that do not contribute to this goal. Fourth, MPOs can use the flexibility in federal funding programs to steer regional efforts.

Even so, MPOs are still reliant on support from both state and local agencies (Goetz et al., 2002; Goldman & Deakin, 2000), as well as internal support in the form of strong MPO leadership coupled with technical competence and credibility (Handy, 2010; Handy & McCann, 2011). In sum, its ability to play a leadership role in the region depends not only on the assets the MPO itself brings to the process but also on its relationships with the state DOT and local governments. Keys to success may be summarized as:

- **Top-down support** from the state DOT and state policy more generally to enable the MPO in facilitating its regional planning efforts. Examples include additional suballocation of federal funding to the MPO, assistance with technical analysis, and enabling legislation for regionally scaled local-option taxes or other funding mechanisms. For instance, the State of California took a significant step to devolve transportation funding to MPOs and other county-level entities in 1997. The state also granted the San Francisco Bay Area's MPO the authority to levy a regional gas tax, though it has yet to exercise it.

- **Bottom-up support** from local governments (cities, counties, and others) in the form of consensus on the definition of regional problems and their solutions and a willingness to balance regional solutions with local interests, with backing from local interest groups. For example, Clark County, Nevada, voters approved a new way to fund Las Vegas transportation improvements with regional revenues, allowing the Las Vegas MPO to borrow the county's taxing authority to support projects consistent with the regional plan (Sciara & Wachs, 2007).

- **Internal support** in the form of strong leadership from the MPO board and its director, coupled with strong staff, with technical competence and credibility. Strong MPO leadership can help to foster state and local support where it is lacking (Goetz et al., 2002), and some evidence suggests that MPOs dominated by nonelected managers produce more regionally oriented transportation policies than those dominated by local elected officials (Gerber & Gibson, 2009). Strong board leadership allowed the Lee County, Florida, MPO to face down pressure in 2006 to amend its LRTP to add a new interchange project. Though never part of the regional plan, the project was awarded a $10 million congressional earmark, supported by local real estate developers, builders, and other businesses. Unswayed by the earmark or potential congressional retaliation, the MPO board used its leverage over LRTP approval to reject the patently parochial project (Sciara, 2009).

A number of MPOs have presided over successful regional transportation planning efforts, in which a regional consensus emerged on a new direction for the transportation system. In addition, with encouragement from the federal government and some states, MPOs are increasingly exploring how alternative land use scenarios might address travel needs and reduce transportation's economic, environmental, and social externalities. For instance, California State Law SB375 asks MPOs to develop a regional land use vision that, when paired with supportive transportation investment, will reduce automobile reliance and associated greenhouse gas emissions. As required, each MPO in the state has adopted a "sustainable communities strategy" identifying a set of investments and other actions that, as forecasted by travel demand models, is likely to achieve the reduction targets. The state's four largest MPOs have prioritized infill development in areas served by transit along with investments in alternatives to driving.

Whether regional transportation planning is easier in areas with strong regional planning in other domains is not clear. The Portland, Oregon, example suggests that regional planning efforts can be synergistic: with its regionwide, directly elected council, state-granted authority over the establishment of a growth boundary for the region, as well as the regional transportation planning powers conferred by federal policy, Portland Metro has succeeded more than most regions in achieving effective coordination of transportation and land use strategies (Abbott, 1997; Seltzer, 2004). More generally, regional transportation planning may have a greater chance of success if it builds on previous collaborations at the regional scale (Deyle & Wiedenman, 2014). On the other hand, some research suggests that ongoing collaboration in regional transportation planning may in fact be negatively affected by existing cooperative processes in other local policy arenas, such as watershed planning, other resource management, estuary planning, housing, economic development, and social service delivery (Lubell, Henry, & McCoy, 2010). The limited resources that agencies can devote to collaborative planning efforts may explain this counterintuitive finding. Each local government must weigh its own costs of collaborating—staff time and other resources—against the potential benefits of collaborating (Gerber, Henry, & Lubell, 2013).

An MPO's chances of success in creating and implementing a regional vision may also rest on circumstances beyond its control, including underlying demographic and sociopolitical dynamics in the region it serves. For instance, collaboration appears to be more likely in regions where local governments are more homogeneous with respect to political attitudes, socioeconomic characteristics, and size and growth patterns. Geographic proximity also promotes collaboration by increasing interdependencies, on the one hand, and opportunities for and ease of interacting, on the other hand (Gerber et al., 2013). That said, an MPO can maximize its chances through careful design of the planning process and selection of the key participants (Deyle & Wiedenman, 2014). Devoting adequate resources and dedicated staff to the process, ensuring access to high-quality and credible technical information, and encouraging ongoing participation of individuals who can make commitments on behalf of their organizations can help to overcome any unfavorable starting conditions.

REGIONAL TRANSPORTATION PLANNING IN THE FUTURE

Regional transportation planning in the United States has evolved since the mid-20th century as an intergovernmental process for making plans and decisions for investments in the region's transportation system. It depends on relationships between the state and the MPO region, between the MPO as a whole and the individual local governments that comprise it, between the MPO and transit agencies, between the MPO and the public, all within the context of federal requirements. Although this decidedly intergovernmental process results in notable checks on MPOs and their authority to assert a regional vision, many MPOs have succeeded in doing so by using the powers they do have in effective and creative ways. Federal requirements, which traditionally emphasize process, have recently added a new outcome-based dimension to evaluation of MPOs accomplishments, through performance-based planning. A number of trends suggest that in the coming decades the regional scale will become increasingly important in planning the transportation system to meet societal needs while also addressing its economic, environmental, and social impacts. At the same time, regional planners are likely to face many new challenges and uncertainties, including:

• **Increasing connectivity.** Metropolitan regions are becoming increasingly interconnected, with extensively overlapping commute sheds and interdependent economies. The concept of the "megaregion" is used to describe parts of the United States where economic interactions spread far beyond the conventional boundaries of metropolitan regions (Ross, 2009; Wheeler, 2009). Examples include the Boston-to-Washington corridor on the East Coast, and the Los Angeles–San Diego and Bay Area–Sacramento conurbations in California. What does a half-century of experience with regional planning reveal about how best to address "megaregional" problems?

• **Technology.** While telecommuters are not likely to overtake drive-alone commuters as a share of total daily travel (see Chapter 4), other technological innovations may rearrange daily travel patterns in unpredictable ways. What will driverless cars, for example, mean for the regional transportation system? Crowd-sourced traffic information and routing guidance (e.g., Waze) could improve the efficiency of the system, but could they also generate unanticipated consequences? And will new technology-enabled rideshare services alter patterns of vehicle ownership or transit usage?

• **Equity.** Who benefits is another ongoing question for regional transportation planners. With long-standing environmental justice requirements, has the planning process in fact become more inclusive? Will the move toward performance-based planning help to ensure that the needs of all segments of society are well served? How can regional planners find an acceptable balance between satisfying regional needs and ensuring an equitable distribution of costs and benefits at the local level, particularly with respect to vulnerable communities?

• **Environmental Sustainability.** As California has acknowledged in state legislation, regional transportation planning is imperative for achieving needed reductions in greenhouse gas emissions. While technological solutions can accomplish much, analysis shows that reductions in vehicle miles traveled (VMT) are also essential to meeting reduction targets. Is the recent flattening of heretofore long-standing increases in VMT per capita a lasting trend? If not, what strategies can accomplish VMT reductions while also meeting needs for access? In the meantime, regional planners must also identify strategies for adapting transportation systems to already inevitable climate impacts. Adaptation requires answers to a host of questions: What are the critical system components, what is their vulnerability to unavoidable and expected climate impacts, and what options exist to provide system redundancy?

As new challenges and discoveries continue to unfold with respect to the connectivity, technology, equity, and sustainability of urban transportation, the question for metropolitan regions is: How best to plan for the changes to come? Answering that question is what regional transportation planning is all about.

REFERENCES

Abbott, C. (1997). The Portland region: Where city and suburbs talk to each other and often agree. *Housing Policy Debate, 8*(1), 11–51.

Advisory Commission on Intergovernmental Relations. (1972). *Multistate regionalism.* Washington, DC: Author.

Altshuler, A. A., & Luberoff, D. E. (2003). *Mega-projects: The changing politics of urban public investment.* Washington, DC: Brookings Institution.

Ames, S. C. (1993). *A guide to community visioning: Hands-on information for local communities.* Prepared for the Oregon Chapter of the American Planning Association.

Banfield, E. C. (1961). *Political influence: A new theory of urban politics.* New York: Free Press.

Banfield, E. C., & Grodzins, M. (1958). *Government and housing in metropolitan areas.* New York: McGraw-Hill.

Beimborn, E., & Kennedy, R. (1996). *Inside the black box: Making transportation models work for livable communities.* Washington, DC: Citizens for a Better Environment and the Environmental Defense Fund.

Bloom, M. S., & Bennett, N. (1998, September–October). U.S. highway financing: Historical perspective and national priorities. *TR News,* pp. 3–7, 43.

Bond, A., & Kramer, J. (2010). Governance of metropolitan planning organizations: Board size, composition, and voting rights. *Transportation Research Record,* No. 2174, 19–24.

Bond, A., & Kramer, J. (2011). Administrative structure and hosting of metropolitan planning organizations. *Transportation Research Record,* No. 2244, 69–75.

Bregman, S., & Watkins, K. E. (Eds.). (2013). *Best practices for transportation agency use of social media.* Boca Raton, FL: CRC Press.

Burnes, D., Neumark, D., & White, M. J. (2014). Fiscal zoning and sales taxes: Do higher sales taxes lead to more retailing and less manufacturing? *National Tax Journal, 67*(1), 7.

Castiglione, J., Hiatt, R., Chang, T., & Charlton, B. (2006). Application of a travel demand microsimulation model for equity analysis. *Transportation Research Record: Journal of the Transportation Research Board, 1977,* 35–42.

Deyle, R. E., & Wiedenman, R. E. (2014). Collaborative planning by metropolitan planning organizations: A test of causal theory. *Journal of Planning Education and Research, 34*(3), 257–275.

Dong, X., Ben-Akiva, M. E., Bowman, J. L., & Walker, J. L. (2006). Moving from trip-based to activity-based measures of accessibility. *Transportation Research Part A: Policy and Practice, 40*(2), 163–180.

Edner, S., & McDowell, B. D. (2002). Surface-transportation funding in a new century: Assessing one slice of the federal marble cake. *Publius, 32*(1), 7–24.

Federal Highway Administration. (2000). An overview of transportation and environmental justice. Retrieved June 3, 2015, from *www.fhwa.dot.gov/environment/environmental_justice/overview/ej2000.pdf.*

Federal Highway Administration. (2007). *The transportation planning process: Key issues—A briefing book for transportation decisionmakers, officials, and staff* (Publication No. FHWA-HEP-07-039). Retrieved June 3, 2015, from *www.planning.dot.gov/documents/BriefingBook/BBook.htm.*

Fishman, R. (2000). The death and life of American regional planning. In B. Katz (Ed.), *Reflections on regionalism* (pp. 107–123). Washington, DC: Brookings Institution.

Frick, K. (2013). The actions of discontent: Tea Party and property rights activists pushing back against regional planning. *Journal of the American Planning Association, 79*(3), 190–200.

Gage, R. W. (1988). Regional councils of governments at the crossroads: Implications for intergovernmental management and networks. *International Journal of Public Administration, 11*(4), 467–501.

Gage, R. W., & McDowell, B. D. (1995). ISTEA and the role of MPOs in the new transportation environment: A midterm assessment. *Publius, 24*(3), 133–154.

Gerber, E. R., & Gibson, C. C. (2009). Balancing regionalism and localism: How institutions and incentives shape American transportation policy. *American Journal of Political Science, 53*(3), 633–648.

Gerber, E. R., Henry, A. D., & Lubell, M. (2013). Political homophily and collaboration in regional planning networks. *American Journal of Political Science, 57*(3), 598–610.

Goetz, A., Dempsey, P. S., & Larson, C. (2002). Metropolitan planning organizations: Findings and recommendations for improving transportation planning. *Publius, 32*(1), 87–105.

Goldman, T., & Deakin, E. (2000). Regionalism through partnerships?: Metropolitan planning since ISTEA. *Berkeley Planning Journal, 14,* 46–75.

Goldman, T., & Wachs, M. (2003). A quiet revolution in transportation finance: The rise of local option transportation taxes. *Transportation Quarterly, 57*(1), 19–32.

Grant, M., McKeeman, A., Bowen, B., Bond, A., Bauer, J., LaSut, L., et al. (2014). *Model long range transportation plans: A guide for incorporating performance-based planning.* Washington, DC: Federal Highway Administration.

Handy, S. (2008). Regional transportation planning in the US: An examination of changes in technical aspects of the planning process in response to changing goals. *Transport Policy, 15*(2), 113–126.

Handy, S. (2010). *Transportation—land use coordination in the Austin region: Keys to making it happen.* A joint initiative of the Central Texas Sustainability Indicators Project and the Center for Sustainable Development funded by the Snell Endowment.

Handy, S., & McCann, B. (2011). The regional response to federal funding for bicycle and pedestrian projects. *Journal of the American Planning Association, 77*(1), 23–38.

Hartgen, D. T., & Neumann, L. A. (2002). Performance (A TQ point/counterpoint exchange with David T. Hargten and Lance Neumann). *Transportation Quarterly, 56*(1), 5–19.

Hill, E. W., Geyer, B. K., & Puentes, R. (2003). *Slanted pavement: How Ohio's highway spending shortchanges cities and suburbs.* Washington, DC: Center on Urban and Metropolitan Policy and Brookings Institution.

Holmes, E. H. (1973). The state-of-the-art in urban transportation planning, or how we got here. *Transportation, 1*(4), 379–401.

Hoover, J., McDowell, B. D., & Sciara, G.-C. (2004). *Transit at the table: A guide to participation in metropolitan decisionmaking.* Washington, DC: Federal Transit Administration.

Howard/Stein–Hudson Associates. (2011). *Public participation strategies for transit: A synthesis of transit practice.* Washington, DC: Transit Cooperative Research Program, National Academy of Sciences.

Howard/Stein–Hudson Associates & Parsons Brinckerhoff Quade and Douglas. (1996). *Public involvement techniques for transportation decision-making.* Washington, DC: Federal Highway Administration.

Johnston, R. A. (2004). The urban transportation planning process. In S. Hanson & G. Giuliano (Eds.), *The geography of urban transportation* (3rd ed., pp. 115–140). New York: Guilford Press.

Jones, D. (1985). *Urban transit policy: An economic and political history.* Englewood Cliffs, NJ: Prentice-Hall.

Leavitt, H. (1970). *Superhighway—Super hoax.* Garden City, NY: Doubleday.

Lewis, P. G. (2001). Retail politics: Local sales taxes and the fiscalization of land use. *Economic Development Quarterly, 15*(1), 21–35.

Lewis, T. (1999). *Divided highways: Building the interstate highways, transforming American life.* Harmondsworth, Middlesex, UK: Penguin Books.

Lubell, M., Henry, A. D., & McCoy, M. (2010). Collaborative institutions in an ecology of games. *American Journal of Political Science, 54*(2), 287–300.

Lupo, A., Colcord, F., & Fowler, E. P. (1971). *Rites of way: The politics of transportation in Boston and the U.S. city.* Boston: Little, Brown.

Mallett, W. J. (2010). *Metropolitan transportation planning* (Report No. R41068). Washington, DC: Congressional Research Service.

Matute, J., & Pincetl, S. (2013). *Unraveling ties to petroleum: How policy drives California's demand for oil.* Los Angeles: California Center for Sustainable Communities, UCLA, for the Next10 Foundation. Retrieved June 3, 2015, from *http://next10.org/unraveling-petroleum.*

McDowell, B. D. (1999). *Improving regional transportation decisions: MPOs and certification.* Washington, DC: Brookings Institution.

Meyer, M. D., & Miller, E. J. (2001). *Urban transportation planning: A decision-oriented approach* (2nd ed.). New York: McGraw-Hill.

Mladinov, J. K. (1963). Federal aid for planning and intergovernmental cooperation. *Traffic Quarterly, 17*(1), 79–94.

Mohl, R. A. (2008). The interstates and the cities: The U.S. Department of Transportation and the freeway revolt, 1966–1973. *Journal of Policy History, 20*(2), 193–226.

Morris, A., & Fragala, L. (2010). *Effective public involvement using limited resources* (NCHRP Synthesis 407). Washington, DC: National Cooperative Highway Research Program, National Academy of Sciences.

National Committee on Urban Transportation. (1958). *Better transportation for your city.* Chicago: Public Administration Service.

Nelson, A. C., Sanchez, T. W., Wolf, J. F., & Farquhar, M. B. (2004). Metropolitan planning organization voting structure and transit investment bias: Preliminary analysis with social equity implications. *Transportation Research Record,* No. 1895, 1–7.

Neshkova, M. I., & Guo, H. D. (2012). Public participation and organizational performance: Evidence from state agencies. *Journal of Public Administration Research, 22,* 267–288.

O'Toole, R. (2008). *Roadmap to gridlock: The failure of long-range metropolitan transportation planning policy analysis* (Vol. 617). Washington, DC: Cato Institute.

Pickrell, S., & Neumann, L. (2001). Use of performance measures in transportation decision making. In Transportation Research Board (Eds.), *Performance measures to improve transportation systems and operation* (Conference Proceedings 26). Washington, DC: National Academy Press.

Polzin, S. E., & Pisarski, A. E. (2015). *Commuting in America 2013: The National Report on Commuting Patterns and Trends.* Washington, DC: American Association of State Highway and Transportation Officials.

Poorman, J. (2001). Implementing performance measures. In Transportation Research Board (Eds.), *Performance measures to improve transportation systems and operation* (Conference Proceedings 26). Washington, DC: National Academy Press.

Post, Buckley, Schuh & Jernigan, Inc. (2006). *How to engage low-literacy and limited-English-proficiency populations in transportation decisionmaking.* Washington, DC: Federal Highway Administration.

Rosenbloom, S. (1988). Transportation planning. In F. S. So & J. Getzels (Eds.), *The practice of local government planning.* Washington, DC: International City Management Association.

Ross, C. (Ed.). (2009). *Megaregions: Planning for global competitiveness.* Washington, DC: Island Press.

Sanchez, T. W. (2006). *An inherent bias?: Geographic and racial–ethnic patterns of metropolitan planning organization boards: Metropolitan policy program.* Washington, DC: Brookings Institution.

Sciara, G.-C. (2009). *Planners and the pork barrel: Metropolitan engagement in and resistance to congressional transportation earmarking.* Unpublished doctoral dissertation, University of California, Berkeley, CA. Retrieved June 3, 2015, from *www.uctc.net/research/diss166.pdf.*

Sciara, G.-C. (2012). Planning for unplanned pork: The consequences of congressional earmarking for regional transportation planning. *Journal of the American Planning Association, 78*(3), 239–255.

Sciara, G.-C., & Handy, S. L. (2013). *Cultivating cooperation without control: California's MPO-driven Smart Growth Programs.* Davis: Institute of Transportation Studies, University of California.

Sciara, G.-C., & Wachs, M. (2007). Metropolitan transportation funding: Prospects, progress, and practical considerations. *Public Works Management and Policy, 12*(1), 378–394.

Seltzer, E. (2004). It's not an experiment: Regional planning at Metro, 1990 to the present. In C. P. Ozawa (Ed.), *The Portland edge: Challenges and successes in growing communities* (pp. 35–60). Washington, DC: Island Press.

Transportation Research Board. (2000). *Highway capacity manual 2000.* Washington, DC: National Academy Press.

Transportation Research Board. (2001). *Performance measures to improve transportation systems and operation* (Conference Proceedings 26). Washington, DC: National Academy Press.

Turnbull, K. T. (2007). *The metropolitan planning organization, present and future* (Conference Proceedings 39). Washington, DC: Transportation Research Board.

U.S. Department of Transportation. (2012). *Highway statistics 2012.* Washington, DC: Office of Highway Policy Information. Retrieved June 3, 2015, from *www.fhwa.dot.gov/policyinformation/statistics/2012.*

U.S. Government Accountability Office. (2009a). *Metropolitan planning organizations: Options exist to enhance transportation capacity and federal oversight.* Washington, DC: Author.

U.S. Government Accountability Office. (2009b, September). *Survey of metropolitan planning organizations* (GAO-09-867SP, September 2009), an e-supplement to GAO-09-868. Retrieved June 3, 2015, from *www.gao.gov/special.pubs/gao-09-867sp.*

Wachs, M. (1987). Forecasts in urban transportation planning: Uses, methods, and dilemmas. *Climate Change, 11,* 61–80.

Wachs, M. (2004). Reflections on the planning process. In S. Hanson & G. Giuliano (Eds.), *The geography of urban transportation* (3rd ed., pp. 141–162). New York: Guilford Press.

Wassmer, R. W. (2006). The influence of local urban containment policies and statewide growth management on the size of United States urban areas. *Journal of Regional Science, 46*(1), 25–65.

Weir, M. (2000). Coalition building for regionalism. In B. Katz (Ed.), *Reflections on regionalism* (pp. 127–153). Washington, DC: Brookings Institution.

Wheeler, S. (2009). Regions, megaregions, and sustainability. *Regional Studies, 43*(6), 863–876.

Wolf, J. F., & Farquhar, M. B. (2005). Assessing progress: The state of metropolitan planning organizations under ISTEA and TEA-21. *International Journal of Public Administration, 28,* 1057–1079.

CHAPTER 7

Land Use, Travel Behavior, and Disaggregate Travel Data

MARLON G. BOARNET

Interest in neighborhood-oriented transportation planning has grown dramatically over the past three decades. That neighborhood context has brought land use–travel behavior interactions to the forefront of transportation policy questions. A focus on land use and travel behavior requires understanding disaggregate flows, which are the microlevel travel patterns of individual persons and households.

If planners and developers built denser, more mixed-use, transit-oriented cities, would persons travel differently, for example, by driving less and walking or taking transit more? In other words, can city building be a tool to achieve transportation goals? To understand those questions, and to understand whether, for example, Smart Growth neighborhoods or transit-oriented developments change travel behavior, it is necessary to understand how persons travel within cities. Hence the need to understand the disaggregate travel patterns of individual persons or households (see also the discussion of disaggregate modeling approaches and disaggregate data sources in Chapter 5).

This chapter proceeds in several sections. We first discuss why planners and policymakers increasingly need to understand disaggregate travel flows and then turn to the methods and data sources for analyzing how land use influences travel behavior. After that, we discuss what we currently know about the influence of land use on travel, and then the conclusion highlights next steps and future research directions.

WHY STUDY DISAGGREGATE TRAVEL FLOWS?

From Prediction to Prescription in Land Use–Travel Interactions

The decades leading up to and immediately following the passage of the 1956 National Defense and Interstate Highway Act shaped much transportation planning in the United States, and methods for understanding how land use influences travel are no exception. The four-step model (described in Chapter 5) was developed to analyze how much highway capacity was needed and where to locate highway routes in the national network. For those purposes, it was not necessary to understand how individuals traveled, although four-step models were refined to work with classes or groups of commuters. The model was and is inherently highly aggregated, predicting aggregate commute flows rather than individual behaviors.

In the last three to four decades, transportation policy questions have gone well beyond predicting aggregate commute flows. In successive waves, several theories tied land use to travel in ways that were intended not merely to predict travel flows but to apply land use planning as a transportation policy tool. The New Urbanism, transit-oriented development, and Smart Growth all proposed to employ land use as a policy tool to change driving in ways that would, for example, reduce traffic congestion or automobile emissions.

Relatedly, economists had long theorized that urban areas could not build their way out of congestion by adding more highway capacity, because increases in travel supply will induce more driving (see Downs, 1962, for a classic article on the topic.) By the 1970s, attention had begun to move from increasing the supply of transportation (usually highway) infrastructure to managing travel demand.

The shift from building infrastructure to managing travel was also a shift from prediction to policy prescription, which requires a deeper understanding of behavior than that provided by aggregate approaches. What if, for example, persons who would otherwise ride transit move to transit-oriented developments (TODs)? Or what if, in response to reduced congestion, persons move farther from their workplace and drive longer distances on uncongested suburban routes in the same amount of time?[1] Efforts that seek to use land use and neighborhood-oriented policies to change travel behavior require a refined understanding of travel behavior that, in turn, necessitates disaggregate analysis. Yet that is only part of the story. At the same time that land use–travel interactions have moved to the forefront of planning debates, variations in travel have moved neighborhood-oriented transportation planning, which inherently highlights interactions between land use and travel, to the forefront of many urban policy discussions.

Variations in Travel Behavior

The best data available on travel behavior in the United States indicate three things: (1) Most travel in the United States is by car. (2) Despite the "car dominance" that national data indicate, other modes are important in particular places. (3) Car travel may have peaked, and there are hints that travel by non-auto modes may be increasing.

On net, these trends paint a picture of a shift from a "one size fits all" auto-oriented transportation planning to a planning framework that can be characterized as an auto-oriented theme with meaningful noncar variations. In the United States, the 2009 National Household Travel Survey shows that 83.4% of all trips are by private vehicle (Santos, McGucken, Nakamoto, Grey, & Liss, 2011.) Yet national data can be deceptive. There is important variation in travel modes across places.

Table 7.1 lists the top 10 transit commuting, walking commuting, and bicycle commuting metropolitan areas, with the "top 10" defined by the metropolitan areas with the highest share for each mode. Public transportation carries over 10% of all commute trips in the Boston, Chicago, New York, San Francisco, and Washington, DC consolidated metropolitan statistical areas. Smaller metropolitan areas, often college towns, have mode splits for walking that in some cases exceed 10% and bicycling commute shares that in some cases approach 10%. Note that walking and bicycling are likely more common for trips other than commuting.

The variations in travel are even more dramatic when looking within urban areas. Table 7.2 shows walking mode split (walk trips as a percent of all trips) for selected regional study areas within the Los Angeles metropolitan area.[2] Walking comprises 22% of all trips in the Santa Monica Bay regional study area, 29% of all trips in the urbanized mid-Los Angeles region, and over half of all trips for downtown residents. Three suburban regions—greater Riverside, Greater Fullerton, and Simi Valley (in the far north of Los Angeles's San Fernando Valley) are shown for comparison, and in each of those walking is less than 10% of all trips.

Plate 7.1 shows a map of vehicle miles traveled (VMT) in the urbanized portion of the Los Angeles metropolitan area. The map shows household VMT from the 2001 Southern California Association of Governments travel survey and is based on VMT for 10,630 households, matched to the location of their residence. VMT is then interpolated between household locations, creating a smooth surface that represents VMT across the Los Angeles region based on the 10,630 surveyed households. Areas in the urbanized portion of Los Angeles, near downtown and toward the urbanized Westside, have low VMT, as does the relatively urbanized and accessible area in central Orange County called South Coast Metro (near Irvine.) High VMT areas, in red, are largely but by no means exclusively on the exurban fringe.

It would be tempting to conclude that the travel variations shown in Table 7.2 and Plate 7.1 reflect common distinctions between cities and suburbs, but that is only part of the story. As an example, consider the results from a recent study of travel in eight small (approximately half-mile radius) areas in the inner suburban South Bay region of Los Angeles County, described in Boarnet, Joh, Siembab, Fulton, and Nguyen (2011). The South Bay is home to approximately a million persons, many in communities near the beach, and the region is mostly classic auto-oriented suburbs. Boarnet, Joh, and colleagues conducted a 1-day travel diary of 2,125 South Bay residents in three survey waves from 2005 through 2007. The survey indicated that 86% of trips in the region were by car, almost identical to the national average of 83.4% of all trips traveling by private vehicle from the 2009 NHTS. But even in this relatively homogenous suburban region, averages are deceiving.

The survey respondents were drawn from eight neighborhoods, each of which had residences surrounding commercial development, but the physical arrangement of the commercial development varied. In four neighborhoods, the commercial zone was in the center of the study area, with residential development radiating outward (called "centers"), while in the other four neighborhoods the commercial activity was arrayed linearly along a major boulevard (called "corridors"; see Plate 7.2 in the color insert). In the four centers, persons took an average of 0.19 walking trips per day and 2.00 driving trips per day, compared to 0.07 daily walking trips and 2.3 daily

TABLE 7.1.—Top 10 Transit, Walk, and Bicycle Commuting Metropolitan Areas, 2009

Top 10 Transit Commuting Metropolitan Areas		Top 10 Walk Commuting Metropolitan Areas		Top 10 Bicycle Commuting Metropolitan Areas	
CMSA	% commute trips by public transportation	CMSA	% commute trips by walking	CMSA	% commute trips by bicycle
New York–Northern New Jersey–Long Island, NY–NJ–PA	30.50%	Ithaca, NY	15.10%	Corvallis, OR	9.30%
San Francisco–Oakland–Fremont, CA	14.60%	Corvallis, OR	11.20%	Eugene–Springfield, OR	6.00%
Washington–Arlington–Alexandria, DC–VA–MD–WV	14.10%	Ames, IA	10.40%	Fort Collins–Loveland, CO	5.60%
Boston–Cambridge–Quincy, MA–NH	12.20%	Champaign–Urbana, IL	9.00%	Boulder, CO	5.40%
Chicago–Naperville–Joliet, IL–IN–WI	11.50%	Manhattan, KS	8.50%	Missoula, MT	5.00%
Philadelphia–Camden–Wilmington, PA–NJ–DE–MD	9.30%	Ocean City, NJ	8.40%	Santa Barbara–Santa Maria–Goleta, CA	4.00%
Seattle–Tacoma–Bellevue, WA	8.70%	Iowa City, IA	8.20%	Gainesville, FL	3.30%
Baltimore–Towson, MD	6.20%	Hinesville–Fort Stewart, GA	8.20%	Logan, UT–ID	3.30%
Los Angeles–Long Beach–Santa Ana, CA	6.20%	Jacksonville, NC	8.10%	Chico, CA	3.00%
Portland–Vancouver–Beaverton, OR–WA	6.10%	State College, PA	8.00%	Bellingham, WA	3.00%

[a]Among the largest 50 metropolitan statistical areas.

Source: 2009 American Community Survey as summarized in McKenzie, 2010, Table 2, pp. 5–6, and McKenzie and Rupino, 2011, Table 4, p. 10 and Table 3, p. 10.

Notes: CMSA, consolidated metropolitan area statistical area. The top 10 transit commuting metropolitan areas are from among the 50 largest metropolitan areas. It is unlikely that smaller metropolitan areas would have higher transit commute mode splits.

driving trips for corridor residents.[3] Possibly more striking, 47% of center residents said they usually walk when traveling to their centers while only 12% of corridor residents usually walked when traveling to the commercial corridor in the middle of their neighborhood (Boarnet, Joh, et al., 2011). Clearly, significant differences in urban spatial structures and in travel behavior exist among suburbs, not just between city and suburb.

The study of land use and travel behavior is, by and large, an effort to explain these localized differences, and to understand how such differences inform popular planning practices such as transit-oriented development and Smart Growth. Looking beyond variations in locations, a recent literature has focused on the question of peak driving, asking whether per capita driving has begun to drop in developed countries (see the discussion in Chapter 1, and, e.g., Goodwin, 2012). Disaggregate travel flows are necessary to understanding localized variations in travel across locations and over time, and the behavioral foundations of any such variation. Furthermore, disaggregate travel flows are increasingly important in helping us understand how personal travel responds to policies that might, for example, charge for on-street parking, allow solo drivers to travel on carpool lanes by paying a toll, or encourage employers to promote alternatives to single occupancy automobile commuting. The next sections discuss the methods and results of disaggregate travel behavior analysis, with a focus on land use–travel behavior interactions.

TABLE 7.2. Walking as Percentage of All Trips, Select Los Angeles Subregions, 2009

	% trips by walking	Number households surveyed in subregion	Subregion population, 2010 census	Subregion population density (in persons per square mile), 2010 census
Urbanized subregions				
Los Angeles CBD	54.8%	26	151,950	24,638
Santa Monica Bay	22.3%	142	328,480	4,683
Mid Los Angeles	28.9%	330	1,172,079	12,160
Suburban subregions				
Greater Riverside	9.7%	239	708,664	1,923
Greater Fullerton	9.6%	109	222,643	5,992
Simi Valley	6.9%	77	175,631	1,183

Source: Analysis of 2009 NHTS data, by 55 regional study areas (the subregions) defined by the Southern California Association of Governments, as reported in Joh et al. (2014).

STUDYING DISAGGREGATE TRAVEL BEHAVIOR

Theory

Travel behavior has long been regarded as a derived demand. With the exception of recreational travel (possibly important for nonmotorized travel), derived demand theory posits that persons do not travel for its own sake but rather to consume goods and services that require travel. This viewpoint is informed by neoclassical microeconomics (see, e.g., Domencich & McFadden, 1975, for a detailed discussion). As with any product, the demand for travel should depend on the price of travel and the traveler's income, among other variables.

Land use—the pattern of possible trip origins, destinations, and the ease and speed of moving between those origins and destinations—affects the price (most simply, the time cost) of travel. This leads to a clear theoretical link between land use and travel—a link that has been recognized and that has been part of travel demand modeling for decades. Mitchell and Rapkin (1954) titled one of the early and influential academic monographs on travel modeling *Urban Traffic: A Function of Land Use*. Yet the land use–travel behavior relationship, while clear at a simple level, yields few theoretically unambiguous answers to today's policy questions.

Persons consume more of goods that are less expensive. Hence shortening the distance or travel time between, for example, your home and the neighborhood grocery store might cause you to make more trips to the grocery store, not fewer. Crane (1996) notes that land use plans that shorten travel distance or travel time can lead to more car trips unless the change in access is large enough to encourage a mode shift (e.g., from driving to using transit or nonmotorized modes). In urban areas with multiple travel modes, whether land use changes will result in more or less travel by specific modes cannot be settled with a priori theory; it requires empirical study.

Complicating matters further, the derived demand view of travel focuses attention on economic factors—travel costs and the traveler's income—while other factors are also likely to be important for travel. Psychology, preferences, and habit might influence how persons travel (see, e.g., McFadden, 2007). Recent studies have measured attitudes toward the environment or safety, and those studies typically find that travelers' attitudes are associated with their travel behavior (see, e.g., Anable, 2005).

The results of the empirical literature are summarized later in this chapter, but for now note that individual characteristics, most notably income and family structure (such as whether children are present in the household and the ages of children), have a large impact on travel behavior. Empirical results suggest that the role of land use variables is often smaller than the role of individual/household characteristics (see, e.g., Hanson, 1982, for an early demonstration of this result), but policies cannot easily change the characteristics of individuals and households, so these are not considered policy variables. Comparing land use changes to policies that might include congestion tolls or travel demand management programs, the impact of comprehensively changing several land use variables at once may be comparable to, or only slightly smaller than, the role of changing price variables. These differences are discussed later in this chapter.

Methods

The usual method for studying disaggregate travel behavior is regression analysis, with observations for individuals or households. The data are drawn from travel diary surveys, which ask persons to log all their travel for a survey period (usually 1 or 2 days, although the length of time can vary). Household members often divvy up responsibilities for travel and other duties. For example, a wife with a long commute can ask her husband to do more travel for day-to-day errands, or vice versa.[4] Research questions vary, sometimes seeking to understand household travel and sometimes studying within-household travel tradeoffs, and hence the analysis might be at the level of either households or the individuals within households. A typical travel behavior regression is shown below:

$$\text{Travel Behavior Variable} = \beta_0 + \textit{Land-Use-Variables}\boldsymbol{\beta}_1$$
$$+ \textit{Sociodemographic-Variables}\boldsymbol{\beta}_2 + u$$

where Travel Behavior Variable = a measure of individual or household travel; *Land-Use-Variables* = a vector of built environment variables, usually measured near the traveler's or household's place of residence, but the variables can be measured near travel destinations or along routes; *Sociodemographic-Variables* = a vector of individual characteristics; the β terms are scalars or vectors of coefficients to be estimated (with coefficient column vectors $\boldsymbol{\beta}_1$ and $\boldsymbol{\beta}_2$ shown in bold); and u = the regression error term.

This method of regression analysis of individual travel dates to early work by Hanson and Hanson (1981) and Vickerman (1972), both of whom used early travel diaries to explain disaggregate travel in a regression model. Since then, with the increasing availability of disaggregate travel data and GIS-based land use information, the literature has grown substantially. The method outlined in the equation above is, in some ways, conceptually simple. Characteristics of the traveler (an individual or, more commonly now, a household unit), and characteristics of land uses (near places of residence, key destinations such as places of employment, or along travel routes and nontraveled alternative routes) should explain travel behavior. This is the essence of disaggregate travel analysis. Within that simple framework, the modeler must make several choices and resolve a number of methodological modeling issues.

Dependent and Independent Variables

Dependent variables for disaggregate travel (the travel behavior variables in the equation above) include measures of trip generation (number of trips, sometimes by mode), mode choice, and distance or time traveled. Depending on the policy focus, distance traveled may be measured by VMT or time spent in active travel (walking or bicycling.) The dependent variable is, and should be, tailored to the policy questions being asked. Concerns about greenhouse gas emissions imply a focus on VMT, while questions of access suggest measures of total travel, either measured by number of trips or distance traveled. Given the timeframe for travel diaries, often 1 or 2 days, the dependent variable is typically daily travel—for example, measures such as VMT per household per day or nonwork car trips per household or per person per day are common.

There are two groups of independent variables for a disaggregate land use–travel behavior regression: variables that describe the sociodemographic characteristics of the individual or household, and measures of land use or, more broadly, the travel environment. Since the early days of the disaggregate travel behavior literature, studies have found that individual demographic characteristics are at least as important, and often more important, in determining travel behavior than land use. See, for example, Hanson (1982, Table 7), who used a 35-day travel diary from Uppsala, Sweden, to demonstrate that individual demographics (age, gender, household size, household income) were at least as influential as the number of nearby establishments when explaining trip generation and trip distance in a regression analysis of travel behavior. The literature has consistently found that persons with higher incomes travel more (e.g., Crane & Guo, 2011). Several other characteristics influence household travel, including the number of persons in the household, the ages of persons in the household (particularly the number of adults and their ages and number of children who are not of driving age), and the vehicles available in the household. Other variables that are typically in the set of sociodemographic characteristics in land use–travel regressions include education levels for the adults in the household (or the individual, if individual travel is being studied), gender (which is more typical when the focus is individual travel but studies that include the gender of the household head are also common), and race and ethnicity.

Land use variables have commonly included direct measures of land use such as density or land use mix and the supply of transportation options, such as distance from nearby transit stations or highway on-ramps. Land use variables are most commonly measured by the "D variables"—a phrase first coined by Cervero and Kockelman (1997). In Cervero and Kockelman's original formulation, they modeled travel behavior as a function of density, diversity, and design near survey respondents' place of residence. Density is typically measured as residential density (persons or, less commonly, households divided by land area.) Diversity is land use mix, and is typically measured by an entropy index or a dissimilarity index. See Krizek (2003) or Hess, Moudon, and Logsdon (2001) for assessments of the relative advantages and disadvantages of entropy and dissimilarity indices. Design is typically the street design, juxtaposing the grid-oriented street patterns advocated by New Urbanists with more curvilinear street patterns indicative of post-World War II suburban development. Street design elements are commonly measured by the fraction of street intersections that are four-way (as opposed to three-way "T" intersections or one-way cul-de-sacs) and by the number of intersections or street blocks per unit of land area (e.g., Frank, Schmid, Sallis, Chapman, & Saelens, 2005).

Additional D variables have been added since Cervero and Kockelman's (1997) research, the most common of which are distance to transit (usually straight line or street-network measures from a survey respondent's residence to the nearest transit station) and destination access (typically a gravity variable that measures accessibility to employment or jobs, both to proxy possible commuting destinations and because places that employ persons are also the stores, movie theaters, and schools that are nonwork travel destinations.) For examples, see Bento, Cropper, Mobarak, and Vinha (2005) and Zegras (2010). Measures of safety, such as crime rates or indicators of objective or subjective feelings of safety in the environment, are associated with travel, and the evidence suggests that such associations might be stronger for women than for men (Loukaitou-Sideris, 2005; Lynch & Atkins, 1988).

An important emerging literature asks whether individual perceptions of the built environment, distinct from objective measures, influence travel behavior. McGinn, Evenson, Herring, Huston, and Rodriguez (2007) find evidence that perceptions of the built environment (measured by study subjects' answers to survey questions) had an independent effect on travel behavior after controlling for objective built environment measures of the sort captured by the "D variables." Hawthorne and Kwan (2012) discuss how individual perceptions can be used to improve access measures that rely only on distance. Here we focus on objective measures of the built environment, noting though that perceptions about the built environment can also influence travel.

When measuring land use near a survey respondent, two questions are important. How near is near—that is, what geographic scale should be used—and where is the relevant location—the survey respondent's residence, workplace, or other activity locations? Typically the land use variables are measured near survey subjects' residences, but that is mostly a matter of convenience (most travel diaries have information on the subject's place of residence), and focusing on land use at trip destinations is an important and understudied topic (for studies that find that land use near work locations is a determinant of travel behavior, see Chatman, 2003; Hanson & Schwab, 1987). The question of geographic scale largely defaulted to quarter-mile areas around the survey respondent's home, based on studies that argued that most persons would not walk more than a quarter-mile in most urban settings (see Untermann, 1984). Yet even early results had hints that differing scales of geography were important in measuring how land use is associated with travel behavior (e.g., Boarnet & Sarmiento, 1998; Handy, 1995). More recently, there is evidence that the influence of rail transit access on VMT can be detected at distances up to 2 miles from rail stations, although the effect dampens with distance (Bailey, Mokhtarian, & Little, 2008), and gravity measures of access to employment, which extend well beyond a quarter-mile (although also with an effect that declines with distance) are often more important predictors of household VMT than localized measures of density, land use mix, or street design (see, e.g., the results in Ewing & Cervero, 2010; Salon, Boarnet, Handy, Spears, & Tal, 2012).

Lastly, there is a question about how to measure land use. Early efforts studying land use and travel did not have access to computerized georeferenced information available in current GIS databases, and instead had to geocode locations manually, limiting the size of data sets that could be analyzed. Yet the insights from these studies presaged and were often similar to the findings from later work (see, e.g., Hanson, 1982; Hanson & Schwab, 1987). Georeferenced data available via GIS allow land use to be measured around hundreds or thousands of survey respondents.[5] As of the early 2000s, GIS databases were often limited to street networks and parcel-level land use, with limited information about sidewalks or aesthetics that may be unimportant for driving but possibly influential for active travel. As a response, researchers developed environmental audit tools, which require researchers to walk neighborhoods to measure aspects of the built environment. The environmental audit tools are used to gather information about sidewalk coverage and quality, street speed limits, traffic calming measures, and aesthetic factors such as street trees or graffiti that are rarely, if ever, available from GIS. For a review of environmental audit tools, see Brownson, Hoehner, Day, Forsyth, and Sallis (2009). More recently, researchers have used environmental audit tools with images from Google StreetView or other tools, which eliminates the need to physically visit a study location (e.g., Clarke, Ailshire, Melendez, & Morenoff, 2010), although Google StreetView images are sometimes not completely up to date, which can be a disadvantage compared to in-person audit tools in locations where the built environment has changed recently.

The early environmental audit tools measured a large number of characteristics; the Irvine Minnesota Inventory (IMI), a popular audit tool, measured 162 built environment elements. Research has established that only a subset of those characteristics are consistently associated with active travel. Boarnet, Forsyth, and colleagues (2011) studied associations between IMI variables and walking travel using data from the Minneapolis–St. Paul metropolitan area. Walking was associated most commonly with sidewalks, curb cuts, stop signs, low posted vehicle speed limits (25 miles per hour), and the availability of shops on the street. The quality of the sidewalk network, including sidewalk completeness on the block and good maintenance, was also associated with walking travel. Aesthetic characteristics, such as street trees, street amenities, nature elements, and building characteristics were not associated with walking travel. See Boarnet, Forsyth, and colleagues (2011, Tables 4–6) for details.

Estimating Travel Behavior Regressions—The Need to Move beyond Ordinary Least Squares

The dependent variable in Equation (1) is often nonlinear, requiring the use of regression methods that are more advanced than ordinary least squares (OLS). The number of trips that an individual or household takes is what econometricians call "count data." Households might take zero, one, or more car trips in a day, but an individual household does not take a half or a third of a car trip. In other cases, for example, when studying the amount of VMT or walking, the dependent variable is left-censored: travel diary data of VMT or walking or bicycling minutes will have many observations censored at zero, meaning that large portions of the data set have a value of zero for the dependent variable. As an example, approximately 24% of the 2000–2001 California Statewide Travel Survey households reported no driving trips on the survey day (Heres del Valle & Niemeier, 2011). All of these cases—count data, left-censored data, and also mode choice models where persons choose from among a small number of alternatives—require regression techniques that go beyond OLS. For examples of econometric techniques when OLS is not appropriate, see, for example, Boarnet, Joh, and colleagues (2011) for an application of methods to estimate count data (number of trips), Heres del Valle and Niemeier (2011) for an application of methods to estimate a left-censored variable (VMT), and Ben-Akiva and Lerman (1985) for a detailed discussion of methods to estimate discrete choice models, including mode choice.

Unraveling Causality

Methodological debates in the land use travel literature have grown increasingly important, and in some ways increasingly heated, over the past two decades. Briefly stated, the debate is between persons who argue that the body of evidence on land use–travel associations illuminates a causal impact that can be used for policy purposes, and those who do not believe that causality has been sufficiently demonstrated.

Likely the most studied and debated methodological concern in this literature is the residential selection problem. Certainly persons drive less and walk more in downtowns than in the suburbs (see, e.g., the data presented earlier in this chapter), but that might simply reflect the fact that persons who seek to travel by walking and transit choose to live in transit-oriented locations. An early hint of this is work by Cervero on travel patterns of residents in transit-oriented developments. Cervero (1994) surveyed residents of transit-oriented developments (TODs) in California and found that among those TOD residents who commuted by rail, 42.5% had been rail commuters before they moved to their existing residences. Cervero stated, "Part of the high incidence of rail usage among station-area residents, then, could be due to the fact they have a proclivity to patronize rail transit, whether due to habit, personal taste, or happenstance" (p. 177). A simple regression of transit ridership on variables that include access to rail transit, following the specification in Equation (1), would find a correlation, but the effect of transit-proximity may not be causal—persons who ride transit might simply have moved to housing near the rail station.

This concern about residential selection has been studied extensively over the past two decades. See Cao, Mokhtarian, and Handy (2009) for a review. As summarized by Cao and colleagues (2009), the association between land use characteristics and travel behavior reflects partly direct causality (persons do appear to change their travel based on land use characteristics) and partly residential selection. Yet the debate goes deeper.

People make many choices that influence the relationship between land use patterns and travel outcomes. Persons can choose to travel at different times of day, or by different routes, avoiding or reducing traffic delays. Such choices are constrained at times (your boss might prefer that you arrive at work at 9:00 A.M., not 2:00 P.M.), but most persons have some discretion over travel time and path. As already mentioned, persons might move their residence based on their desired travel pattern. Individual attitudes might influence travel choices (inherent in the idea that some persons might prefer transit, for example, and hence move to transit-oriented developments), and it is unclear whether those preferences are relatively fixed or can change over time (see, e.g., the discussion in McFadden, 2007). Furthermore, in the long run, the businesses, government offices, recreation locations, and educational institutions that are travel destinations might relocate in response to changing residential locations, creating a feedback loop between short-run land use–travel behavior links and long-run patterns of urban development. Such feedback loops are illustrated in historical changes in urban spatial structure (see Chapter 3) and in the contemporary context are modeled in advanced urban simulation programs (e.g., Waddell, 2002).

Most empirical studies in the social sciences, studies of travel behavior included, are nonexperimental; that is, transportation scholars cannot randomly select a sample of individuals or households and make one randomly chosen part of the sample live in a transit-oriented development and the remaining part live in the far suburbs so as to compare the behaviors of the two groups. Sometimes, though, the world provides circumstances that approximate a true experiment. A new transit line might open in a neighborhood, or several schools might introduce improvements to the pedestrian and bicycle infrastructure. If those projects were built independent of the preferences of persons living in the neighborhoods, researchers could compare how travel behavior among persons living near the new infrastructure project changed relative to travel changes among similar persons who were not located near the new transit station, pedestrian improvements, or other infrastructure. These studies can mimic the experimental design of clinical trials in the medical sciences. For examples, see the experimental-control group studies of travel behavior change around new light rail lines in Salt Lake City (Brown & Werner, 2008) or Charlotte (MacDonald, Stokes, Cohen, Kofner, & Ridgeway 2010), or studies of changes in school travel after implementation of California Safe Routes to School infrastructure improvements (Boarnet, Anderson, Day, McMillan, & Alfonzo, 2005).

While using new projects (such as new transit stations or infrastructure improvements) to create a "natural experiment" is a powerful way to determine causality, some researchers debate this approach. The experiment might have affected an idiosyncratic group of persons or places, and there will always be debates about how well any one such natural experiment can generalize to a broader population. Possibly the communities that received a new transportation project lobbied hard for that funding and so the project serves persons who wanted that infrastructure more than others and hence who are primed to use the new rail station or sidewalk. Absent good knowledge of the underlying behavioral relationships, questions will remain about how well any experimental evaluation generalizes to a broader population. Having said that, experimental approaches have been woefully underapplied in travel behavior studies, and much can be learned from carefully studying travel behavior changes before and after new projects are constructed, comparing persons affected by the project (an experimental group) and persons less affected by the project (a control group).

Data Sources

The data for land use–travel studies come primarily from two sources: survey or observational data of travel behavior, which can be used to form the dependent variables, and land use data from either georeferenced GIS data bases or direct observation using environmental audit tools. Data on sociodemographic characteristics— income, ages, employment status, and other characteristics of travelers—are typically obtained from the same travel surveys that provide information on the dependent (travel) variables.

Traditional Travel Surveys

Travel diary surveys are conducted by most large metropolitan planning organizations and national governments in the U.S., and often in other countries. Survey respondents are asked to record details of their travel, including all trip origins and destinations, trip purposes, modes, and the times of travel, for a specified period, most commonly one or two days.

The U.S. National Household Travel Survey (NHTS) is a good example of a large, modern, travel diary. The NHTS, and its predecessor the Nationwide Personal Transportation Survey (NPTS), were conducted by the U.S. Department of Transportation in 1969, 1977, 1983, 1990, 1995, 2001, and 2009. The 2009 NHTS surveyed approximately 150,000 households from April 2008 through April 2009, asking respondents to record details of the travel of all household members and also asking a large number of questions about the demographics and summary travel information of household members. Similar travel diary surveys have been developed by some states and larger metropolitan planning organizations, often after U.S. census years, to provide data for regional transportation planning models. Traditional land use–travel behavior studies obtain data about household residence locations, usually using street addresses geocoded to latitude–longitude coordinates, to get data on land use near those households using GIS software.

Electronic Travel Data Collection

As mentioned in Chapter 5, electronic data collection, including GPS tracking, has the potential to provide a wealth of information on travel behavior. Some studies have used GPS as an analog to larger travel diary surveys, partly to quality-check the self-reported diary data (e.g., Bricka, Zmud, Wolf, & Freedman, 2009). Many smart phone applications have the ability to track the user's location, although access to those data depends on the smart phone platform. Using such detailed locational data would require research protocols with careful subject confidentiality safeguards. In some cases, such as GPS data from smart phones, the traditional sociodemographic information recorded in diary surveys does not exist, which could limit the ability to control for the individual and household characteristics that are known correlates of travel behavior. Using GPS data to track travel requires considerable postprocessing, to infer travel speed, mode, and trip origin and destination from the GPS location data. Such approaches are possible, and in the future surveys are likely to continue to refine combinations of GPS data and self-reported information, both to obtain sociodemographic information and to obtain self-reported information on trip mode or trip purpose to supplement the detailed route and speed data available from GPS.

WHAT DO WE KNOW?

Land Use

Several recent studies have summarized the literature on the effect of land use on travel behavior. One approach is meta-analysis, which combines the results from a large number of studies to form an average or, at times, a weighted average impact. The most commonly cited meta-analysis of land use–travel behavior research is Ewing and Cervero (2010). An alternative approach is to summarize the impacts from a small number of methodologically sound studies, as was done in Salon and colleagues (2012). For the most part, the results from Ewing and Cervero (2010) and Salon and colleagues are similar.

Both Salon and colleagues (2012) and Ewing and Cervero (2010) converted coefficient estimates from land use–travel studies—typically with specifications of the sort shown in Equation (1)—into elasticities. *Elasticities* are ratios of percentage changes. As an example, the elasticity of household VMT with respect to population density near the household's residence shows how VMT will change, in percentage terms, for a 100% change in density. Expressing results as elasticities allows easy comparisons of the magnitude of the impact across several different variables.

Table 7.3 shows the range of elasticities from Salon and colleagues (2012) and Ewing and Cervero (2010) for the effect of selected land use variables on household VMT. The elasticities, individually, are often small. The range for the population density elasticity of VMT is from −0.08 to −0.19, land mix elasticities range from −0.02 to −0.10, and the range for the regional accessibility elasticities is −0.03 to −0.25.[6] For example, the elasticity of VMT with respect to population density, ranging from −0.08 to −0.19, means that doubling population density (a 100% increase) will reduce household VMT by between 8 and 19%. This highlights one of the criticisms of land use as a transportation policy tool: the impact of most land use variables is small. Doubling population density in many communities would not be feasible, and even such a large policy change would reduce VMT in that community by at most 19%, according to the results in Table 7.3. Yet it is important to recognize that the elasticities in Table 7.3 reflect the impact of manipulating one variable by itself, and land use policy often changes more than one of the variables in Table 7.3.

Many Smart Growth planning interventions aim to increase density, improve land use mix, create more grid-oriented street patterns, improve access to jobs (and hence to a range of travel destinations), and provide alternatives to driving in the form of transit or pedestrian or bicycle infrastructure. Those policies are holistic—changing many variables rather than manipulating, for example, only density, or only land use mix. A more realistic test of the impact of land use policy would be to ask about the cumulative impact on travel of changing several land use variables at once. Bento and colleagues (2005) examined the effect of taking a representative household and changing all land use variables for that household from Atlanta's values to Boston's values, using coefficient estimates from their regression study. The purpose was not to imply that one could change Atlanta's land use to mimic Boston's, but rather to examine the cumulative impact of changes to several different land use variables. Bento and colleagues found that simulating a hypothetical move of a representative

TABLE 7.3. Ranges for Elasticity of VMT with Respect to Land Use Variables

Land use variable	Elasticity range from Salon et al. (2012)	Elasticity estimate from Ewing and Cervero (2010)
Population density	−0.08 to −0.19	−0.04
Land use mix	−0.02 to −0.10	−0.09
Intersection density	0 to −0.19	−0.12
Regional accessibility to jobs	−0.03 to −0.25	−0.20

Note. All land use variables are measured near the survey household's residence, usually within ¼ or ½ mile or in geographies corresponding to census block groups or tracts.

household from Atlanta to Boston reduced household VMT by 25%. The impacts of individual variables in Bento's study were smaller; for example, they estimate an elasticity of VMT with respect to population density of −0.07 or less. Bento and colleagues suggest that changing several land use variables at once can have a larger effect than changing one variable singly.

Pricing and Other Policies

Researchers have examined the relationship between travel behavior and congestion pricing (i.e., charging higher tolls in places and at times of traffic congestion), but those studies typically examine traffic flows on tolled links or within areas subject to cordon pricing, rather than total VMT. Because traffic flows on a link or within a cordoned area can be diverted elsewhere, traffic flows are not the same as VMT. The elasticity of traffic flows with respect to link tolls ranges from −0.1 to −0.45, and the elasticity of traffic crossing cordons with respect to the cordon price ranges from −0.2 to −0.3 (Salon et al., 2012). Leape (2006) finds that in the case of the London congestion cordon, about one-quarter of the reduced trips were diverted outside of the cordon zone.

Theory suggests that pricing and land use policies should be complementary. Persons living in neighborhoods with travel options, or where density or land use mix allow easier walking or transit travel, would find it easier to reduce driving in response to congestion pricing or mileage fees. Similarly, pricing policies would provide additional incentive for persons living in dense or mixed-use neighborhoods to adopt travel patterns that reduce the use of personal vehicles. Guo, Agrawal, and Dill (2011) found empirical support for this proposition in their analysis of the results of a vehicle pricing experiment in Portland, Oregon, conducted in 2006 and 2007. The experiment enrolled 183 households, tracking all household vehicle travel with GPS devices. Household travel was tracked for a 5-month baseline period (June–October, 2006), after which half of the households were charged a flat, per-mile, driving fee, and the other half of the households were charged a mileage fee that was higher during peak travel periods. Household travel was then tracked again for another 5 months with the VMT fee in place, from November 2006 through March 2007. For both groups of households, the VMT fee was calculated to replace the state gasoline tax, which participants did not pay during the experiment. Guo and colleagues found that households who were charged peak pricing reduced their VMT more if they lived in neighborhoods with greater land use mix, supporting the idea that land use and congestion pricing policies are complements.

Context Sensitivity

Very few studies have examined how land use–travel behavior relationships vary across different places. Virtually all research in this area has measured average effects for metropolitan areas or larger geographies, and little is known about whether that average effect masks important variation across different land use contexts. Might land use changes have the largest effect on travel in relatively low-density areas, or in highly urbanized locales, or in places in between? Two recent studies give insights.

Boarnet, Houston, and colleagues (2011) used data from a 2001 travel diary in the five-county Los Angeles metropolitan area to study the effect of land use on household VMT. They divided the households into quintiles based on gravity measures of employment accessibility centered on each household's residence and then used regression analysis to examine the relationship between VMT and sociodemographic and land use variables, allowing the effect of employment accessibility on VMT to vary across the five quintiles. The authors found that the effect of employment accessibility on VMT was not statistically significant for households in the top and bottom quintiles of employment accessibility, but the effect was statistically significant for the second through fourth quintiles. In those "middle accessibility" locations the implied elasticity of VMT with respect to the employment gravity variable was larger than is typical in the literature; elasticities ranged as high as −0.8 for the fourth quintile. The results suggest that improving employment accessibility, for example, by adding residences to employment centers or bringing job locations closer to existing concentrations of residential development, might have the greatest impact on reducing VMT for places "in between"—neither the most highly accessible urban core locations nor places on the urban fringe.

Instead, efforts to improve employment accessibility might reduce VMT the most in the near-urban or more densely developed inner-ring suburban locales. Salon (2013) found a similar result using statewide data for California; the effect of job accessibility within 5 miles of a household's residence was statistically significant only for "in between" locations that were neither central downtowns nor exurbs, although the implied elasticities of VMT with respect to employment access from Salon's research, not larger in magnitude than −0.12, were lower than those in Boarnet, Houston, and colleagues (2011).

Employment accessibility influences both work and nonwork VMT. Using the underlying data from Boarnet, Houston, and colleagues (2011), average work VMT drops from 28 to 20 miles per household per day moving from the second to the fourth quintile of accessibility (a move from outlying to more central locations in the Los Angeles metropolitan area) and average total VMT drops from 57 to 45 miles per household per day from the second to the fourth quintile of employment accessibility. The fraction of household VMT from travel to work is relatively stable across the middle three accessibility quintiles, from 48% in the second accessibility quintile to 44% in the fourth accessibility quintile, implying that the effect of employment accessibility on travel does not flow predominantly through travel to work but instead that employment accessibility proxies a broad range of work and nonwork trip destinations. In the most accessible locations, the fifth quintile of employment accessibility, work travel is 40% of total household travel.[7]

NEXT STEPS AND FUTURE RESEARCH DIRECTIONS

Experimental Evaluation

Urban transportation has entered an era of experimentation. The once staid practice of designing and managing bus or rail transit systems, and debating transit- versus automobile-oriented policies, has given way to an array of different policy approaches and market innovations. San Francisco and Los Angeles are implementing real-time parking pricing programs that adjust parking prices to equilibrate demand and available supply. Carsharing, once fringe, is becoming mainstream with applications and web-based companies that help match riders and drivers. Bicycle sharing programs have been implemented in cities as diverse as Minneapolis, Austin, and New York. New transportation options, often promoted by new companies, have sprung up seemingly overnight. Examples include Uber and Lyft, which provide on-demand ride-matching service via social media, and Bolt Bus and Megabus, which provide intercity bus service that combines luxury attributes such as wireless Internet with social media reservation systems. Metropolitan areas are implementing bus rapid transit, light rail transit, and heavy rail. Traffic calming, complete streets, and pedestrian and bicycle projects are becoming broadly accepted. This wave of experimentation is driven by social media technology that allows more efficient matching of consumers and providers, increased segmentation of the transportation market, and the variations in urban travel described earlier in this chapter. The increasingly diverse range of public-sector and private-sector transportation services increases both the possibility and the need for experimental evaluation in transportation.

Policy interventions can be treated as experiments, and researchers could enroll persons who are exposed to the intervention and persons less exposed to the new policy or project and then mimic experimental–control group evaluations. There are still only a few such efforts, but momentum is building. The ability to collect travel data via the use of mobile GPS devices or survey tools opens the door to experimental evaluations that can more clearly establish causal links than can cross-sectional regression analyses. Recent experimental evaluations of light rail transit (Boarnet, Houston, & Spears, 2013; Brown & Werner, 2008; MacDonald et al., 2010; Spears, Boarnet, & Houston, 2016) illustrate methods that can be used to evaluate a broad range of transportation projects and policies, and Guo and colleagues' (2011) analysis of experimental–control group data illustrate how experimental programs can inform policy implementation.

Real-Time Information and Traveler Preferences

Recent research suggests that attitudes toward travel—including perceptions of safety, the convenience of the car versus transit, and the importance of environmental protection—affect travel choices (see, e.g., Handy, Cao, & Mokhtarian, 2006; Spears, Houston, & Boarnet, 2013). Some of these attitudes likely combine traveler preferences and aspects of the policy environment, such as perceptions about the convenience of car versus transit travel. But other attitudinal measures in recent studies appear to be unrelated to aspects of the transportation system, and instead may reflect underlying concerns about, for example, time availability or environmental protection or personal safety. This psychological aspect of travel behavior has long been studied separately from land use issues, but in recent research land use, policy interventions, and traveler attitudes are all being used to explain disaggregate flows.

Looking forward, refined research can help answer whether the attitudinal measures being used in transportation research relate to psychological constructs among travelers and whether attitudes are stable across the life course or whether attitudes change as people age or move to different environmental contexts. As transportation agencies increasingly provide information to travelers about system performance, congestion, and wait times (often in real time), understanding whether traveler attitudes can be changed and if so how such marketing efforts might be adapted to transportation policy will also be important.

Context Sensitivity

Context sensitivity is broader than variations in effect sizes across different land use patterns. Might travelers with different attitudes respond differently to land use or pricing policies, with some persons changing their travel quickly while others stay relatively fixed in their routines? Might information flows be adjusted to be complementary to land use and pricing policy, advising travelers about alternative options and tailored in ways that address key concerns about service reliability or fear of crime? How might real-time communication with travelers help public agencies and private providers manage both day-to-day operations and exceptional circumstances that occur during unique events such as the Olympics or World Cup competitions or when a disaster strikes? The availability of real-time communication, in the form of smart phones, and real-time data about transportation performance makes research into such topics timely.

Moving forward, disaggregate travel flows will certainly be an increasing part of transportation analysis. Technology allows disaggregate flows to be measured and analyzed more easily, and more often, than was the case in the past. The questions being asked of transportation—from ties to land use and policy initiatives to the segmentation of travelers by type and in different contexts—require a disaggregate analysis. The use of disaggregate flows is part of the increasing trend to view transportation as a behavioral phenomenon.

NOTES

1. See, e.g., Cervero (1994) on the possibility that transit-oriented development residents rode transit before they moved to TODs, and Gordon, Richardson, and Jun (1991) on the fact that commuting times remained stable even as suburbs expanded in the 1980s.

2. The data in Table 7.2 include walking trips for access to or egress from public transit. If transit access/egress trips are excluded, the percentage of all trips that are walking in the three urbanized subregions is 37.2% for Los Angeles CBD, 19.8% for Santa Monica Bay, and 19.4% for Mid Los Angeles. Because the number of households in the Los Angeles CBD subregion is small, Joh, Chakrabarti, Boarnet, and Woo (2014) compared the percentage walk trips in 2009 to similar data on walking from the 2001 Southern California Association of Governments (SCAG) travel survey. For 2001, there were 144 surveyed households in the Los Angeles CBD sub-region, and the percentage of trips by walking in that sub-region was 35.5 percent excluding trips to access/egress transit and 46.7 percent including trips to access/egress transit in 2001.

3. The survey included 1-day travel diaries for 813 respondents who lived in centers and 1,312 respondents who lived in corridors. For details, see Boarnet, Joh, and colleagues (2011).

4. See, e.g., Hanson and Hanson (1981) for information about how travel responsibilities are traded off within households and the gendered nature of the division of those duties.

5. For example, the 2009 NHTS has over 150,000 surveyed households, but likely due to difficulty obtaining consistent, fine-scaled land use GIS maps and data nationwide, there are still relatively few studies of disaggregate land use–travel behavior interactions using national data. An exception is Bento and colleagues (2005).

6. All estimates were statistically significantly different from zero in the underlying original studies, with the exception of one insignificant estimate for street intersection density; hence that range includes zero (see Salon et al., 2012, Table 2). The regional accessibility variable is typically a gravity measure of access to jobs, weighting jobs by the inverse of the distance from the household to those jobs, or at times a power function of the inverse of the distance (see, e.g., the accessibility equations in Chapter 1).

7. Based on author's calculations from the travel diary data used in Boarnet, Houston, and colleagues (2011).

REFERENCES

Anable, J. (2005). "Complacent car addicts" or "aspiring environmentalists"?: Identifying travel behaviour segments using attitude theory. *Transport Policy, 12,* 65–78.

Bailey, L., Mokhtarian, P. L., & Little, A. (2008). The broader connection between public transportation, energy conservation and greenhouse gas reduction. Fairfax, VA: ICF International.

Ben-Akiva, M., & Lerman, S. (1985). *Discrete choice analysis: Theory and applications to travel demand.* Cambridge, MA: MIT Press.

Bento, A. M., Cropper, M. L., Mobarak, A. M., & Vinha, K. (2005). The effects of urban spatial structure on travel demand in the United States. *Review of Economics and Statistics, 87*(3), 466–478.

Boarnet, M. G., Anderson, C., Day, K., McMillan, T., & Alfonzo, M. (2005). Evaluation of the California Safe Routes to School legislation: Urban form changes and children's active transportation to school. *American Journal of Preventive Medicine, 28*(2, Suppl. 2), 134–140.

Boarnet, M. G., Forsyth, A., Day, K., & Oakes, M. (2011). The street level built environment and physical activity and walking: Results of a predictive validity study for the Irvine Minnesota Inventory. *Environment and Behavior, 43*(6), 735–775.

Boarnet, M. G., Houston, D., Ferguson, G., & Spears, S. (2011). Land use and vehicle miles of travel in the climate change debate: Getting smarter than your average bear. In Y.-H. Hong & G. Ingram (Eds.), *Climate change and land policies* (pp. 151–187). Cambridge, MA: Lincoln Institute of Land Policy.

Boarnet, M. G., Houston, D., & Spears, S. (2013, October). *The Exposition Light Rail Line Study: A before-and-after study of the impact of new light rail transit service.* Unpublished study prepared for the Haynes Foundation, Los Angeles, CA.

Boarnet, M. G., Joh, K., Siembab, W., Fulton, W., & Nguyen, M. (2011). Retrofitting the suburbs to increase walking: Evidence from a land use–travel study. *Urban Studies, 48*(1), 129–159.

Boarnet, M. G., & Sarmiento, S. (1998). Can land use policy really affect travel behavior? *Urban Studies, 35*(7), 1155–1169.

Bricka, S., Zmud, J., Wolf, J., & Freedman, J. (2009). Household travel surveys with GPS: An experiment. *Transportation Research Record: Journal of the Transportation Research Board,* No. 2105, 51–56.

Brown, B. B., & Werner, C. M. (2008). Before and after a new light rail stop: Resident attitudes, travel behavior, and obesity. *Journal of the American Planning Association, 75*(1), 5–12.

Brownson, R. C., Hoehner, C. M., Day, K., Forsyth, A., & Sallis, J. F. (2009). Measuring the built environment for physical activity: State of the science. *American Journal of Preventive Medicine, 36*(4), S99–S123.

Cao, X., Mokhtarian, P., & Handy, S. L. (2009). Examining the impacts of residential self-selection on travel behavior: A focus on empirical findings. *Transport Reviews, 29*(3), 359–395.

Cervero, R. (1994). Transit-based housing in California: Evidence on ridership impacts. *Transport Policy, 1*(3), 174–183.

Cervero, R., & Kockelman, K. (1997). Travel demand and the 3Ds: Density, diversity and design. *Transportation Research D, 2*(3), 199–219.

Chatman, D. (2003). How workplace land use affects personal commercial travel and commute mode choice. *Transportation Research Record: Journal of the Transportation Research Board,* No. 1831, 193–201.

Clarke, P., Ailshire, J., Melendez, R., & Morenoff, J. (2010). Using Google Earth to conduct a neighborhood audit: Reliability of a virtual audit instrument. *Health and Place, 16*(6), 1224–1229.

Crane, R. (1996). On form versus function: Will the new urbanism reduce traffic, or increase it? *Journal of Planning Education and Research, 15,* 117–126.

Crane, R., & Zhan, G. (2011). Toward a second generation of land-use/travel studies: Theoretical and empirical foundations. In N. Brooks, K. Donaghy, & G. Knaap (Eds.), *Oxford handbook of urban economics and planning* (pp. 522–543). New York: Oxford University Press.

Domencich, T. A., & McFadden, D. (1975). *Urban travel demand: A behavioral analysis.* Amsterdam, The Netherlands: North Holland.

Downs, A. (1962). The law of peak hour expressway congestion. *Traffic Quarterly, 16*(3), 393–409.

Ewing, R., & Cervero, R. (2010). Travel and the built environment: A meta-analysis. *Journal of the American Planning Association, 76*(3), 265–294.

Frank, L. D., Schmid, T. L., Sallis, J. F., Chapman, J. E., & Saelens, B. E. (2005). Linking objectively measured physical activity with objectively measured urban form: Findings from SMARTRAQ. *American Journal of Preventative Medicine, 28*(2, Suppl. 2), 117–125.

Goodwin, P. (2012). *Peak travel, peak car, and the future of mobility: Evidence, unresolved issues, policy implications, and a research agenda* (Discussion Paper No. 2012-13). Unpublished paper prepared for the International Transport Forum, Organisation of Economic Co-operation and Development. Retrieved from *www.itf-oecd.org/sites/default/files/docs/dp201213.pdf.*

Gordon, P., Richardson, H. W., & Jun, M.-J. (1991). The commuting paradox: Evidence from the top twenty. *Journal of the American Planning Association, 57*(4), 416–420.

Guo, Z., Agrawal, A. W., & Dill, J. (2011). Are land use planning and congestion pricing mutually supportive? *Journal of the American Planning Association, 77*(3), 232–250.

Handy, S. L. (1993). Regional versus local accessibility: Implications for nonwork travel. *Transportation Research Record,* No. 1400, 58–66.

Handy, S. L., Cao, X., & Mokhtarian, P. (2006). Self-selection in the relationship between the built environment and walking: Empirical evidence from Northern California. *Journal of the American Planning Association, 72*(1), 55–74.

Hanson, S. (1982). The determinants of daily travel-activity patterns: Relative location and sociodemographic factors. *Urban Geography, 3*(3), 179–202.

Hanson, S., & Hanson, P. (1981). The impact of married women's employment on household travel patterns: A Swedish example. *Transportation, 10*(2), 165–183.

Hanson, S., & Schwab, M. (1987). Accessibility and interurban travel. *Environment and Planning A, 19,* 735–748.

Hawthorne, T. L., & Kwan, M.-P. (2012). Using GIS and perceived distance to understand the unequal geographies of healthcare in lower-income urban neighborhoods. *Geographical Journal, 178*(1), 18–30.

Heres del Valle, D., & Niemeier, D. (2011). CO_2 emissions: Are land use changes enough for California to reduce VMT? Specification of a two-part model with instrumental variables. *Transportation Research Part B, 45*(1), 150–161.

Hess, P. M., Moudon, A. V., & Logsdon, M. G. (2001). Measuring land use patterns for transportation research. *Transportation Research Record,* No. 1780, 17–24.

Joh, K., Chakrabarti, S., Boarnet, M. G., & Woo, A. (2014, June). *The walking renaissance: Insights from the Greater Los Angeles area.* Unpublished working paper. College Station, TX: Department of Urban and Regional Planning.

Krizek, K. J. (2003). Residential relocation and changes in urban travel: Does neighborhood scale urban form matter? *Journal of the American Planning Association, 69*(3), 265–281.

Leape, J. (2006). The London congestion charge. *Journal of Economic Perspectives, 20*, 157–176.

Loukaitou-Sideris, A. (2005). Is it safe to walk here?: Design and policy responses to women's fear of victimization in public places. In *Research on Women's Issues in Transportation: Report of a Conference, Vol. 2: Technical Papers, Transportation Research Board Conference Proceeding, 35* (pp. 102–110). Washington, DC: National Research Council.

Lynch, G., & Atkins, S. (1988). The influence of personal security fears on women's travel patterns. *Transportation, 15*(3), 257–277.

MacDonald, J. M., Stokes, R. J., Cohen, D. A., Kofner, A., & Ridgeway, G. K. (2010). The effect of light rail transit on body mass index and physical activity. *American Journal of Preventive Medicine, 39*(2), 105–112.

McFadden, D. (2007). The behavioral science of transportation. *Transport Policy, 14*, 269–274.

McGinn, A. P., Evenson, K. R., Herring, A. H., Huston, S. L., & Rodriguez, D. A. (2007). Exploring associations between physical activity and perceived and objective measures of the built environment. *Journal of Urban Health, 84*(2), 162–184.

McKenzie, B. (2010, October). Public transportation usage among U.S. workers, 2008 and 2009. *American Community Survey Briefs*. Washington, DC: U.S. Census Bureau. Available at *www.census.gov/prod/2010pubs/acsbr09-5.pdf*.

McKenzie, B., & Rupino, M. (2011, September). *Commuting in the United States: 2009. American Community Survey Reports*. Washington, DC: U.S. Census Bureau. Available at *www.census.gov/prod/2011pubs/acs-15.pdf*.

Mitchell, R. B., & Rapkin, C. (1954). *Urban traffic: A function of land use*. New York: Columbia University Press.

Salon, D. (2013, July). *Quantifying the effect of local government actions on VMT*. Draft of the Final Report to California Air Resources Board, Sacramento, CA.

Salon, D., Boarnet, M. G., Handy, S., Spears, S., & Tal, G. (2012). How do local actions affect VMT?: A critical review of the empirical evidence. *Transportation Research Part D, 17*(7), 495–508.

Santos, A., McGucken, N., Nakamoto, H. Y., Grey, D., & Liss, S. (2011). *Summary of travel trends: 2009 National Household Travel Survey*. Washington, DC: U.S. Department of Transportation. Available at *http://nhts.ornl.gov/2009/pub/stt.pdf*.

Spears, S., Boarnet, M. G., & Houston, D. (2016). Driving reduction after the introduction of light rail transit: Evidence from an experimental-control group evaluation of the Los Angeles Expo Line. *Urban Studies*. [Epub ahead of print]

Spears, S., Houston, D., & Boarnet, M. G. (2013). Illuminating the unseen in transit use: A framework for examining the effect of attitudes and perceptions on travel behavior. *Transportation Research Part A, 58*, 40–53.

Untermann, R. K. (1984). *Accommodating the pedestrian: Adapting towns and neighborhoods for walking and bicycling*. New York: Van Nostrand Reinhold.

Vickerman, R. W. (1972). The demand for non-work travel. *Journal of Transport Economics and Policy, 6*(2), 176–210.

Waddell, P. (2002). UrbanSim: Modeling urban development for land use, transportation and environmental planning. *Journal of the American Planning Association, 68*(3), 297–314.

Zegras, P. C. (2010). The built environment and motor vehicle ownership and use: Evidence from Santiago de Chile. *Urban Studies, 47*(8), 1793–1817.

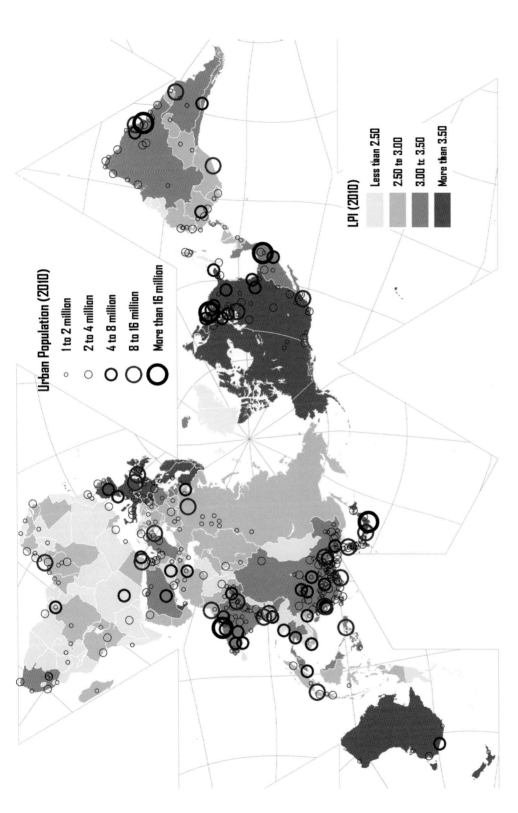

Urban Population (2010)

- ∘ 1 to 2 million
- ◦ 2 to 4 million
- ○ 4 to 8 million
- ◯ 8 to 16 million
- ◯ More than 16 million

LPI (2010)

- Less than 2.50
- 2.50 to 3.00
- 3.00 to 3.50
- More than 3.50

PLATE 2.1. National urban populations and the Logistics Performance Index. The labels give the countries' urban populations in millions, which is also denoted by the size of the circles.

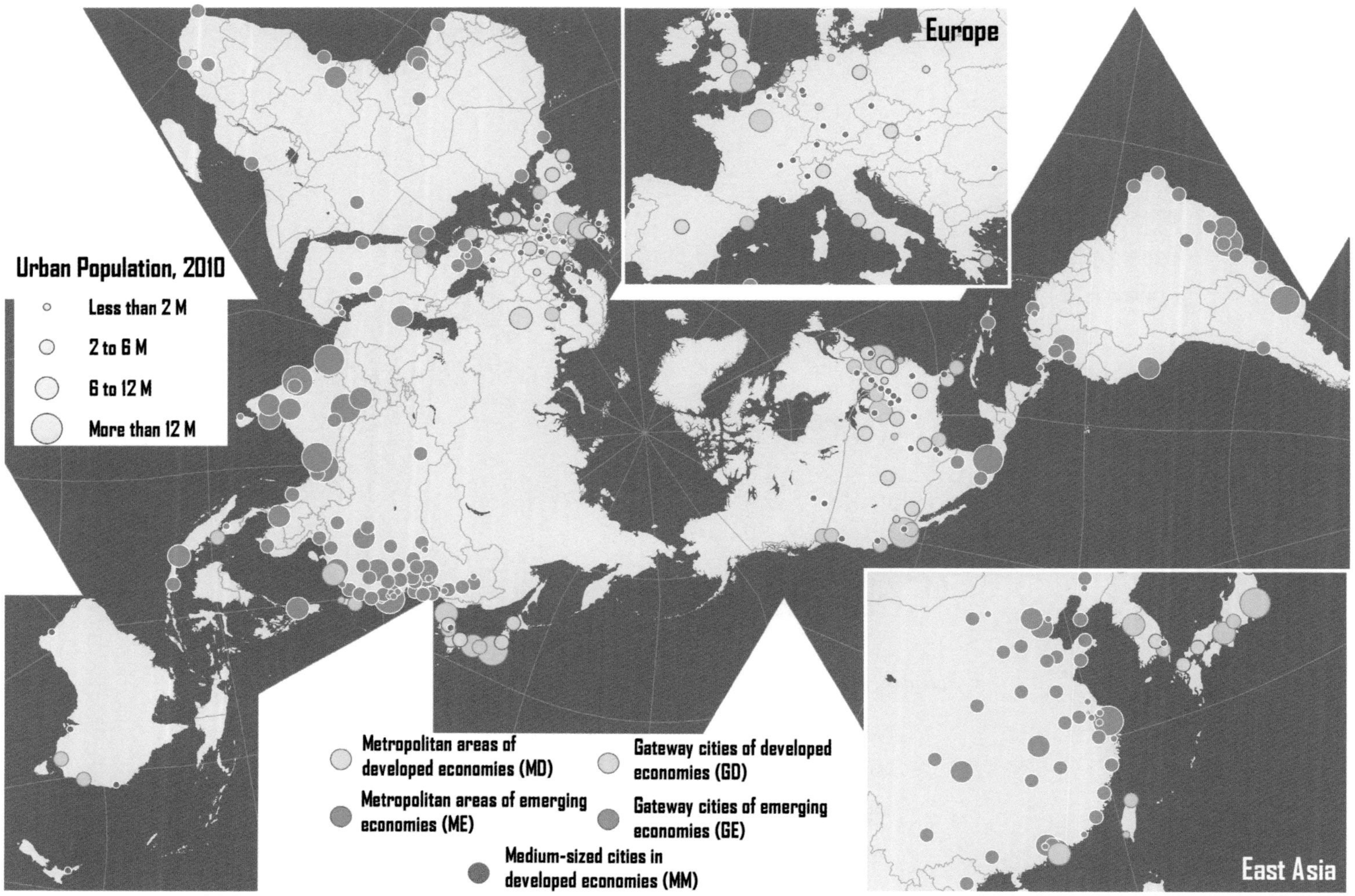

Europe

East Asia

Urban Population, 2010

Less than 2 M

2 to 6 M

6 to 12 M

More than 12 M

Metropolitan areas of developed economies (MD)

Metropolitan areas of emerging economies (ME)

Gateway cities of developed economies (GD)

Gateway cities of emerging economies (GE)

Medium-sized cities in developed economies (MM)

PLATE 2.2. Global city logistics typology.

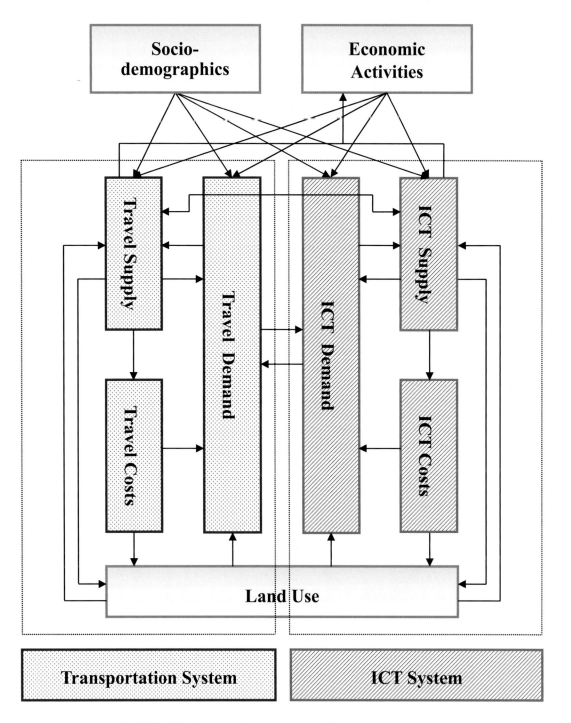

PLATE 4.1. Relationships between ICT and transportation.

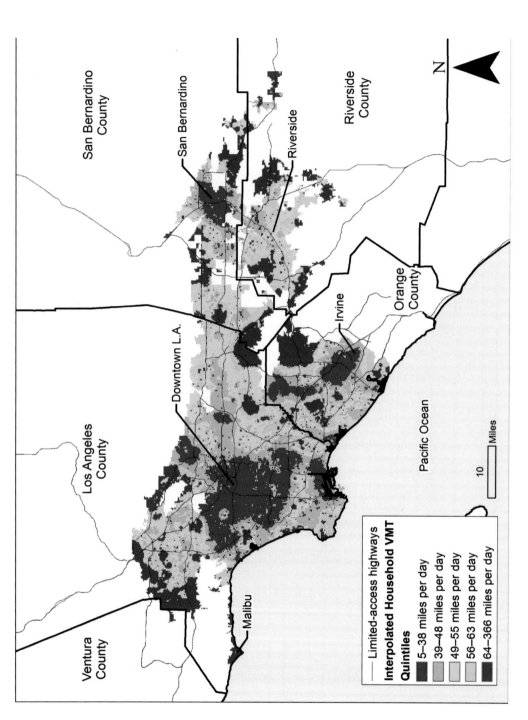

PLATE 7.1. Map of VMT in the urbanized portion of the Los Angeles metropolitan area.

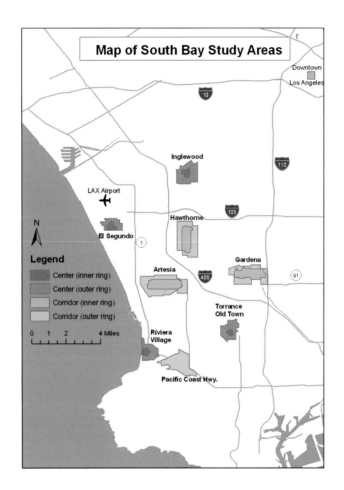

PLATE 7.2. Map of South Bay Study areas.

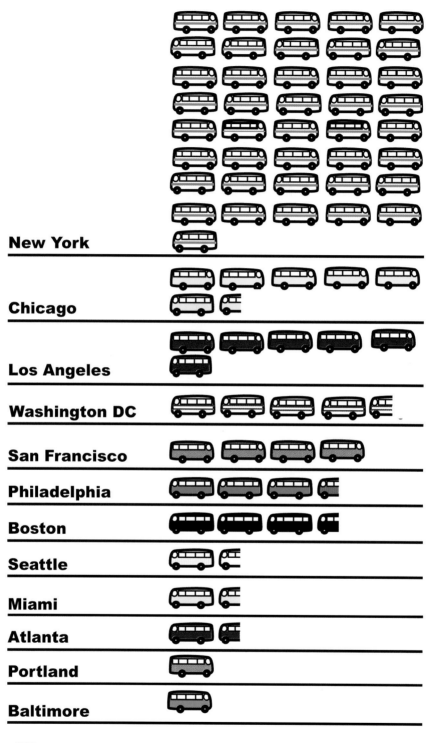

New York

Chicago

Los Angeles

Washington DC

San Francisco

Philadelphia

Boston

Seattle

Miami

Atlanta

Portland

Baltimore

= 100,000,000 unlinked trips

PLATE 8.1. Unlinked passenger trips in selected U.S. regions, FY 2012.

PLATE 8.2. Light rail transit operating at street level.

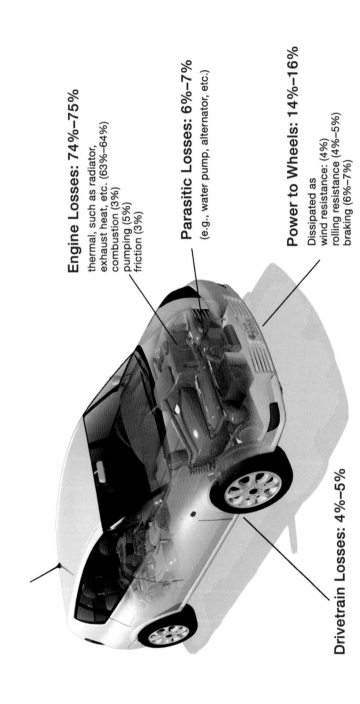

Engine Losses: 74%–75%

thermal, such as radiator,
exhaust heat, etc. (63%–64%)
combustion (3%)
pumping (5%)
friction (3%)

Parasitic Losses: 6%–7%

(e.g., water pump, alternator, etc.)

Power to Wheels: 14%–16%

Dissipated as
wind resistance: (4%)
rolling resistance (4%–5%)
braking (6%–7%)

Drivetrain Losses: 4%–5%

Idle Losses: 6%

In this figure, they are accounted for as part of the engine and parasitic losses.

PLATE 12.1. Energy losses in a modern ICE vehicle: combined U.S. test cycles. Source: *www.fueleconomy.gov.*

PART III

POLICY ISSUES

CHAPTER 8

Mass Transit

LISA SCHWEITZER

Transit advocates have high hopes for what transit can accomplish. As a result, most writing on transit lauds its potential benefits: Transit cleans up the air! It alleviates congestion! It raises property values! It revitalizes retail! It combats global warming! It makes people thinner! But transit can do few of those wonderful things if it does not first serve passengers, and high-quality passenger service requires both vision and prudent management. Transit service quality depends on many factors, including service frequencies, information, geographic placement, security, driver courtesy, station design, and fares. Does a train that comes every 10 minutes provide better service than a bus that comes every 5 minutes, simply because the former is a train and the latter a bus? Some argue yes; others argue no. Should transit operators try to serve every far-flung suburb in a region, or should cities focus on small, geographically limited networks surrounding particular activity centers, like downtown? Again, good arguments appear on both sides. This chapter presents an introduction to what transit is, how it works, and the dilemmas that transit operators face as they both navigate and shape urban geographies. We shall assume throughout that transit is important to cities, thus leaving aside advocacy in favor of exploring the field's many challenges.

TRANSIT BASICS

Transit, like other fields in transport, has its own unique vocabulary around operations and service. *Transit,* defined loosely, is passenger transport where individuals share vehicles at a scale larger than carpooling or ridesharing. *Mass transit* refers to moving large numbers of people in shared vehicles. *Public transit* refers to government-owned systems.

Measuring Service and Use

Some preliminary concepts simply measure service. *Service miles* refer to the miles of route served, unrelated to the number of vehicles. *Vehicle miles* are the number of miles a vehicle travels, and this number can be summed across routes and vehicles. Vehicle miles do not necessarily tell us much about *ridership,* which is the number of trips taken during a given time period. Empty vehicles running along long routes can accumulate many vehicle miles. As a result, the most common measures of service use reflect passenger numbers in some way.

Simple passenger counts suffer from the same problem as simply toting up miles: carrying 50 passengers 1 mile is not the same task for a transit provider as carrying 50 people 10 miles. *Passenger-miles* sum the distances each passenger rides. For example, 10 passengers that ride for 10 miles equal 100 passenger miles. Table 8.1 summarizes the passenger miles for 10 transit providers in major U.S. regions. While most people would probably guess that New York is the biggest transit system in the United States, the passenger miles listed in Table 8.1 show just how much more transit service gets consumed in New York than everywhere else, including high-quality transit environments like Boston, Chicago, and San Francisco.

Passenger trips, too, can be a misleading measure, depending on how we count trips. A *linked passenger trip* is a passenger's entire trip, from origin to destination, on the transit system. Information on full trips can be difficult for transit companies to gather because most agencies do not track transfers. *Transfers* refer to passengers disembarking from one vehicle to board another as part of what is, for them, one single trip. Some stations or stops do not serve as transfer points. At those, we can assume that passengers boarding are at their trip origin, and those alighting are at their destination. Many stops and stations, however, serve as transfer points. At those, providers generally do not have information on whether a person is at his or her origin or destination when he or she gets on or off a vehicle, or whether she has a midtrip transfer.

TABLE 8.1. Passenger Miles Consumed on Selected U.S. Metro Systems, 2011

Provider	Passenger Miles (Thousands)
MTA New York City Transit (NYCT)	12,170,490
Los Angeles County Metropolitan Transportation (Metro)	2,189,194
Chicago Transit Authority (CTA)	2,121,635
Washington Metropolitan Area Transit Authority (WMATA)	2,032,395
Massachusetts Bay Transportation Authority (MBTA)	1,814,253
Southeastern Pennsylvania Transportation Authority (SEPTA)	1,625,946
San Francisco Bay Area Rapid Transit District (BART)	1,442,860
Metropolitan Atlanta Rapid Transit Authority (MARTA)	731,062
Tri-County Metropolitan Transportation District of Oregon (Tri-Met)	452,363

Source: American Public Transportation Association (2013), Table 3; compiled by the author.

Unlinked trips simply count the people entering and leaving a vehicle (Plate 8.1 shows the dominance of New York for unlinked passenger trips). Unlinked trips are (1) the most readily available information about service use that agencies are likely to have and (2) quite uncertain measures for large transit service areas with frequent transfers. The measurement problems associated with unlinked transit trips make evaluating service changes difficult. If an agency reconfigures routes or opens a new line, increases in unlinked trips can result from any of multiple factors that are hard to differentiate from each other. Higher numbers of unlinked trips may result from higher service use due to lower transit journey times (i.e., an actual service improvement) or because the service changes induced more transfers. New service may prompt patrons to take on new transfers as part of their journey, as it opens up new trip possibilities. Also, if transfers are timed well—a key service quality factor—the measurement problem can worsen as more patrons may be willing to undertake trips with multiple transfers (Iseki & Taylor, 2009).

With some farecard reader technologies, agencies can trace passengers' movements through the system, but that strikes many as a violation of patrons' privacy. In addition, transfers often occur across transit agencies. In many metropolitan regions, one agency may operate bus service and another may operate rail service, which requires an interagency transfer as well as an intermodal transfer. To get data on full passenger trips for those types of transfers would require tracking passengers across agencies as well as across vehicles, something that few agencies can take on. With social media and geolocator services, it may be possible to derive some trip information, but social media users' transit choices may not mirror those of transit users as a whole group.

If planners and analysts want to gather linked trip information, agencies can conduct surveys and travel diaries for a sample of their patrons, but those surveys are usually expensive to do. *Passenger loads* (the number of people on a vehicle at given time) and *passenger counts* are easier and less expensive for transit agencies to gather than linked trips. As a result, most transit trip data count unlinked trips or passengers. These are displayed by U.S. metro region in Plate 8.1. The number of unlinked trips on all U.S. transit systems exceeded 10 billion in 2012. Again, most transit trips in the United States occur in the New York metro area, whose various providers served 4.1 billion of those trips.

Mode Share

Trips, passengers, and miles all cover the "what" of public transit. Mode covers the "how." *Mode* describes travel by different means: driving alone, riding the bus, walking, or driving a motorcycle. *Mode share* describes the breakdown of all travel according to mode, usually given in percentages, and usually at the city level. Table 8.2 shows the mode share for India and the United States, and for a large selection of cities that have trip data available by mode for all trips. National data on mode shares mask many different and conflicting factors about modes; the United States and India, for example, are two populous countries with large cities but also large rural areas. Because car travel is nearly ubiquitous in rural and suburban areas of the

United States, the aggregate mode share for cars in the United States, at 83% of all trips, does not reflect the car mode share in cities like New York, Chicago, Philadelphia, and San Francisco, where cars serve a much lower portion of trips than for the United States as a whole. In India, an estimated 13% of all trips occur by car, but the percentages in urban areas like Chandigarh and Bangalore are nearly twice that because car acquisition in India has occurred more among affluent, urban residents in midsize cities than in rural regions or the largest cities.

TABLE 8.2. Mode Share for Selected Places

	Pop (millions)	Land area (km²)	Rail (%)	Bus (%)	Transit overall (%)	Walk (%)	Bike (%)	Private car (%)
U.S. (National)[1]	313	9+ million	0.6	3.3	4	10	1	83
India (National)[2]	1,120	3+ million	—	—	27	28	11	13
Bejing[3]	11.7	1,368	2	21	23	21	32	20
Chicago[4]	2.7	589	5	11	16	9	1	63
Delhi[2]	11	431	—	—	42	21	12	19
London[5]	7.8	1,579	12	15	27	30	2	40
Madrid[6]	3.1	606	—	—	34	36	—	29
Melbourne[7]	4.1	1,566	4	3	7	13	2	77
Mumbai[2]	12.5	603	—	—	45	27	6	15
New York[8]	8.2	790	12	10	23	39		33
Paris[9]	6.5	762	—	—	62	4	1	32
Philadelphia[2]	5.2	347	—	—	25	9	2	60
Singapore[3]	5.1	712	19	25	44	22	1	29
Sydney[10]	4.6	1,580	6	5	11	18		69
Taipei[3]	2.6	272	14	18	32	15	4	46
Tokyo[3]	8.8	622	48	3	51	23	14	12
Vienna[6]	1.6	415	—	—	36	28	5	31

[1]U.S. Department of Transportation, Federal Highway Administration, National Household Travel Survey, Table 3.3.

[2]Ministry of Urban Development, 2008, Study on Traffic and Transportation Policies and Strategies in Urban Areas in India, Table 2.19.

[3]Land Transport Authority Singapore, 2011. *Journeys: Passenger Transport Mode Shares in World Cities,* reference figures.

[4]Chicago Metropolitan Agency for Planning, 2010. Chicago Regional Household Travel Inventory, Table 8.

[5]Transport for London, 2011, Travel in London, Supplementary Report: London Travel Demand Survey, Table 5.1.

[6]Urban Transport Benchmarking Initiative, 2006. Year Three Final Report.

[7]Department of Transport, 2009, Victoria Integrated Survey of Travel and Activity, 2007, Figure 2.1.

[8]New York Department of Transportation, 2011, 2009 NHTS Tabular Summary and Graphics.

[9]Enquête Nationale Transports et Déplacements, 2013, *Chiffres Cles Due,* multiple tables.

[10]New South Wales, Bureau of Transportation Statistics, *2009/10 Household Travel Survey Key Indicators for Sydney.*

Some locations break their mode share estimates out by rail and bus; others only report by mass transit. While rail is a higher capacity mode, buses carry a lot of trips both in the United States and in cities around the world. The exception to that rule concerns Japanese cities, where rail transit captures about 35% of trips in Tokyo and a similar share in Osaka. All modes of transit combined, including rail, bus, and ferry where applicable, serve between 7% of trips in Melbourne to 62% in Paris.

Mode shift describes a change in the percentage of trips between the various modes. Concerns about pollution, traffic, and sprawl have made mode shift away from driving alone to other modes, such as walking, biking, or transit, a central concern for urban planning and policy in the United States. Cities with a high transit mode share usually have a high walking mode share as well. A reduction in the car mode share does not automatically mean an increase in transit mode share or in the overall number of trips taken by transit. The outcomes depend on where work, home, and other opportunities are located, and how transit and walking work in tandem in the urban geography.

Table 8.3 presents the modes shares for *commuting trips* (just trips to work) for four of the largest regions. About 31% of commuters (nearly 3 million people) in New York City take mass transit daily; compare this with the lower percentage of 23% for all trips. The difference? Large sections of New York City have plenty of nonwork destinations that residents can walk to. Few people take the subway or a bus to go to the dry cleaner or a grab a sandwich because the city has dry cleaners and sandwich shops on just about every corner. But work locations relative to home locations are much less plentiful, and it can be very difficult to find an affordable place to live within walking distance of your employer. As a result, transit for work trips gets proportionally more patronage than for all types of travel, and walking captures more proportionally for all travel.

Mode share is a common way to discuss how people travel, but percentages can mask important information. Driving and cars have a much lower mode share in New York with 50% (compared to 70% in other U.S. regions), but that percentage covers many people driving alone. On an average day, about 4,420,600 million people are driving in the New York region—that means more drivers on the roads in that region alone than men, women, and children living in Nebraska and New Mexico combined. Population scale matters.

TABLE 8.3. Commuting Mode Shares for Large U.S. Regions

	Commuters (Estimated)	Drive alone	Carpool	Transit	Other
New York	8,807,184	50%	8%	32%	10%
Los Angeles	5,954,970	74%	11%	6%	9%
Chicago	4,368,216	71%	9%	12%	9%
Washington, DC	3,130,006	66%	10%	15%	9%

Source: *American Community Survey Data 2011,* Table S0804; compiled by the author; the daily commuting figures are estimates.

Modes of Transit

Modes in transit also break down into specific service types. Most transit services are *fixed route*, in that they operate on a repetitive, preset schedule along a specific route, with vehicles that stop at designated stops or stations. Unlike fixed route services, *paratransit* services, in general, are on-demand services scheduled by appointment for specific passengers to their specified destinations. Paratransit provides point-to-point service, much like biking, walking, or driving a car. Paratransit can be quite costly for transit agencies to operate; nonetheless, some jurisdictions require transit agencies to provide paratransit services by law as a supplement to fixed route transit for passengers who may have disabilities that make it impractical for them to use fixed route services.

Fixed route transit make up a system of different technologies that vary by operating characteristics such as speed and capacity. *Capacity* is the number of passengers a service can carry per some unit time. In general, capacity is a function of vehicle size, speed, and *frequency*, which is how often the service runs. The larger the vehicle size, the more passengers it is possible to carry. The faster the service, the more frequent the trips, and the larger the cars operated, the more passengers per hour a transit mode can serve. Speed, in addition to contributing to capacity, is a service quality characteristic: faster is better in transit, as it is for most transport modes. Also, the more frequent the service—whether vehicles serve the stops along the route every 10 minutes or every 30 minutes—the more passengers per hour it is possible to serve. Just as with speed, frequency contributes to service quality as well as capacity. A bus that comes every 5 minutes is much more useful to riders than one that comes only once an hour. Frequency also aids passengers if they transfer from one part of the system to another.

Design factors also can improve both speed and safety. *Rights of way* is the physical space where transit vehicles operate. *Exclusive rights of way* mean that the vehicles operate on space they share with no other vehicles or services. Subways and commuter trains usually operate on exclusive rights of way. Vehicles operating on *shared rights of way* have to use the space along with other vehicles. Buses operating in regular traffic use shared rights of way, for instance. Exclusive right of way increases speeds because it frees transit vehicles from congestion. *Grade separation*, an exclusive right of way that is built either above or below ground, can allow transit vehicles to operate at higher speeds more safely, and thus serve greater numbers of passengers. Grade-separated, exclusive rights of way enable transit vehicles to avoid any traffic congestion on the surface streets, and they eliminate the need for street crossings where transit vehicles can conflict with pedestrians, cars and trucks, or other transit vehicles.

Many in transit argue that rail investment offers the highest standard for service because rail often (though not always, as we shall see) operates on grade-separated, exclusive rights of way. Transit systems can mix all different types of design decisions to fit context and service area. Despite the variety, I shall cluster transit modes by their general design and, with caution, by capacities in order to provide an introduction. Table 8.4 summarizes the different modes going, roughly, from the smallest vehicle employed to the largest.

TABLE 8.4. Typical Traits of Basic Modes of Transit

Bus
- Operates on surface streets, mixed traffic
- 70 to 150 person vehicles
- Rubber tire
- Powered by many different fuels

Trolley/Streetcar
- Operates on surface streets, mixed traffic
- 70 to 150 person vehicles
- Steel wheel
- Electrical power
- Circuit or limited route

Light Rail
- Surface streets, but often some grade separation
- 150 to 200 person cars
- Electrical power
- Steel wheel

Heavy rail
- Exclusive right of way
- Longer trains, larger cars (200+ people)
- Third rail, electrical power or diesel
- Steel wheel
- Longer distances define commuter rail

Before launching into the differences among modes, I should point out that just because some modes can carry more passengers does not mean that they carry the most passengers in total. Buses are, in general, the lowest capacity transit mode since the vehicles have to be relatively small in order to operate on surface streets, but they still do a lot of work in U.S. transit. Over the past two decades, buses effected roughly 50% of passenger trips, in contrast to heavy rail at about 35% and light rail and commuter rail with 5% each (Public Transportation Association, 2013; statistics compiled by the author). Lower capacity modes play a large role in supplying transit, even in cities with mature, large-capacity rail systems.

Buses

Transit experts and advocates disagree over the usefulness of buses: bus routing customizes well to different contexts, which some view as an advantage and others as a disadvantage. Because bus routes are easily altered, agencies can experiment with bus service by adding or taking away stops or altering route alignments in response to shifting settlement or employment patterns. Buses can be operated in their own exclusive rights of way in a bus lane. Bus lanes can be converted back into shared rights of way, or they can provide the right of way for future railway development. Some see that flexibility as a flaw rather than a benefit. Rail investment is usually permanent, and it indicates a long-term commitment to service, one that can act as an anchor for residential, commercial, and employment development that buses may not in the absence of higher investment (Cervero, 1998). Buses are ideal for distributing passengers via *local service,* a term that describes service over short distances, with frequent stops and low speeds, and for conveying passengers in a local area and to and from higher capacity parts of the systems, like commuter rail.

Bus Rapid Transit

Recent innovations have led to improvements that can increase bus usefulness beyond simple distributor functions. Bus rapid transit (BRT) can get buses moving through regions much faster. The highest standard of BRT implementation can mean grade-separated, exclusive right of way, and multiple other system improvements. BRT systems use *express* or *skip-stop service,* which means the service runs on a regular route, but serves only select stops on that route. The longer distances between stops speed up service along the route. However, BRT can operate faster than simple skip-stop service if buses enjoy *priority signalization,* technology that allows the vehicle to coordinate with traffic signals along the corridor. That coordination allows the bus either to receive the green signal early, or to extend the green cycle, in order to get buses through intersections faster.

The idea sounds simple enough, and when it functions, it can significantly speed up service. Unfortunately, problems can arise. Transit companies normally do not control traffic operations; cities do. In fragmented regions with multiple cities, all the cities may not agree to allow priority at their signals. If some jurisdictions agree and some do not, those sections of the route where the bus has to operate without priority can limit operations. Even with priority signalization, traffic can grow bad enough that the speed gains from signalization erode over time.

These problems can become serious enough that some authors, like transit researcher Vulcan Vuchic (2007), argue that BRT should really be considered semirapid transit, as buses can never attain the speeds or capacities that rail operating on an exclusive right of way can. Some rail advocates push the point further, and suggest that BRT simply deludes those who want to avoid spending money to invest in rail into thinking that buses can somehow be configured to provide the same levels of service as rail for less capital investment. Bus advocates counter that BRT can, in fact, provide service comparable to that of much more capital-intensive, and therefore more costly, light rail service, once traffic variables are brought under control.

Innovative transit agencies have provided amenities and large platforms for high-demand bus stops, and then, in some cases, transformed those lines to rail service when ridership proved out. For all the dogma put out among advocates about whether bus is better than rail or vice versa, in practice agencies design systems based on how the budget, expected demand, and politics come together for each investment (Brown & Thompson, 2009; Levine, 2013).

Trolleys, Streetcars, Funiculars

Perhaps nothing reflects the role that preferences play in investment more than the trolley renaissance in the United States in the past two decades. Investments in funiculars, trolleys, streetcars, and cable cars tend not to be made just to improve mobility.

While advocates for these investments often claim broad environmental benefits, constituency groups as varied as downtown real estate interests and artists' organizations have pursued downtown trolley projects in many cities simply to anchor place development. Trolleys and streetcars are fun to ride, and they can help distinguish a neighborhood within a wider metro region. Because these systems tend to benefit local landowners and residents the most, trolleys can be controversial investments as their proponents seek funding for start-up capital from broader regional, state, and federal funding sources. Opponents argue that local property owners and residents capture both the amenity and property value benefits from the investment, and thus should pay for the system themselves (see Chapter 9). In some instances, this controversy has prompted trolley advocates to pursue, and sometimes win, local ballot-box self-taxation measures for their local trolley investment.

Light Rail Transit (LRT)

Trolleys in general are the shortest route, lowest capacity light rail transit (LRT). Light rail transit operates on fixed guideways. Many LRT systems usually operate on surface streets and interact with traffic, although many LRT run on exclusive rights of way for a portion of their routes (see Plate 8.2). LRT systems around the world have many different configurations and operating strategies, but most LRT cars tend to carry around 150 passengers or more, and they usually operate on steel tracks using electricity. As a result of the limited power available, LRT trains tend to have smaller cars, with trains of fewer cars and shorter station platforms, than heavy rail.

Criticisms of LRT center on the match between the level of service it provides and its associated capital costs. Skeptics often argue that LRT offers few, if any, operating improvements over buses, but tends to be costlier to build. Light rail proponents counter that midsized cities such as Portland, Oregon, Charlotte, North Carolina, and San Diego, California, have used LRT to build up transit patronage and development opportunities far beyond what would have been possible using buses alone long before their populations would have been able to support heavy rail investments.

Heavy Rail

Heavy rail investments tend to be costlier than LRT and are best reserved for the places within metro regions where population and employment densities are highest. The New York City Subway, the London Underground ("the Tube"), and the Paris Metro are examples of the world's most iconic heavy rail systems. Much heavy rail is grade-separated like subways, with all the costs that building underground entails. They operate on exclusive rights of way with larger cars (200 passengers or more) and longer trains, requiring more energy to power than LRT systems. Heavy rail includes commuter rail systems that connect suburbs to downtown, and heavy rail capacities, depending on frequencies and train lengths, can approach 60,000 people per hour.

What type of investment to make—on heavy rail versus light rail, light rail versus bus—hinges on many factors in the urban geography, and despite volumes of research and advocacy, no hard-and-fast rules seem credible. The important issue for service involves matching investment levels to the existing and future potential transit demand along the corridor. Since transit service supply and transit-oriented development can themselves alter those settlement and activity patterns, predicting how many riders an investment might attract is one of the greatest challenges in transit planning. I revisit this issue later in the chapter.

THE SPACE–TIME GEOGRAPHY OF TRANSIT

Networks come in many geographic shapes and sizes, and can be optimal for different goals, such as speed or spatial coverage. *Radial service* refers to networks that connect outlying areas to a central business district, but a radial network may not be optimal in regions with multiple large centers of employment, shopping, or population densities.

Spatiotemporal Factors in Service Quality

Service area describes the geography that a particular transit agency covers, usually coincident with administrative and political units. In cases where metro regions transcend political boundaries, service areas may extend across jurisdictions via operating agreements, or the area may have a regional transit provider. In either case, the distribution of service across the geography usually prompts some political wrangling about which parts of the service area should receive higher levels of investment.

Service span describes the hours of the day when transit companies offer service, such as from 5:00 A.M. until midnight, or 24-hour service. Twenty-four-hour service is very expensive for transit agencies to provide, but it can be a lifeline for workers on late-night shifts.

Throughout the service span and service areas, transit trips have different time elements. *In-vehicle time* refers to the time a passenger spends riding on vehicles. *Waiting time* refers to the time that patrons spend at a stop waiting for service. As a general rule, passengers dislike waiting, and one service goal is to minimize the time spent for the next vehicle to arrive (Hess, Brown, & Shoup, 2009). Frequent service helps minimize waiting time. If a route is operating on 20-minute *headways* (the time between vehicles), the service frequency is three times an hour. If the passenger just misses the vehicle, then the headway time becomes his waiting time, and the additional 20 minutes of waiting time adds to the 20 minutes of in-vehicle time for a total trip time of 40 minutes, assuming he has a short walk to and from the stations. With long headways, transit trip times can take much longer than taking the same trips by car, which poses a problem if transit is to compete with other modes for patrons (Walker, 2012). Transit commutes in the United States are much longer, on average, than commutes undertaken by other modes, and have been since the earliest studies on the subject (Taylor & Ong, 1995). In 2009, commutes by car took on average about 23 minutes and by transit 53 minutes. These averages mask considerable variation, however.

Whether a transit operator can support high frequency service depends on how much potential ridership exists along a transit corridor. *Transit corridors* refer to the geographic area surrounding the route. The area around a station or a stop where transit customers are likely to be coming and going is called the *service capture area*. Transit planners have used a quarter-mile to a half-mile as a general rule for service capture area for decades. Individuals vary, however, in how much they are willing to walk, and urban environments vary in how nice they are to walk in, which influences whether and how far people are willing to walk (Lee, 2013). Biking or driving to a station can also increase its service capture area, so just about any assumption will be approximate. In any case, high densities of either homes or businesses within those corridors yield more patrons than sparsely populated areas, and more population in service capture areas can support higher capacity service.

High frequencies and short headways can vastly improve service, but they are not the only possible solution. New technologies can track vehicles and alert patrons via phone or text when the vehicles arrive in real time, so that passengers can minimize their wait time and still make sure they arrive at their departure point on time. Whether transit vehicles adhere to their schedules is also important. If the bus is late, patrons may miss a later transfer connection and have to spend even longer completing the trip.

Scheduling from the Operator's Perspective

Frequencies, transfers, service coverage, waiting, and total trip time cover much, but not all, of what patrons look for in service quality. How to deliver service from a transit agency's perspective differs markedly from what patrons see as they consume the service. Some terms of art can help illustrate the basic challenges related to transit operations. A *revenue trip* refers to trips when vehicles are out serving customers and generating revenue. *Revenue hours* refer to the total time that a vehicle operates for revenue; *revenue miles* are an analogous measure for the miles in which a vehicle is in service. The *running time* refers to the amount of time that a vehicle operates for service, usually measured as time per route segment.

Patrons often refer to the schedule, which is a bit of a misnomer. Patrons and the public have access to what is more precisely called the *timetable,* which simply lists the times for stops at each location served. The *schedule,* by contrast, is an agency document that contains detailed operating plans and instructions for each revenue trip to help drivers, dispatchers, and road supervisors do their jobs. *Time points* on the schedule mark the geographic locations where a vehicle should arrive before, but not leave earlier than, the stated time according to schedule. Stations and stops serve as time points, but schedules often include additional time points to help drivers time their routes as they go.

Delays occur for many reasons, particularly for bus or light rail service. Traffic or traffic signal problems can delay both these modes and throw them off schedule. Missed trips can confound both the passengers and the agency, and those occur for many prosaic reasons such as driver illness, road calls for vehicle problems, accidents, or dispatch errors. Passenger loads can also delay service. The *load factor* is the ratio of passengers on the vehicles to the vehicle's total passenger seating capacity. Load factors below 1 mean that seats should be available. A load factor of greater than 1 means that passengers are standing in the vehicle, which via slang are *standees* or *strap-hangers. Crush loads* refer to vehicles really jammed with standees. An Indian railroad official recently coined the term "superdense crush load" for load factors higher than 2 on their trains.

Maximum load points are those stops on a given route where the most passengers board, and understanding these locations is important to helping the agency plan the route's time points accordingly. Maximum load points can shift by time of day, season (during the school year and out), special events, missed trips by previous vehicles on the same or different routes, and other factors. Long routes in highly patronized, large regional systems are likely to have multiple locations with high patronage.

Transit agencies face an enormous challenge in trying to provide on-time service. And on-time service is among the most basic aspects of transit performance and customer service. As a result, transit operations ranks among the most complex and rewarding fields in transportation, and it employs professionals with a wide range in skills from marketing to industrial engineering and operations research.

THE FINANCIAL STRUCTURE OF THE TRANSIT INDUSTRY

Private companies developed the earliest omnibus, horse cart, and, eventually, streetcar systems primarily as a service to support their residential real estate developments in growing metropolitan areas (see Chapter 3). New York's early subway investments resulted from groups of investors who wished to open up opportunities for high-end home sales farther from New York's teeming business and industrial districts (King, 2011). This early model continues to inspire advocates for transit-oriented development (TOD) today. Like those who built toll roads before them, business owners in cities after the Industrial Revolution invested in omnibuses and streetcars in urban areas because transit supported development directly by opening new land up on the urban fringe. It also helped commerce by providing relatively low-cost mobility for workers, and by attracting pass-by customers at station areas. What happened to all those private transit companies holds some lessons about the financial problems transit faces as an industry.

One of the most common anecdotes about public transit concerns the great General Motors conspiracy from a U.S. Senate antitrust attorney, Bradford Snell (1974). The conspiracy goes something like this: holding companies began buying up streetcar companies throughout the late 1930s and early 1940s in the United States. These holding companies pursued these buy-outs and takeovers with the financial backing of General Motors (GM), along with what conspiracy advocates refer to as "rubber tire and big oil" interests—Standard Oil, Firestone Tires, Goodyear, Chevron—among others eventually tried for anticompetitive behavior. These holding companies replaced the streetcars with buses. In some instances, the holding companies attempted hostile takeovers of small-city transit lines, like the Key System in Oakland, California. Some conspiracy believers argue that the holding companies were an attempt by big oil and big auto to foster auto dependency via destroying streetcar systems and replacing them with buses.

But the structure of the transit industry may be more at fault for the demise of private streetcar companies across the United States and Europe. Hostile takeover bids and selling out among smaller streetcar firms was hardly unusual by the late

1930s through the 1950s by the time GM and the rest got into holding companies. Industry consolidation like this occurs in virtually all transportation markets, for passenger and goods movement alike, even today. With network goods, transport firms gain a geographic business advantage when they merge their network with the network of another firm—which again, routinely occurs with logistics and airlines companies today. Mass transit service has high capital requirements in order to operate. At the very least, it requires *rolling stock*, which are the vehicles in the transit company's inventory, and streetcar operations required in-road rail and (in many instances) electricity generation and distribution capital. "Big auto" and "big oil" may prompt modern-day disdain, but they were able to enter transit markets without the coinciding real estate development profits precisely because they already were big corporate concerns that could leverage financing for the high capital costs required for transit services they took over. The only other entities capable of doing so either at the time or since have been municipal governments backed by taxpayers, or private franchisers who operate services after taxpayers front the capital.

Along with the high up-front costs of capital, operating mass transit still generally proves to be unprofitable unless the operator or company (1) serves a small, densely populated island (Japan, Singapore, Hong Kong); (2) has a way to leverage its capital for things other than simply selling tickets (fares) to passengers, such as through advertising or station-area concessions, or real estate, like the early streetcar operators; or (3) provides a monopoly service, like a ferry or some small-scale, specialized paratransit services. Just about every other fixed-route transit provider loses money running service. The *farebox recovery ratio* refers to the amount of money gained in fare revenue divided by the operating costs. Table 8.5 displays a selection of farebox recovery ratios reported from transit operators around the United States. A recovery ratio of 1 would mean that an operator is breaking even; above 1 would indicate profit, and below 1 indicates operating losses.

It is clear from Table 8.5 that passengers pay a small amount of the day-to-day operating expenses incurred to supply public transit—on average about 30% across all systems. When capital costs are included, the recovery at the farebox falls to around 20% on average across all systems in the country. The gap between operating revenues and costs is the *operating subsidy*. *Capital subsidies* come from taxpayers at the federal, state, and local levels to help transit agencies pursue capital projects and improvements.

For advocates, public investment in rail makes sense based on the environmental and other secondary benefits that come from encouraging transit use instead of car use. Based on these benefits, transit proponents argue that taxpayers should recognize that transit is a merit good that will require both capital and operating financial support from the public. But, while auto pollution, crashes, and congestion all create real and sizable costs, preventing those costs does not directly result in revenue that transit agencies can then use to pay their bills. Instead, those gains to public welfare are often captured or reflected in private income and property, and a tax-averse populace or high demand for public investment in other needs, like schools or health, may redistribute those gains to other public endeavors.

The idea that taxpayers should just expect to pay for transit provision also has its problems. Transit may lose money for many reasons, including poor management,

TABLE 8.5. Farebox Recovery, 2012, for Selected U.S. Transit Operators

Name	Service Area	Recovery Ratio (Break even = 1.0)
Urban heavy rail/subways		**0.36**
MTA New York City Transit (NYCT)	New York–Newark, NY–NJ–CT	0.78
Southeastern Pennsylvania Transportation Authority (SEPTA)	Philadelphia, PA–NJ–DE–MD	0.52
Washington Metropolitan Area Transit Authority (WMATA)	Washington, DC–VA–MD	0.68
Massachusetts Bay Transportation Authority (MBTA)	Boston, MA–NH–RI	0.51
Miami–Dade Transit (MDT)	Miami, FL	0.24
Staten Island Rapid Transit Operating Authority	New York–Newark, NY–NJ–CT	0.22
Alternativa de Transporte Integrado–ATI (PRHTA)	San Juan, PR	0.13
Commuter rail		**0.42**
Southeastern Pennsylvania Transportation Authority (SEPTA)	Philadelphia, PA–NJ–DE–MD	0.57
Peninsula Corridor Joint Powers Board Caltrain (PCJPB)	San Francisco–Oakland, CA	0.50
Metro-North Commuter Railroad Company	New York–Newark, NY–NJ–CT	0.62
Metro Transit	Minneapolis–St. Paul, MN–WI	0.17
Connecticut Department of Transportation (CDOT)	Hartford, CT	0.08
Denton County Transportation Authority (DCTA)	Denton–Lewisville, TX	0.03
Light rail		**0.23**
San Diego Metropolitan Transit System (MTS)	San Diego, CA	0.57
Massachusetts Bay Transportation Authority (MBTA)	Boston, MA–NH–RI	0.49
Metropolitan Transit Authority of Harris County, Texas (Metro)	Houston, TX	0.32
Dallas Area Rapid Transit (DART)	Dallas–Fort Worth–Arlington, TX	0.12
Port Authority of Allegheny County (Port Authority)	Pittsburgh, PA	0.18
Central Arkansas Transit Authority (CATA)	Little Rock, AR	0.08
Streetcar		**0.16**
New Orleans Regional Transit Authority (NORTA)	New Orleans, LA	0.25
Memphis Area Transit Authority (MATA)	Memphis, TN–MS–AR	0.13
Hillsborough Area Regional Transit Authority (HART)	Tampa–St. Petersburg, FL	0.28
Kenosha Transit (KT)	Kenosha, WI–IL	0.10

Source: American Public Transportation Association, 2014, Table 26; compiled by the author.

ill-advised expansions of service, and costly capital investments. Should we continue to expand transit, even if the marginal revenue is always likely to be negative? In the absence of new revenue sources for operations, new construction projects could cause a transit agency to lose money faster than it was before and jeopardize service quality across the region.

It might be helpful to examine just how serious shortfalls can be. The New York City transit system is the envy of many cities, and millions of people rely on it every day. But in 2010, the U.S. economy experienced a downturn and exposed the Metropolitan Transit Authority's (MTA) serious financial problems. The MTA had at the time nearly a $12 billion budget; its shortfall was $1 billion, or about 9% of its total budget (Office of the State Comptroller, 2010). Growing debt from building new projects can simply doom an agency like the MTA to keep cutting services and raising fares unless the agency finds new revenue sources. In subsequent years, the MTA budget recovered, but the recession revealed just how unstable the agency's finances can be.

Contemporary privatization advocates suggest that the private sector could vastly improve service, capital costs, and farebox recovery numbers through better management. Some of that advocacy probably boils down to simple ideology: the belief that governments are always less efficient than private markets. But privatization advocates also point out, correctly, that public ownership of transit leads to many perverse incentives for transit agencies and patrons.

Reprivatizing Transit—A Question for the Future

As Table 8.5 shows, some transit agencies receive as little as a fifth to a third of their revenue from serving their patrons. If two-thirds of the budget comes from the voters and the politicians who represent them, then the many decisions that public transit managers have to make about fare levels, service geographies, frequencies, and capital investments target the desires of voters and politicians at least as much, and perhaps even more, than the desires of their transit patrons. This problem is especially acute in metro regions where transit captures a small percentage of commuters. Even when transit riders are a sufficiently large group that they can influence democratic politics, their influence may also lead to political pressures that prompt short-term gains for passengers at the expense of long-term sustainability for the public provider of the service.

Fare Policy under Public and Private Transit Management

For instance, some privatization advocates argue that market-rate fares would be an improvement over the artificially low fares that public agencies offer. Governments provide transit in part so that residents without cars can still access social and economic opportunities within regions (Sanchez, 1999). For impoverished people within metro areas, transit may be the only way they have of getting around (see Chapter 13). As a result, fare increases are highly controversial. Low fares were among the first things municipal governments demanded of private streetcar companies in early franchise agreements for this reason. Privatization advocates counter that overex-tended transit agencies losing money on low fares provide poor service that limits the job and educational opportunities passengers realistically can use. Cheap fares leave transit riders with perpetually poor-quality service, privatization advocates argue, because general taxpayers will agree to pay only so much for tax revenue going to transit they will never use themselves because of its poor quality. If these advocates are correct, transit always remains a last-resort, second-rate service rather than an amenity most people are willing to pay market rate fares to patronize. Private tran-sit companies, accountable to passengers, shareholders, and donors, would be free to offer a variety of transit services at different price points so that passengers can choose what suits them rather than the one-size-fits-all of government-run service delivered at low-quality due to pressures to keep fares low.

Investment Strategies in Public and Private Transit Management

Because public agencies have to focus on pleasing voters or politicians rather than customers, transit agencies market the service to voters based on secondary benefits to nonpatrons, like relieving auto congestion and alleviating air quality. Government funding for capital investment decisions means that politics can drive where invest-ment occurs more than demand (see Chapter 9 for more information on this issue). This problem may be particularly acute in places that use ballot-box initiatives to seek taxes for funding transit (Schweitzer, 2012; Sellers, Grosvirt-Dramen, & Ohenesian, 2009). Such ballot-box initiatives may have to put together project lists that spread the money around, so voters in each district get their cut of the money. Spreading investments around in this way can mean overinvesting in places with few potential riders (but many voters) and underinvesting in places with higher demand. Similar concerns exist about federal funding for transit. Political wrangling over funding at the federal level leads to the sort of deals where transit funding gets spread across states, rather than targeted to the regions where the investments would be most pro-ductive (Brown, 2005). Private companies, unlike government, are more likely to seek out investments only in potentially lucrative new markets, and then to market and price the service as needed in order to attract new passengers, rather than scramble trying to appease voters and influential politicians.

Information and Market-Matching Problems in Transit Investment

Researchers have identified a persistent problem of public agencies' forecasting of capital costs and ridership (Siemiatycki, 2009). The most common means for analyz-ing public investments is cost–benefit analysis. If the projected benefits, measured in dollars, of a project outweigh its costs, then the project is worth investing in accord-ing to social welfare criteria. Research demonstrates that many cost–benefit analy-ses in transportation—from highways to transit—appear to distort benefit and cost calculations to make projects look better to decision makers and the public than those projects really are. That is, the analyses show that agencies systematically underestimate how much new capital investments in transit will cost while overestimating how many passengers will use the new facility. Table 8.6 shows the projected costs, the as-built costs, and the costs per kilometer of recent transit projects built in the United States. Only two came in at or under projected cost, while the rest were over budget, costing up to nearly 80% more than projected cost.

The bias in forecasting has generated considerable debate. One explanation is that transit agencies and politicians want a project in their district, and forecasters, most likely those working in consulting firms paid by the public agency, generate analyses that show these projects in the most positive light possible. Research on planning ethics has found that agency demands and political pressures whipsaw forecasters between clients' demands for positive results and their professional integrity (Wachs, 1990). Project proponents face few consequences when projections prove false, so long as they can get approval to begin construction. Critics refer to this phenomenon as a "commitment trap." Few politicians or public sector managers will stop work on construction and abandon a half-constructed project. Instead, agencies simply pay the extra costs. The commitment trap also extends to poor ridership forecasting. Once a project is finished, agencies will not generally shut it down if ridership is below expectations.

Other explanations focus on the many challenges to accurate transit forecasting, most of which bias ridership forecasts upward. One method to forecast demand is to ask residents living and working near proposed new transit facilities whether they would use the new service. Focus groups and surveys are two methods, but asking people about future projects can lead to misleading forecasts. Because residents seldom pay the full costs of new service, they have little reason to say no to the new

TABLE 8.6. Projected and Actual Costs for Individual Transit Projects ($Millions)

Mode	Region	Agency	Projected cost	Total cost	Cost per km
BRT	Cleveland	Greater Cleveland Regional Transit Authority	168.4	197.2	14.8
Rail	Minneapolis	Metropolitan Council	$265.0	$317.0	$4.1
Rail	San Juan	Departamento de Transportación y Obras Públicas	$1,250.0	$2,250.0	$131.4
Rail	South Florida	South Florida Regional Transportation Authority	$327.0	$345.6	$4.9
LRT	Oceanside	North Coast Transit District	$213.7	$477.6	$13.6
LRT	San Francisco	San Francisco Muni	$667.0	$787.0	$74.4
LRT	Dallas	Dallas Area Rapid Transit	$988.0	$1,700.0	$75.9
LRT	Los Angeles	Expo Authority/LA Metro	$640.0	$930.0	$67.6
LRT	Portland	TriMet	$283.0	$350.0	$37.7
LRT	Charlotte	Charlotte Area Transit System	$331.1	$462.8	$30.1
LRT	Salt Lake City	Utah Transit Authority	$89.3	$83.6	$34.15

Source: Office of Planning and Environment Federal Transit Administration, U.S. Department of Transportation, Before-and-After Studies of New Starts Projects, Report to Congress for the years 2006 to 2011. Retrieved June 22, 2013, from www.transit.dot.gov/sites/fta.dot.gov/files/docs/FY2013_Before_and_After_Studies_Report_to_Congress_Final.pdf.

amenity, or they may simply not know for sure if they will use the service until they experience what the service is really like.

Forecasters can also examine other similar transit services in similar neighborhoods and examine ridership there. This process, however, is also likely to overstate ridership on new service because of *residential self-selection*. People who want transit access choose to live where such access is available, and thus they self-select into neighborhoods with transit. Some people will not even consider moving to a neighborhood unless they have transit access—call them Group A. Others may place transit access far down their list of things they need in a neighborhood, behind access to high-quality schools, for example. But they might, if they had more access to transit, begin to use it if the service is good enough to meet their needs. We can call them Group B. Other people, in Group C, may be simply unwilling to use transit, no matter what. For an area that has little to no existing service, we do not know how many people really fall into each group, B or C. We only know there are no members of Group A there. Using ridership figures from existing transit services attributes transit use levels of people in Group A to people in Groups B and C, and thus is likely to overestimate their willingness to use transit.

These groups can be flexible. It is possible that people in Groups B and C will join Group A if energy prices, parking charges, or other costs of driving a car go up, or if transit service is good enough. Nonetheless, skeptics counter that we ignore residential self-selection at our own peril. By building transit where people have demonstrated they do not value it enough to pay higher rents to live near it, public agencies will oversupply the service. Our history of transit investment demonstrates that sometimes advocates are correct, in that redevelopment around stations can help to increase ridership, particularly over time as more people from Group A move next to new service, while other times skeptics were right and ridership never really thrives.

Private companies, in theory, have more reason to be prudent with managing risk around capital planning and finance than public agencies, which can pass the fiscal consequences of poor cost-and-demand forecasting onto taxpayers. As long as there are tax revenues to pay off the bonds governments issue to cover transit agencies' capital costs, bond investors do not care how well the agency has projected costs or revenues from potential new riders. Private transit providers, however, who would have to obtain financing from venture capital or other private funds, face market discipline vis-à-vis unrealistic cost and revenue projections.

Privatization and Operating Cost Savings

In addition to market fares and better risk management, privatization advocates argue that private transit companies, driven by the profit motive or specific philanthropic goals, rather than numerous ill-defined and conflicting public goals, would be free to pursue cost savings in ways that government agencies are not. A particularly contentious issue for public agencies concerns unionized labor, which could be one source of cost savings in private industry. Labor is among the highest costs in operations for public sector agencies, and in general, unions can command higher wages, benefits, and protections for their members than what would be paid to a nonunionized labor force.

Split shifts and the use of part-time labor exemplify some of the problems with flexibility and costs that prompt conflict between transit management and labor. Transit, like just about every other type of transportation service, suffers from a peaking problem. Work schedules prompt many people to commute in the morning and the late afternoon—the *peak*. Demand for transit is no different, which explains why transit service uses larger vehicles and runs every 5 to 10 minutes during rush hours. However, that demand structure creates a problem for transit agencies because no single peak period time corresponds with a driver's 8-hour shift. In supplying service, a manager would like to be able to split some of the drivers' shifts into about 4 hours for the morning peak and 4 hours for the evening peak. Alternatives include hiring part-time drivers to put in 4-hour shifts here and there or contracting out to privately owned contractors to supplement the full-time drivers. But split shifts are inconvenient for labor, and part-time or contracted labor to cover those peaks pose a threat to full-time, unionized workers' job and wage security.

Privatization advocates argue that labor is only one area where competition might prompt private transit providers to cut costs or innovate. Other ways the private sector might innovate include technology adoption and amenity provision to attract customers.

The Privatization Record So Far

Research in privatization and mass transit shows mixed results for all the benefits claimed on behalf of privatization. Contracting out bus operations in the United Kingdom during the 1980s, for example, resulted in lower subsidies and more frequent service with the introduction of minibuses. But that experience contrasts with the public–private partnership (PPP) forged around the London Underground (public) and Metronet (private) beginning in 1998. The deal came about in part because the transit provider, Transport for London, was financially stretched and its capital stock decayed. Metronet ran into the capital cost overruns that plague public agencies and went into insolvency. An audit cited Metronet's poor management as the primary reason for the failed partnership (National Auditor's Office, 2009).

These problems with privatization happen with toll roads and other infrastructure as well. Governments really are the lowest cost borrowers, again because they can fall back on taxpayers, and the costs of transferring ownership from public to private entities, let alone maintaining the relationships and contracts, add up. Thus these deals gamble on private-sector productivity gains being large enough to offset higher borrowing and transactions costs.

Less dramatically, but perhaps more importantly, a structural problem plagues privatization deals. For private entities seeking to operate a portion of the transit network, the rational strategy is to bid on those parts of the network where revenues are greatest—the lowest cost, highest ridership portions of the network. But if transit agencies give up revenues from the highest revenue parts of their system, they are left with lower revenues to run the rest of their system. By using money from high-demand locations to support service on lower-traffic areas, the whole network becomes more useful and financially stable. Allowing private companies to cream off the best parts of a network increases the subsidy burden required to run the rest of the system. One way around this issue is to allow both positive and negative bids on contracts. Private contractors can bid positive amounts for the parts of the system that do not require subsidy, and they can bid negative amounts (what they would need to be paid to run a low-demand route) on other parts of the networks (Levinson, 2014). Alternatively, the high-demand and low-demand parts of the network can be bundled, requiring contractors to run both if they want access to the profitable route.

Despite the many unanswered questions, privatization remains a lively field of debate in transit planning, largely because the financial problems of the industry persist and threaten the service's sustainability. Low farebox recovery numbers (Table 8.5) mean that, in general, transit operations lose money. That situation is not financially sustainable, and fixing it requires some combination of lower costs, greater revenues, or a reliable source of operating subsidies year in and year out. In theory, privatization might provide the first two, which would lessen the need for taxpayer subsidy for operations. In practice, however, the gains to privatization have proven more elusive than advocates suggest. Other possible solutions to financial stability include cost-saving measures such as saving labor costs through driverless vehicle technology or by scaling back the system. Still others suggest that revenues from "green" taxes or tolls on cars should be used to support public transit. There are many potential answers to the financial problems of the industry other than privatization.

It is worth pointing out that public and private services support each other as well as compete. New York's subway system operates with the help of taxis on surface streets that fill inevitable gaps in the system. Jitneys, though often unlicensed, operate in a similar manner. Privatization advocates may be correct when they argue that monopoly franchise laws are not needed to protect public transit agencies and that these laws ultimately hurt passengers by reducing service choices. Finally, the private sector already supplies quite a bit of public transit, mainly through simple contract arrangements for everything from construction to janitorial services to station-area concessions.

THE TRANSIT MARKET: WHO CURRENTLY RIDES TRANSIT

Transit markets vary a great deal from place to place. This section contrasts three U.S. transit markets: Atlanta, Los Angeles, and New York.[1] The figures on demographics throughout this section show percentages. The "All" category is the percentage of the total population each demographic characteristic represents in the city. For each mode, by contrast, the percentages represent how many people riding that particular mode (of all transit riders) possess that particular characteristic. If each mode had exactly the same percentage of each group as the population as a whole, all people, regardless of their characteristics, would be equally distributed throughout the modes. But that is not the case, as we shall see.

Age

Transit use by age and mode are presented in Figure 8.1. The comparison across three cities shows a couple of important points about age and transit markets. First,

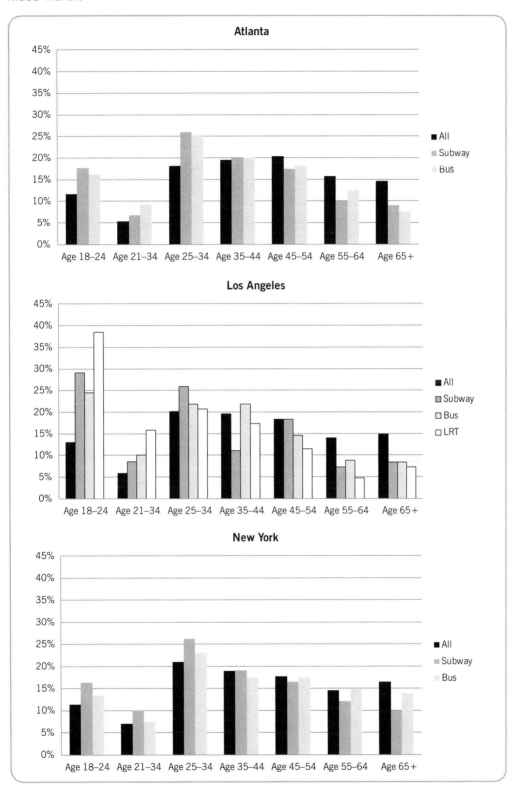

FIGURE 8.1. Transit use by mode and age.

Los Angeles and New York have younger population bases than Atlanta. Atlanta's population base is bimodal, with comparatively large groups of young people and a big percentage of residents (nearly 30%) who are over 65. Los Angeles and New York have relatively high percentages of their populations under 35. Second, and more significantly, older people tend not to use transit to the same degree as younger people in any of these markets. In each city, transit users are disproportionately younger, and transit usage starts to drop off after the age of 45, and declines substantially after the age of 55.

In Atlanta, bus riders are much more likely to be young (18–24) than represented in the overall population. The same group is overrepresented among light rail riders in Los Angeles. In New York, transit riders mostly mirror the general demographics in terms of age, until the older age group. At that point, even in New York, older residents are disproportionately less likely to take subways and buses, though the percentage of transit users on the subway declines as well.

Although these data are for only three regions, the age demographics of transit riders tend to show similar drops as people age, and have done so for decades (Rosenbloom, 2001). Even in the best transit market in the United States, New York, older residents tend not to use transit to the same degree as younger riders do. Multiple reasons may explain the decline. First, as we saw with the commuting data, transit is a mode of choice for many commuting trips, particularly in New York City. As people age and retire, they do not commute. Not commuting does not mean that their overall level of daily travel necessarily goes down, but these noncommuting trips may be easier to serve with walking or carpools than with transit.

Income and Ethnicity

Income differences among the users of different transit modes have been the focus of debate for some time. Figure 8.2 shows the percentage of transit riders in each mode falling into each income group relative to the entire urban population. Note that the top two income categories overlap in that households that earn over $100,000 a year are included in both of the last two subgroups. Bus riders in Los Angeles are disproportionately low income, and the same is true, if less severe, in Atlanta. Atlanta's interesting feature shows up on the other side of the income distribution. Subway riders in Atlanta at the highest two income groupings almost exactly mirror Atlanta as a whole, but people in those income groups do not ride buses. In New York, subway riders are disproportionately affluent because of the concentration of both wealth and subway service on and near Manhattan.

What to make of these differences has bothered travel behavior and transit researchers for some time, along with community activists. Bus riders in Los Angeles and Atlanta, as well as elsewhere, have organized to challenge the spending and investment patterns of regional transit providers who pour money into capital projects like trains that serve primarily wealthy communities while letting bus service languish (Grengs, 2002). Some argue that affluent people willing to take trains will not use buses because trains are fast and comfortable, while buses are slow, dirty, and crowded. Ergo, the argument goes, transit agencies should invest in trains to meet the demands of the fussier passengers, thereby raising service quality for all passengers.

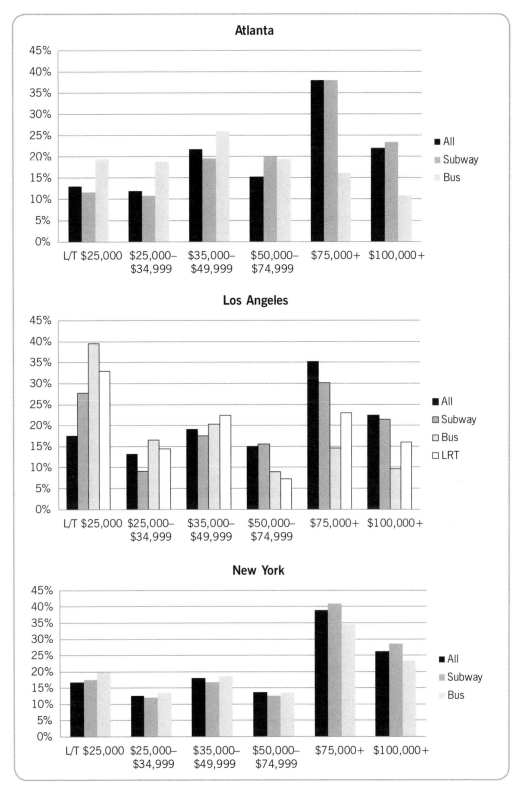

FIGURE 8.2. Transit use by mode and income.

That rationale is hard to accept when we examine the numbers from the five boroughs of New York City and take ethnicity into account (Figure 8.3). Transit riders in New York of all income levels and all ethnic groups take the bus. The idea that white passengers will only use a particular mode is refuted by these data—the data show that people use all the modes available in a robust transit market like New York.

Residential segregation by race and its effects on urban politics likely explain differences in income and ethnicity in rail patronage more than anything inherent to the mode. In places where transit is ubiquitously supplied, transit demographics across modes mirror those of the urban population generally. In places where transit investment is more spread out, politically influential communities, like those that are white and affluent in the United States, are likely to get expensive capital improvement projects, like trains, sooner than other groups, and the ridership on those services will reflect those political outcomes. Transit access is an amenity: people who are affluent are likely to be able to rent and buy housing closer to stations in places where high-quality transit supply is scarce, pushing less affluent and less privileged home-seekers out to places with less transit. As more and more communities throughout metro regions demand a share of capital projects, these differences may change, and the regional transit demographics may begin to look like the general population demographics, as with New York.

Transit policy and planning, however, cannot ignore these differences among groups right now in favor of hoping that it will all work out in the future. The fact that train riders in Los Angeles and Atlanta (and other regions) are likely to be more affluent than bus riders has equity consequences for changes in service, particularly given the fiscal problems of transit agencies. In many regions the solution to funding problems tends to be to cut service. Cutting service and raising fares directly affects the subgroups who use that mode the most. Allowing bus service quality to languish in favor of supporting train service hurts low-income patrons and some ethnic groups more than others. Those choices need to be part of the policy and planning discussion around service equity (see Chapter 13).

SECONDARY BENEFITS AND MERIT GOOD ARGUMENTS FOR TRANSIT

Now that we have covered the basics of modes, scheduling, networks, finance, and service consumption, we can return to where many transit advocates start: the major, nonmobility selling points of transit. Advocates of public transit for cities have been coming up with rationales for public investment in transit for decades, but transit seems to deliver some promised benefits better than others. First are the *direct mobility benefits* discussed throughout the chapter. Transit benefits cities by helping people get where they want to go. When transit is supplied well, people living in cities can avoid having a car and its attendant expenses, hassles, and accident mortality. Transit can help those either unwilling or unable to drive to gain social and economic citizenship within regions. Even for people who do have cars or bikes, transit can provide an option in case the car goes on the fritz, when fuel prices spike, or if a person becomes unable to operate a car or walk for whatever reasons.

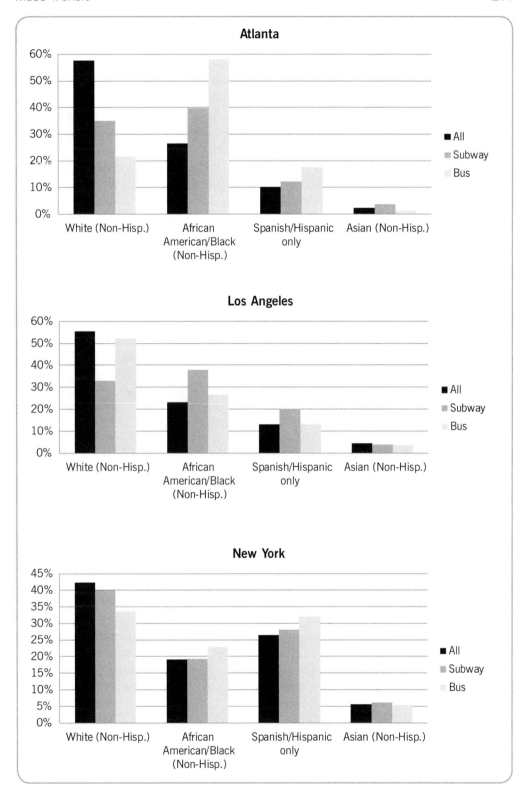

FIGURE 8.3. Transit use by mode and ethnicity.

Place-based benefits are the second category of benefits. Place-based benefits are environmental or economic or both. Transit allows denser land use development than does auto-oriented development, which requires acres of parking (see Chapter 9). More intensive land development via transit can help reduce the habitat destruction and ecological disruption of urban development. In cities that have robust real estate markets, transit stops and stations can support commercial, light industrial, and residential housing development. Transit allows for other place-based benefits of agglomeration, like the economic opportunities that arise via density, and these benefits may be sizable (Chatman & Noland 2013). Transit access is an amenity for all types of land uses, and the value of the mobility it provides gets incorporated into the price of land surrounding rail stations. Transit stations or stops can be lively, well-designed amenities that enhance the public spaces available for people to use. All these are potential place-based benefits.

The third category of benefits, *mode-shift benefits,* are those that happen because people use transit rather than cars for trips. Cars cause major problems in urban areas, from crash deaths to mortality and morbidity from pollution. They are also a nuisance. These benefits, however, can be the hardest to obtain and demonstrate. The mobility and place-based benefits of transit can occur no matter what happens with auto travel. In the market for travel, trips by all modes can go up simultaneously with economic and income growth. Even if more cars are on the road than ever before, increases in transit usage reflect mobility benefits, and, in all likelihood, increases in land values and development opportunities near stops and stations. The modal shift benefits, however, hinge on transit ridership. If people are taking transit *instead of* driving, then transit can help reduce the negative environmental and safety problems associated with cars. These problems include air, noise, climate, and water pollution and all their attendant human and ecological health effects (see Chapters 1, 11, and 12). Consequently, getting cars off the road has been a policy goal in the United States since the early 1970s, and supplying transit has been a key part of that goal. But supplying transit alone generally is not likely to take cars off the road. To discourage driving, multiple policy and design factors have to be in place. Some involve the mixed-use development strategies covered in Chapter 7. Other policies directly raise the costs of driving relative to taking transit, like high fuel taxes, price floors for gasoline, road charges, parking charges, car rationing, or carbon taxation (Sperling & Gordon, 2009). Plying travelers with many transit options can help get people to leave their cars, but it is no guarantee.

Absent policies that make cars more expensive or time-consuming to use than transit, transit ridership tends to underperform on public goals for the service. Half-empty trains and buses do not "clear the air" or "fight climate change" as much as full ones do. Even worse, absent policies that restrict driving, new transit-oriented development and swanky new trains can increase the number of cars on road rather than decrease them, as when people drive to new developments to enjoy the restaurants, art walks, and sports venues those places offer, even if *some* people go to those places via transit. For example, consider a place that, prior to a transit-oriented new development, attracted 2,000 trips a day: 80% by car, 2% by transit, and 18% by walking and biking. After a new transit-oriented development, the place attracts 4,000 trips per day, and shifts 20% of visits to transit, 20% to walking and biking, and only 60% to driving. Those are big mode shifts, but with the growth in trips overall, the new development actually increased the number of car trips overall (from around 1,700 to 2,400, an increase about of about 600 car trips per day over the baseline) even as it boosted transit and the other alternatives. The car mode share would need to drop from 80% to 44% to have fewer cars on the road after the new development.

Policy and design can influence that outcome. If, for example, parking at the new development is costly, people will be more likely to leave their cars at home and take transit. The same is true if people face high fuel taxes or cordon tolls. Transit provision is a crucial part of a portfolio of policies to discourage driving not just because transit provides, as it is often described, an "alternative" to the car, but because it can be a much lower cost alternative if and when policies to restrict car use come into play. Voters and politicians will not accept restrictions on cars or driving if there are no legitimate alternatives. And a long, uncomfortable transit ride with two transfers and unreliable service is not really a legitimate alternative.

One oft-sought but grossly misunderstood modal-shift benefit is congestion. Transit service is routinely sold to voters as a means for clearing up traffic congestion. It may be possible to reduce auto traffic by restricting cars or making them more expensive, but traffic congestion in the city is also a function of population density, which transit requires if the service is to be cost-effective. Congestion in cities, unlike air pollution, is not strictly negative. Congestion means that many people want to be in a particular place (Taylor, 2002). By increasing the time and money costs of auto use, traffic congestion can help make transit more attractive. But solving congestion with transit is neither likely nor desirable.

Are all the benefits of transit worth the sometimes billion-dollar price tags we see on transit projects? The answers depend on the individual project and the context, as those determine what type of mobility, place-based, and modal-shift benefits might come into play. High-quality transit service that really serves riders unlocks all the other things we want public transit to accomplish, whether it is helping clear up the air, getting people to walk more in order to be more active for their health, or boosting economic growth.

TRANSIT AND THE FUTURE

This chapter has provided an overview of what transit is and how it operates. Several important points guide the discussion. First, the leading systems around the world supply far more transit than just about all the other systems put together. In the United States, that leading system is New York's, supplying more passenger miles per year than the other top 10 providers in the United States combined. This imbalance is seen across the world, with the transit in mega-cities like London, Mumbai, Tokyo, Mexico City, and Beijing serving far more passengers than other cities.

Second, different modes serve different, but complementary, functions to create a system. Exclusive rights of way contribute to increasing capacity on all modes, and larger cars enable heavy rail transit to carry the most passengers per unit time. But distributor services like buses serve almost half of all U.S. passenger trips. Buses often lead in ridership numbers in the aggregate, even in places with mature rail systems. System planning requires thinking about how, whether, and where investments in the different modes work in tandem.

Third, transit alters urban geography in multiple ways. Infrastructure investment choices are one way; operations and scheduling are another. Overlapping routes, high-frequency service, and preferential treatments like signal prioritization change how much transit access and capacity a particular place has right along with new physical infrastructure.

Fourth, passengers themselves cover only a portion of the funds required to supply public transit. The rest comes from federal, state, and local governments. That financial structure means transit agencies balance the many goals placed on the agency by higher levels of government, political leaders, voters, and passengers' expectations for service. The low levels of passenger contributions lead to multiple, varying opinions on what the problem with transit might be. For many, transit's many external benefits, like decreasing deaths from auto crashes or improving air quality, demonstrate that transit is a merit good that justifies taxpayer support. For others, the gap between the farebox and the costs to supply the service suggests poor public management in need of cutback, reform, or both. Those who advocate for privatizing public transit do not, however, have at hand dozens of transit privatization success stories to tell. Good contracted services and terrific examples of privately run jitneys and taxis have proven out in many places around the world, but strictly private transit services for large-scale systems have not been unqualified successes. Nonetheless, many aspects of privatization remain in play in public policy and management, and funding gaps in service guarantee that privatization debates are here to stay.

So where does transit go from here? Futurists love to produce visions of what transit will look like in the coming decades, usually with streamlined, silvery vehicles that sport everything from cut-your-own locavore salad bars to stations with ziplines. We have many reasons to be optimistic about transit's future, regardless of whether that future is mostly public or private, and despite the lingering questions about its financial sustainability. Growing urban populations around the world make transit service more relevant than ever before. Transit ridership in the United States, one of the markets where the personal automobile has dominated travel choice, has grown steadily from the 1990s onward. Advocates have won, for the most part, the "ideological war" by convincing mayors, real estate interests, and the general public that transit enhances urban environments, a win that may help people contribute to building, running, and riding the systems.

As I have discussed throughout this chapter, the industry's financial concerns, however, make planning for higher transit demand more difficult, even with more patrons, political salience, and private-sector interest. For example, some urbanists have argued that the aging of the North American, European, and Australian populations will result in more seniors than ever before riding transit. With transit companies losing money, however, an uptick in demand may be either a boon or a problem, depending on how much excess capacity currently exists. In addition, older populations may be more tax-averse than younger populations. The result for some agencies may be more demand for service at the same time they receive lower fares per trip and less support from general fund taxation.

A similar problem surrounds transit agencies and energy prices. Many transit advocates argue that rising energy costs will eventually force automobile drivers out of their cars and prompt younger people to avoid getting cars at all. All these changes could happen, but transit providers get hit with higher energy prices right along with drivers. New York Metro's energy costs nearly doubled in recent years, from $291 million in 2009 to a projected $498 million in 2014 (Office of the State Comptroller, 2010). Now that we have glimpsed the complexity of actually running transit, instead of merely advocating for it, we know better than to jump to conclusions about whether any given change will be unambiguously good or bad for transit providers. Chances are, most social and economic changes bring a mix of opportunities and headaches for transit providers to grapple with.

Technology changes, however, are likely to be good for transit. Smartphone, satellite, and individual geolocating technologies have already made transit much easier to use by allowing users to see when the next bus or train is arriving. Other things that new technologies allow may further expand urban travel markets and make cars less useful in favor of transit. App-based ridesharing companies like Uber, Sidecar, and Lyft have already taken advantage of smart phone and geolocating technologies to match those who want rides with those who have cars and who would like to have a little extra cash by giving a ride. These kinds of services flourish in regions where residents leave their cars at home, or where people live without owning a vehicle, and they can supplement transit service just as taxis do.

Driverless vehicle technologies may also radically change the way transit works in cities. While the major focus has been on Google's proposal for driverless cars, the potential cost savings to public agencies of driverless transit vehicles could be substantial, including the possibility of eliminating strikes and labor disputes. The logic about market expansion and carsharing also applies here as well. Driverless cars may be readily deployed as a taxi fleet, and travelers can combine car subscriptions or taxi services with public transit, rather than dealing with the expense and hassle of owning their own car.

The main challenges ahead center on three concurrent priorities: addressing costs, raising revenues, and supplying high-quality service—all the while navigating the difficult politics that surround the mode. As its advocates note, transit can help get people out of their cars, which improves safety and air quality. Transit projects can tie together disparate parts of a metropolitan region, uniting it into one geography, a sum of and yet greater than, its disparate neighborhoods. It can help revitalize a neighborhood (Giuliano & Agarwal, 2010). Transit opens crucial social and economic opportunities for people who do not have cars, and it provides additional, often less expensive, travel options even for people who do have cars. Transit can also be costly, however, with major projects in large urban areas requiring billions of dollars of public investment. With these important issues in play, discussions around public transit become controversial quickly, with opposed and entrenched schools of thought that assume their priorities, preferences, and expertise should settle the question. But the tensions among costs, revenues, and service quality never fully resolve because each project and every service change has consequences for social welfare. Innovations simply open up new possibilities within these long-standing debates about how much transit delivers for what we spend on it.

Mass transit draws professionals from many fields: planning, law, engineering, marketing, lobbying, accounting, and management. Each topic introduced in this chapter represents a complex field of research and practice that goes far beyond what I can present in an overview. The references provide an entry into that research for further study into the complex and fascinating topics that govern the world of mass transit.

NOTE

1. The demographic data for particular markets come from CBSOutdoor (now Outfront Media), which gathers it for marketing purposes.

REFERENCES

American Public Transportation Association. (2013). *Public transportation fact book 2013*. Retrieved June 23, 2014, from *www.apta.com/resources/statistics/Documents/FactBook/2013-APTA-Fact-Book.pdf*.

Brown, J. (2005). Paying for transit in an era of federal policy change. *Journal of Public Transportation, 8*(3), 1–32.

Brown, J., & Thompson, G. (2009). Express bus versus rail transit: How the marriage of mode and mission affects transit performance. *Transportation Research Record: Journal of the Transportation Research Board, 2110*(1), 45–54.

Cervero, R. (1998). *The transit metropolis: A global inquiry*. Washington, DC: Island Press.

Chatman, D., & Noland, R. (2013). Transit service, physical agglomeration and productivity in US metropolitan areas. *Urban Studies, 51*(5), 917–937.

Giuliano, G., & Agarwal, A. (2010). Public transit as a metropolitan growth and development strategy. In N. Pindus, H. Wial, & H. Wolman (Eds.), *Urban and regional policy and its effects* (pp. 205–252). Washington, DC: Brookings Institution.

Grengs, J. (2002). Community-based planning as a source of political change: The transit equity movement of Los Angeles' Bus Riders Union. *Journal of the American Planning Association, 68*(2), 165–178.

Hess, D., Brown, J., & Shoup, D. (2004). Waiting for the bus. *Journal of Public Transportation, 7*(4), 67–84.

Iseki, H., & Taylor, B. D. (2009). Not all transfers are created equal: Toward a framework relating transfer connectivity to travel behavior. *Transport Reviews, 29*(6), 777–800.

King, D. (2011). Developing densely: Estimating the effect of subway growth on New York City. *Journal of Transportation and Land Use, 4*(2), 19–32.

Lee, J. (2013). Perceived neighborhood environment and transit use in low-income populations. *Transportation Research Record: Journal of the Transportation Research Board, 2397*, 125–134.

Levine, J. (2013). Is bus versus rail investment a zero-sum game? *Journal of the American Planning Association, 79*(1), 5–15.

Levinson, D. (2014). How to make mass transit financially sustainable once and for all. *Atlantic Citylab*. Retrieved June 23, 2014, from *www.citylab.com/commute/2014/06/how-to-make-mass-transit-financially-sustainable-once-and-for-all/372209*.

National Auditor's Office. (2009). *The failure of Metronet*. House of Commons, Report by the Comptroller and Auditor General (HC 512 Session 2008–2009). Retrieved June 23, 2014, from *www.nao.org.uk/wp-content/uploads/2009/06/0809512.pdf*.

Office of the State Comptroller. (2010). Financial outlook for the Metropolitan Transportation Authority. Retrieved June 23, 2014, from *www.osc.state.ny.us/reports/mta/mta-rpt-52011.pdf*.

Rosenbloom, S. (2001). Sustainability and the aging of the population: The environmental implications of the automobility of older people. *Transportation, 28,* 375–408.

Sanchez, T. W. (1999). The connection between public transit and employment. *Journal of the American Planning Association, 65*(3), 284–296.

Schweitzer, L. (2012). Ballot box planning for transit. In D. Sloane (Ed.), *Planning Los Angeles* (pp. 171–178). Chicago: American Planning Association.

Sellers, J., Grosvirt-Dramen, D., & Ohanesian, L. (2009). *Metropolitan sources of electoral support for transportation funding.* Final report for the Metrans Transportation Institute, University of Southern California.

Siemiatycki, M. (2009). Academics and auditors: Comparing perspectives on transportation project cost overruns. *Journal of Planning Education and Research, 29*(2), 142–156.

Snell, B. (1974). Statement of Bradford Snell before the United States Senate Subcommittee on Antitrust and Monopoly. *Hearings on the Ground Transportation Industries in Connection with S1167.* Retrieved June 23, 2014, from *http://libraryarchives.metro.net/DPGTL/testimony/1974_statement_bradford_c_snell_s1167.pdf.*

Sperling, D., & Gordon, D. (2009). *Two billion cars: Driving toward sustainability.* Oxford, UK: Oxford University Press.

Taylor, B. D. (2002). Rethinking traffic congestion. *Access.* Retrieved June 23, 2014, from *www.lewis.ucla.edu/wp-content/uploads/sites/2/2015/04/taylor.pdf.*

Taylor, B. D., & Ong, P. M. (1995). Spatial mismatch or automobile mismatch?: An examination of race, residence, and commuting in U.S. metropolitan areas. *Urban Studies, 32*(9), 1453–1473.

U.S. Census Bureau. (2011). S0804: Means of transportation to work by selected characteristics for workplace geography. *2007 –2011 American Community Survey.* U.S. Census Bureau's American Community Survey Office, Washington, DC.

Vuchic, V. R. (2007). *Urban transit systems and technology.* New York: Wiley.

Wachs, M. (1990). Ethics and advocacy in forecasting for public policy. *Business and Professional Ethics Journal, 9*(1–2), 141–157.

Walker, J. (2012). *Human transit: How clearer thinking about public transit can enrich our communities and our lives.* Washington, DC: Island Press.

CHAPTER 9

Land Use Impacts of Transportation Investments

GENEVIEVE GIULIANO
AJAY AGARWAL

The role of the transportation system in fostering growth and affecting urban structure is of great interest to academicians, planners, investors, and politicians. The
notion that transportation services and facilities have had an enormous effect on land
use seems to be self-evident. Indeed, the evolution of urban form appears to be so
clearly linked with transportation technology that Muller formulates his explanation
of urban structure in terms of transportation eras (see Chapter 3).

Transportation investment decisions are the subject of much debate and lobbying among planners and policymakers, particularly at the local government level.
Transportation investments are viewed as a way to generate growth and to shape
urban structure. Public transit investment (usually in rail transit) is promoted as a key
to economic revitalization for central cities. Although the highway building era has
come to an end in the United States, highway capacity expansions are still promoted
as a means to maintain regional competitiveness and attract economic growth by
reducing congestion and increasing access to skilled or low-cost labor markets and
cheap land.

Many urban geographers and policymakers advocate higher density, mixed-use
development as a means for solving urban congestion and environmental problems
and for achieving more sustainable urban development (Calthorpe, 1993; Newman
& Kenworthy, 1998). In this chapter we explore the relationship between transportation and land use: we examine the theoretical basis for expecting transportation
to influence land use, and we determine the extent to which such impacts can be
documented. It should be noted that the primary focus of this chapter is the impact
of transportation on land use, not the impact of land use on travel or transportation,
which is the topic of Chapter 7.

The remainder of the chapter begins with a brief description of factors to consider when examining land use impacts of transportation investments. In the second

section, the conceptual relationship between transportation and land use is described in the context of accessibility. The third section reviews the major theories that have been developed to explain transportation–land use relationships. The fourth section summarizes the empirical evidence, discusses some specific case studies, and draws conclusions based on these findings.

GENERAL CONSIDERATIONS

This section describes the four core questions that must be addressed to examine land use impacts of transportation investment. The section also describes the dynamics of land use–transportation relationships, which makes examination of land use impacts of transportation particularly challenging.

Core Questions

First, what do we mean by "land use impacts"? When metropolitan areas grow, land use patterns change as new development and redevelopment take place. This process may take place with or without new transportation investments. The question is whether land use changes are a result of transportation investments or simply the result of metropolitan growth.

Second, how big is the transportation change, and how big does it have to be to generate an impact on land use? Transport system improvements that represent significant technology change (and hence significant change in service level), such as construction of the electric streetcar systems in the late 19th century, can be expected to have much greater impact than incremental changes, such as bus rapid transit. The scale of the investment relative to the existing system is also important. For example, the first 10 miles of expressway constructed within a region should have greater impact than the last 10 miles, all other things being equal, because the relative impact of those first 10 miles will be so much greater. One example of relative scale is the $2 billion I-105 (Century) Expressway in Los Angeles. The 17-mile segment opened in 1993. This major project constitutes just 1% of the region's freeway miles, despite being a critical artery within its own corridor, and consequently its land use impacts at the regional scale are not likely to be significant.

Third, what is the geographic scale of interest? Impacts may vary substantially across different geographic scales. As in the I-105 example above, even a large project in terms of investment, such as a new rail line or expressway segment, might have a limited impact when viewed in the context of the metropolitan area as a whole. However, the same project could have a substantial impact on land use patterns at the neighborhood level, such as near the transit stations or the expressway ramps.

Fourth, what is the geographical context of the transportation investment? Because the built environment is durable, the vast proportion of building stock that will exist 10 or 20 years from now is already built. The cost of reconstruction in developed areas is usually high, because demolition and site preparation must precede new construction. Consequently, changes in land use occur slowly in already built-up areas, even though the uses of the existing building stock may change much faster.

Also, demand for housing or business locations varies across neighborhoods or communities.

Dynamics

It is particularly difficult to examine the land use impacts of transportation because land use and transportation are mutually dependent. A simple illustration is presented in Figure 9.1. The characteristics of the transportation system influence *accessibility,* or the ease of moving from one place to another. Accessibility in turn affects the *location* of activities, or the land use pattern. The location of activities in space, together with the transportation resources connecting them, affects daily *activity patterns,* which in turn result in travel patterns. These *travel patterns,* expressed as flows on the transportation network, affect the transportation system. Land use and transportation are part of the larger urban system: the collection of people, institutions, and infrastructure that together form the urban space economy.

The model in Figure 9.1 shows that a change in land use will affect transportation, just as transportation affects land use. Note that this model does not imply anything about the strength of these relationships. It serves only to illustrate the interdependence of empirically observed land use and transportation changes. It is difficult to isolate the impact of transportation on land use (or, conversely, land use on transportation), as doing so requires examining one-half of the figure, holding the other half constant. The land use system is dynamic, however, and what we observe in empirical analysis is the totality of relationships expressed in Figure 9.1.

Not only are land use and transportation patterns interdependent, but this interdependency is expressed over long periods of time. People, institutions, infrastructure, and technology all change, making it even more challenging to separate out the role of transportation. As is explained in a later section of this chapter, the durability of the built environment, the high costs of relocation for both residents and firms, and the long lag times of major transportation projects all complicate empirical analysis of land use and transportation relationships.

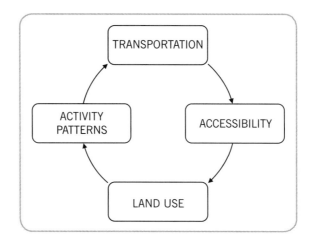

FIGURE 9.1. The urban system.

TRANSPORTATION IN THE CONTEXT OF ACCESSIBILITY

The basic concept underlying the relationship between land use and transportation is accessibility. In its broadest context, "accessibility" refers to the ease of movement between places (see Chapter 1). As movement becomes less costly—either in terms of money or time—between any two places, accessibility increases. The propensity for interaction between any two places increases as the cost of movement between them decreases. Consequently, the structure and capacity of the transportation network affect the level of accessibility within a given area.

How does a change in the transportation network affect accessibility? We illustrate the effect with a simple network example, shown in Figure 9.2. Each node represents a possible origin/destination, and the numbers near each link represent travel times. We begin with the network on the left side of Figure 9.2 (Network K). We then make an improvement such that travel time between nodes B and C is reduced by one-half, as shown on the right side of Figure 9.2 (Network L).

Network accessibility can be measured by calculating the travel time from each node to every other node and summing over each node, as in Table 9.1 (note that this implicitly assumes that each node is equally attractive). Each row in the matrix corresponds to travel times from one to every other node. The row sums are the accessibility measure for each node. Since we have used travel times in the example, lower numbers mean greater accessibility. Comparing the row sums of the two networks, Table 9.1 shows that the network improvement not only increases accessibility between nodes B and C, it also benefits the entire network. This simple example demonstrates how a network improvement (e.g., a new expressway link between two places) affects accessibility not only between the two directly connected places but also within the entire network, and as accessibility increases, the level of spatial interaction increases because travel has, in effect, become less costly.

How does this change affect land use? As more interaction occurs, more activities will locate in those places that have become more accessible. In this example, nodes B, D, and C have benefited most from the improvement, and thus the greater

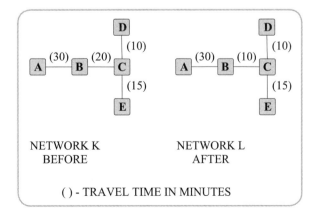

FIGURE 9.2. Transport network accessibility.

TABLE 9.1. Travel Time Accessibility Matrix for Networks Shown in Figure 9.2

| | Network K | | | | | | Network L | | | | | | Change |
From/to	A	B	C	D	E	?	From/to	A	B	C	D	E	?	(%)
A	0	30	50	60	65	205	A	0	30	40	50	55	175	−14
B	30	0	20	30	35	115	B	30	0	10	20	25	85	−26
C	50	20	0	10	15	95	C	40	10	0	10	15	75	−21
D	60	30	10	0	25	125	D	50	20	10	0	25	105	−24
E	65	35	15	25	0	140	E	55	25	15	25	0	120	−14

land use changes would be expected to occur at these nodes. This process may be characterized as regional growth: as population and employment increase, their relative location will be affected by the transportation system. In a theoretical world where capital is completely mobile, then, transportation system changes would result in observable shifts in activity location.

The concept of accessibility can also be used to predict changes resulting from deterioration in the transportation system. Suppose that the link between two places has reached capacity and travel speeds have declined. Accordingly, the level of accessibility declines, generating incentives for activities to shift away from these two places—all other things being equal.

The network example above does not explicitly identify the economic benefits generated by reduced travel costs on the B–C link. Reduced transport costs can lead to greater economic productivity. For example, reduced shipping costs might allow a manufacturer to seek out more distant but lower-cost input suppliers, hence reducing the price of the finished good. Other firms may gather around this highly accessible link, thereby achieving positive externalities from improved access to one another. Reduced transport costs effectively increase a worker's net wage, allowing her or him to spend more on other goods and services. The result of this process is productivity-driven economic growth. Shifts in activity location are part of this process.

Do transportation investments always lead to economic growth? If our sample network was located in a declining city, it is quite likely that the accessibility improvement would not be sufficient to overcome the city's competitive disadvantages that are the underlying cause of its decline. Although shifts in activity location might occur, these shifts could simply be a redistribution of economic activity, leaving other parts of the city worse off, as we shall see in some examples later in this chapter.

THEORIES OF LAND USE

Research on the relationship between land use and transportation has generated numerous theories. Developed by economists or geographers, these theories seek to explain the effect of transportation cost on location choice.

Standard Urban Economic Theory

The standard model of urban spatial structure was developed to explain land value, population and employment distribution, and commuting patterns. Initially developed by Alonso (1964), Mills (1972), and Muth (1969), the theory focuses on residential location choice. Household location is expressed as a utility-maximizing problem in which choice depends on land rent, commuting cost, and the costs of all other goods and services. The standard theory is based on several simplifying assumptions. In addition to the usual assumption of rational behavior, identical preferences, and perfect information, the following assumptions are also made in the simplest version of the theory:

- The total amount of employment is fixed and located at the center of the city.
- Each household has only one worker, so only one work trip is considered.
- Housing is a function of capital and land, and therefore location and lot size are the distinguishing factors.
- Unit transport cost includes both time and monetary costs, and it is constant and uniform in all directions.

Residential location theory gives rise to a city form in which the greatest population density and highest land values are at the center, the point of greatest accessibility, and in which these density and land value (also called land rent) gradients decline with distance from the city center. Figure 9.3 shows the land rent curve: the value per unit of land as a function of distance from the city center. You can see that the land rent curve is nonlinear; its gradient is constantly decreasing in absolute value with distance from the center. The dotted line R_A denotes the nonurban land rent, and the point at which $R(d)$ equals R_A defines the city boundary. The unit cost of housing must decline with distance from the center of the city, since transport (commuting) costs increase with distance. That is, the value of transportation costs savings is reflected in land (and therefore housing) cost. If this were not the case, all households would locate at the center of the city. It therefore follows that more housing will be consumed as distance from the center increases, and therefore population density will also be constantly decreasing. The theory also predicts commuting patterns; the average commute trip length corresponds to the mean distance of total population to the center.[1]

The best location for a given household is the point at which the marginal savings of housing are equal to the marginal cost of transport, or the savings in housing are just offset by the increase in transport cost. If households have identical preferences, they will be indifferent to any given location. If this assumption is relaxed, the particular location of any given household depends on relative preferences between housing and transport. It is generally assumed (and there is some supporting empirical evidence) that housing preferences are the stronger of the two; the higher the household income, the more housing will be consumed, even at the cost of additional commuting. Thus lower-income households would consume less housing and locate closer to the city center, and higher-income households would consume more housing and locate farther from the city center.

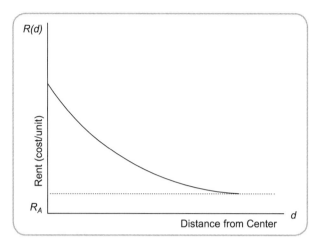

FIGURE 9.3. The relationship between rent and location.

What does the theory predict in response to a change in transport cost? If commuting costs are reduced, the theory predicts movement away from the center, or population decentralization. Figure 9.4 illustrates this process. Curve 1 is the land rent gradient before the transportation improvement; Curve 2 is the land rent gradient after the transportation improvement. The decrease in transport cost reduces rent at the city center, because the location advantage of the center has declined. Consumers take advantage of lower rents by increasing housing consumption and commuting greater distances. As a result, the total amount of land consumed increases, and the city boundary extends. If transport cost is increased, the reverse effect will be observed: households will economize on travel by locating closer to the center and consuming less housing. It is worth noting that population growth will also lead to expansion of the city, because more households will bid for space, raising the entire price gradient (not shown in figure).

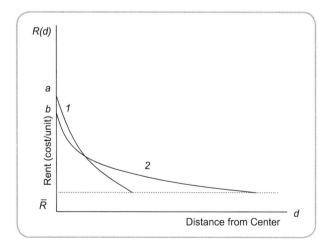

FIGURE 9.4. Response of rent function to a transport cost decline.

In a general way, empirical evidence tends to support the standard theory. Population density does decline with distance from the center, and time series studies document that densities have declined historically in conjunction with transportation system improvements (Anas, Arnott, & Small, 1998; Kim, 2007). Lower-income households are more likely to live near the city center and to have shorter commutes, while higher-income households are more likely to live in the suburbs and to have longer commutes (Glaeser, Kahn, & Jordan, 2008; Mieskowski & Mills, 1993).

Commuting costs can be reduced either by reducing monetary costs (e.g., reducing the price of fuel or transit fares), or by increasing travel speed (e.g., providing an express transit route or adding a lane to an expressway). Different types of travel cost reductions are likely to have differing impacts across income groups. A simple price reduction will affect all income groups in the same way: more income would be available for housing, leading to more housing consumption, and hence to a more distant location choice.

Travel time reductions tend to have a greater effect on higher-income groups, because the value of time is a function of income.[2] Consequently, the decentralizing effect of a travel time reduction is greater for higher-income households. It bears noting that congestion pricing of roads works the same way: the benefits of travel time savings realized from reduced congestion are greatest for higher-income travelers, all else being equal.

Employment Location Theory

In the simplest version of the standard model, employment location is assumed to be fixed in the city center. However, the theory can be applied to employment location (Brueckner, 1978; Koster & Rouwendal, 2013). Another way of expressing this assumption is to say that firms place a higher value on central location than do households. Suppose we allow employment location to vary as a function of land rents, commuter costs, and all other costs. It turns out that employment locations will always be more concentrated than residence locations. Because households will be indifferent between working far from home at a higher wage and working close to home at a commensurately lower wage (as long as the net wage remains constant), the employer's location decision depends on *total transport costs*: the cost of commuting (as reflected in wage rates), transporting inputs of production to the firm, and goods to market. If we assume that the goods import/export node is at the city center, employers will *centralize* to the extent that commuting costs are less than shipping costs, because the central location minimizes goods transport costs. You can see that applying the standard theory in this way implies that as goods transport costs decline relative to commute costs, employers will *decentralize* and locate closer to their workers.

Finally, suppose that the goods produced have no transportation costs, as, for example, the production of information. If the product can be produced and sold anywhere, where will employers locate? In this extreme case, employment would be distributed identically with the population, so long as there are no scale economies of production, meaning as long as the efficiency of producing the good does not depend on the size of the firm. Again, empirical evidence supports theory in a very general

way. Jobs have indeed been decentralizing as transport costs (for both goods and passengers) have declined (Glaeser & Kohlhase, 2004).

Other employment location theories deal with certain types of firms or economic activities. It is useful to distinguish between market-sensitive and non-market-
sensitive activities. *Market-sensitive activities* (such as sales and services) depend on
accessibility to consumers. *Non-market-sensitive activities* include industrial or manufacturing activities that serve national or international markets and hence are not
dependent on a local customer base but must consider instead access to, for example,
source materials and markets for the finished product.

Theoretical Simplicity versus Real-World Complexity

The development of theories and models of complex phenomena requires the use of
simplifying assumptions (see Chapter 5) to make the problem manageable. It would
be practically impossible, for example, to develop a theory of residential location that
takes into account all the factors influencing location choice. Moreover, such a model
would be so complex that the most significant factors might be obscured. In this section we discuss some of the key assumptions associated with the theories described
above and examine how relaxing these assumptions affects our ability to predict land
use–transportation relationships.

The Monocentric City

The standard urban model of location assumes a *monocentric city*: all employment is
assumed to be located at the city center. This assumption is unsupported in contemporary urban areas. Rather, as Muller (Chapter 3) and others have pointed out, large
urban areas today are decentralized and dispersed. Although the downtown central
business district (CBD) generally remains the location of the greatest employment
density and the highest land values within the region, the central city is surrounded
by competing subcenters, with much of the employment and population dispersed
throughout the metropolitan area. Cross-sectional studies of many large U.S. metropolitan areas document the presence of multiple employment centers.[3]

The Los Angeles region illustrates the polycentric urban form of U.S. metro areas.
To document subcenters in Los Angeles, we used different data sources across several decades. We found that spatial structure is remarkably stable: about 40% of all
jobs are located in employment centers, and the large centers (those with more than
100,000 jobs) do not change much even over decades. Figure 9.5 shows centers—
defined as any cluster of employment with at least 10,000 jobs and a density of at
least 10 jobs per acre—for 2005.

The Housing/Transport Trade-off

The standard model relies on the assumption that people decide where to live primarily by selecting a trade-off between commuting cost and land cost. Given a fixed
location of employment, the theory can be used to predict average commute length.
It follows that the observed average commute length can be used as a test of the

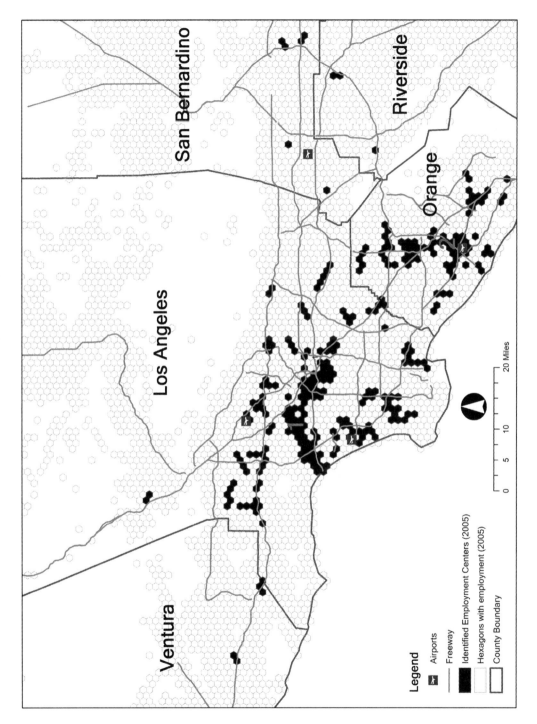

FIGURE 9.5. Employment centers in Los Angeles CMSA in 2005.

Legend

🛧 Airports

Freeway

■ Identified Employment Centers (2005)

Hexagons with employment (2005)

County Boundary

0   5   10   20 Miles

San Bernardino

Riverside

Orange

Los Angeles

Ventura

theory. If households do seek to minimize commute cost, observed commute patterns should reveal an average trip length close to that predicted by the model. But in studies of metropolitan areas around the world, in most cases, the observed average commute length far exceeds that predicted by the standard model, even when model predictions are based on actual employment and population spatial patterns rather than the conventional assumption of monocentricity (Giuliano & Small, 1993; Ma & Banister, 2006).

Why is the average commute much longer than the theory predicts? One explanation is spatial mismatches between workers and jobs. *Mismatches* occur when prices or other characteristics make housing in an area unsuitable or unavailable for workers who hold jobs there. Mismatches may be due to discrimination in housing markets, limiting residential mobility for minorities and the poor, or to a general lack of affordable housing (Ihlandfeldt & Sjoquist, 1998; Quigley & Raphael, 2004). Spatial mismatch leads to longer commutes and poorer job access.

A second explanation for longer-than-expected commutes is related to the job market. Given decentralized employment, households with more than one worker must somehow locate to accommodate multiple job locations. The rise of multiworker households is therefore consistent with longer commutes. Job mobility (i.e., higher rates of job turnover and shorter tenure in any given job) may lead households to locate so as to maximize access to future job opportunities, rather than to a given current job. Finally, the high cost of moving one's residence suggests that households may be willing to trade off a longer commute in order to avoid a residential move (Crane, 1996).

A third explanation for observed commute patterns is that the cost of transportation has declined more rapidly than the cost of housing. One might argue that the cost of transportation (or, more precisely, the price we pay directly for transportation) is so low that households do not attempt to economize on travel. Whereas around 1900 commuting took about 20% of the daily wage (Hershberg, Light, Cox, & Greenfield, 1981), today it's about 4% (Roberto, 2008). On the other hand, as per capita income rises, the time cost of commuting should dominate, limiting the willingness of households to incur ever longer commutes. Hence, even in the largest metropolitan areas, we observe a small share of commuters traveling 1 hour or more: 22.3% for New York, 13.9% for Chicago, 10.8% for Los Angeles.[4]

Household Preferences

The standard urban model assumes identical preferences among households. Preferences are not identical, however, even among households with similar socioeconomic and demographic characteristics. As a result, the demand for housing services (particularly with respect to other goods and services) is not the same for all households. In economic terms, housing demand elasticities are different among households, meaning that relative preferences between transport and housing are not necessarily identical.

Residential location theory defines housing in the broadest of terms as the bundle of all housing-related services, so public services and site amenities can be conceptualized as being contained within this housing bundle. The theory also assumes, however, that housing is identical and uniformly available throughout the region; hence public services and amenities are also uniformly available. This assumption is, of course, unsupported in contemporary urban areas.

Public goods such as parks, schools, and air quality exhibit a high degree of spatial variation. The quality and availability of public goods are important considerations for many households, such that increased commuting costs might willingly be incurred in order to obtain better schools or police services.

A further complication arises from the durability of housing stock. As tastes and per capita income change, so does the new housing stock. For example, the average size of a new single-family dwelling has increased from 1,660 square feet in 1973, to 2,392 square feet in 2010.[5] Since cities grow outward, the largest houses, on average, are located at the periphery. Thus urban areas include a wide variety of housing choices differentiated by age, size, general condition, and the like, with older and smaller units closer to the historic center and newer larger units farther from the center.

A number of hard-to-quantify factors also affect location choice. Examples include ethnic preferences, racial biases, family loyalty to specific neighborhoods, or preferences for specific microenvironments or architectural styles. Last, the financial and psychological costs of moving are so high that considerable inertia is associated with location choice.

These observations imply that location choice is much more complicated than the simple trade-offs among housing commute costs and all other goods that location theory posits. On the contrary, work trip cost is just one among many factors that may be considered. It should be no surprise then that the theory is seldom supported in empirical tests.

ECONOMIC STRUCTURE

Economic restructuring has led to more contract and temporary work, more self-employment, and a resurgence of small enterprises. These shifts are associated with increases in both high-wage professional jobs and low-wage service jobs. Overall, the trend is one of increasing job turnover, which again reduces the relative importance of locating close to any given job.

Restructuring also has implications for employment location. Whereas manufacturing firms typically invest heavily in plant facilities and equipment, making relocation costly, service firms are comparatively more mobile. Office equipment is relatively inexpensive and easy to move, telephone lines are ubiquitous, and high-speed data lines are widely available. Thus the continued shift to service and information-based activities can be expected to increase the mobility of both firms and workers.

The Supply Side

The standard model assumes a perfectly competitive land market, but land markets are affected by many variables. First, land use in the United States is regulated by local jurisdictions. Municipalities influence land use through zoning codes, development requirements and restrictions, and the provision of infrastructure and services.

Second, tax revenues are a major concern for municipalities. Land uses that contribute positively to the tax base, such as big-box retailing, are likely to be encouraged, while uses that add to the public cost burden, such as low-or medium-cost housing, may be discouraged.

Third, municipalities seek to preserve and enhance perceived amenities. Affluent municipalities may wish to preserve a small-town atmosphere, protect environmental amenities or historic districts, or preserve open space. Development fees, growth caps, density restrictions, historic preservation zones land dedications, and infrastructure requirements are some examples of the many policy instruments available to municipalities to accomplish these goals.

Finally, new development or redevelopment typically requires environmental review and local authority approval. Local preferences may therefore play a significant role in determining land use development outcomes, further weakening any systematic relationship between land use and transportation.

Statics versus Dynamics

Finally, the various theories discussed previously are static: they assume instantaneous equilibrium across all markets. Metropolitan areas are, of course, always changing, but changes occur at different rates. Jobs change relatively rapidly, as firms move, or grow, or go out of business. Household composition changes rather quickly as well. But the built environment changes very slowly. A major highway or rail project takes years to build and has an operational life of many decades. Metropolitan areas are never in equilibrium, but rather constantly adjusting to the changing population and employment dynamic.

One might ask whether these simple theories are of any use in understanding the land use–transportation relationship. The answer is an unequivocal "yes": cities have decentralized and become less dense as transport costs have declined; more-accessible places have higher land value than less-accessible places; employment centers in large metropolitan areas are functionally linked; and few workers have 1-hour or more commutes, even in the largest metropolitan areas. Once we go beyond these stylized facts, however, we cannot predict whether real-world land use and transportation changes will conform to theoretical expectations.

EMPIRICAL EVIDENCE OF THE LAND USE IMPACTS OF TRANSPORTATION INVESTMENTS

So far we have discussed land use impacts of transportation investments in generalities, but such investments take many forms: highways, ports, airports, urban transit lines, or high-speed rail lines. Clearly, these investments would have very different land use impacts. The highway system carries the vast majority of both freight and passengers in urban areas, so highway investments should have observable land use impacts. Few major new highways have been built in the United States over the past few decades, and there are few recent studies, but because highways are an essential element of the urban transportation system, they merit our examination. In contrast, investment in urban rail has greatly increased, generating more interest in their impacts, as have airports and ports, which are increasingly important as major nodes of accessibility to national and international markets. In one chapter it is not possible to cover everything; we therefore focus on the two major modes for urban passenger transport: highways and public transit.

Methodological Problems

Studies of specific transport infrastructure projects in the United States have been conducted since as early as 1930. However, there is little consensus on the conclusions to be drawn, mainly because of methodological problems and the challenge of tracing impacts over long periods of time. The ideal research design for establishing causality is the four-way experiment: a "test" group and a "control" group, each monitored before and after the "treatment." Such a design is, of course, impossible in the case of transportation impacts. With enough lead time, a before/after study can be done, but there is no perfect control group with which to compare the before/after results. Let us now explain methodological problems in more detail.

Long-Term Dynamics

Two problems plague attempts to identify land use impacts of transportation investments. First, a change in the transportation system is just one of many changes occurring at the same time, and these changes continue after the investment takes place. Land use changes that may be observed following a given transportation investment could be the result of many other factors.

Second, because of the longevity of capital infrastructure, the market response to changes in transportation may take place only after many years or decades. To complicate matters further, land use changes may appear *before* the project is actually constructed because landowners may act in anticipation of higher land values (Boarnet, 1997; Ko & Cao, 2013). The longer the time span over which these effects take place, the more difficult they are to isolate from all the other changes in economic conditions, regional employment, and population demographics that are occurring at the same time.

Longitudinal versus Cross-Sectional Studies

Longitudinal studies examine changes that occur in one place over a given period of time. If lead time is sufficient and data are available, land use changes within the area of impact (however defined) are traced from before the project took place to sometime after its completion. In order to isolate transportation-related impacts, all other factors thought to be relevant are incorporated into the analysis (e.g., regional housing and commercial space growth rates). The longitudinal approach, however, may not control for land use activities attracted to the area of the transportation investments that may otherwise have located elsewhere in the region. Thus, although significant local changes may take place, there may be no *net* impact on the region. This point is particularly important when decision makers are concerned with the economic growth impacts of transportation investments.

Cross-sectional studies compare the "target" area with nearby similar areas to approximate the "no project" comparison. For example, rail transit impact studies often compare land use changes in corridors with and without rail. This type of analysis has two pitfalls. First, seemingly similar corridors may in fact be quite different. Second, if transportation investments cause a shift in activity location because of improvements in relative accessibility, such a comparison will exaggerate the extent of the impact. What is actually being measured is the combined effect of increased accessibility in the corridor with rail and the decreased accessibility in the nearby corridor without rail.

Finally, neither of these approaches allows us to draw any conclusions on the direction of causality. Even when significant land use changes occur in conjunction with transportation investments, there is no way to determine whether the investment *caused* these changes, or whether anticipated transport investments led to land use changes, which then *created the demand* for the transportation improvements. There is much empirical support for the latter interpretation. The planning and environmental approval process for major projects can span a decade or more, making lags between regional growth and transportation investments quite large.[6] In any event, land use and transportation decisions are so closely tied together that it has been impossible so far to separate their effects.

The Contemporary Context of Urban Highway Investments

The context of urban highway investments can affect the extent of land use impacts. First, any single highway investment is but a part of a much larger urban transportation system. Any new highway investment would add some increment of accessibility to the system. In an area that already enjoys a high level of accessibility, we would not expect a new investment to have much of an impact. In an area with limited accessibility, the same investment should have a much greater impact. Most U.S. urban areas, however, fall into the former category.[7]

Second, land use change is more likely to occur in the form of new construction, and available land is a prerequisite for new construction. All other things being equal, then, the land use impacts of highway investments in areas with vacant land available should be greater than in areas where development has already taken place, because of the higher costs of redevelopment.

Third, even when land is available, local public policy may not be favorable to development. Land use change cannot occur unless local zoning permits it. If local opposition to development is strong, highway investments may have little effect, even if market demand exists.

A fourth factor is the state of the regional economy. Little new investment takes place in a stagnant or declining region. Rather, disinvestment occurs. Residential areas deteriorate, factories close down, and the general level of economic activity declines. Under these circumstances, highway investment should have little or no impact. If it did, it would most likely be at the expense of other areas within the region.

Fifth, the scale of analysis must be considered. Determining whether highway investments have an impact within their immediate vicinity is relatively straightforward. Indeed, any observer of the urban scene can recount changes observed after a new freeway opened or an interchange was constructed (though the careful observer would also be able to point out interchanges where nothing changed). Observation of local changes does not support the hypothesis that highway investments generate land use impacts. In order to test such a hypothesis correctly, we must study the region as a whole. Only then can land use changes taking place in the vicinity of a highway improvement be determined to be significantly different from those occurring outside that vicinity.

Sixth, and finally, careful impact analysis requires that changes in land use patterns be distinguished from general economic growth. That is, land use change will occur in a growing region with or without new highway investment. This point is illustrated in growing regions throughout the United States. Growth in urban travel far exceeded growth in highway capacity from the 1980s through the mid-2000s. Despite negligible additional highway investment, new development was spurred by general economic expansion.

Many of those concerned with the sustainability of our urban areas argue that provision of more highway capacity generates "induced demand," which leads to more auto dependence (see Box 9.1). Because major public investment decisions are being made on the basis of these claims, it is critical to examine land use–transportation relationships as comprehensively as possible.

BOX 9.1
Induced Demand

Many people who support sustainability argue that new highway capacity is poor public policy, because more highway capacity generates more travel. Congestion relief benefits are short term at best, as increased travel demand—demand *induced* by the new capacity—soon results in the same or worse congestion than before, while further increasing auto dependence and its environmental damages. Is it possible that added capacity will generate more travel and hence offset all accessibility benefits?

Consider a congested highway. New lanes are added in each direction. Travel times decline as congestion is reduced. In the short run, travelers using other modes, or taking other routes, or traveling at other times of the day will be attracted to the expanded highway until a new equilibrium across routes, modes, and times is established. The added capacity, by virtue of reducing travel costs, also may attract new or longer trips. Shoppers may choose a bigger mall, farther from home, since the traffic is no longer unbearable. Friends may choose to visit more frequently, hence making new trips. By such choices total travel will indeed increase, and these new or longer trips are induced to the extent that the added capacity reduced travel costs. This is simply supply and demand at work: as the price falls, more of the good is consumed.

In a supply–demand framework, it is obvious that added highway capacity cannot possibly lead to the same or worse conditions, as long as the demand curve remains fixed. If the demand curve shifts to the right (e.g., from population growth), it is possible that congestion will be worse than initial conditions (but not as bad as would have been the case without the added capacity). However, increased travel demand from the added population is not induced.

What if population and jobs were attracted to the area in response to the added capacity? Households and firms residing elsewhere in the region might relocate, redistributing existing economic activity. New households and firms might choose to locate in the area. It is this long-term effect of guiding growth that is at the basis of the highway critics' concerns.

Highway Impacts

Strategies for Measuring Impacts

There are three possible ways for measuring impacts: travel outcomes, property values, or changes in the distribution of population and/or employment. Travel outcomes are the weakest test. We might argue that changes in travel are a necessary but not sufficient condition for demonstrating impact. If the new facility has no significant impact on accessibility, it cannot have an impact on land use, because accessibility is the mechanism by which land use changes are affected. Even if travel shifts are observed, they may not reflect enough of a change in accessibility to promote shifts in the location of households or firms.

Measuring changes in property values is a stronger test of land use impacts. Changes in accessibility generated by a new highway or transit investment should be reflected in land values, per economic theory. As land increases in value, capital (structures) will be substituted for land, leading to more intense development. Hence increased land values can be interpreted as a signal for development. Changes in property values can also be used to measure the physical extent of the impact. For example, we expect that the accessibility–value gradient is steeper for transit than for highways, since access to transit is mainly by foot, and people are willing to walk only short distances.

The third possibility is to measure changes in land use: changes in employment or population density, commercial building, residential units, and so on. This is the most straightforward measure but also is subject to confounding factors, as discussed above.

Summary of Empirical Results

The literature on highway impacts is broad, ranging from national studies of the economic growth effects of highway investment, to regional studies, to highly localized case studies. While earlier US highway investments—made during the 1960s—produced substantial land use changes, subsequent highway investments produced land use changes of comparatively much smaller magnitude. The empirical evidence also suggests that the existing highway network continues to influence land use, particularly employment, although the nature of impacts has become very place-specific.

Early studies of highway impacts used land values as the measure of impact; results showed significant increases in land value near the highway and losses in value in places that became less accessible (Garrison, Berry, Marble, Nystuen, & Morrill, 1959). These early highway investments improved accessibility remarkably, and significant impacts on land values were the result.

By the 1970s, the Interstate Highway System was nearing completion, and expressway systems had been constructed within all the major cities in the United States. The rate of auto ownership had more than doubled, and the shift of both population and employment to the suburbs was well underway (see Muller, Chapter 3, this volume). Hence, accessibility gains from new highway investments were much smaller as compared to those during the earlier period. For example, a comprehensive study of the land use impacts of beltways (circumferential expressways) in 54 U.S.

cities between 1960 and 1977 found mixed results (Payne-Maxie Consultants, 1980). The existence of a beltway had no consistent effect on the distribution of growth: employment and population growth was more rapid in suburban locations with or without the presence of a beltway. The study authors concluded that land use impacts were largely determined by local market conditions.

Subsequent studies have also reported mixed results. Table 9.2 gives summaries of four studies of the regional economic impact of highway investments. These studies were selected for their sound methodology, appropriate data, and state-of-the-art research on this topic. All four studies examine factors associated with economic growth, including highway infrastructure, and three consider distributional impacts (spatial spillover). Boarnet (1998) found a strong link between highways and economic growth within counties, as well as strong spillover effects, meaning that some of the within-county growth was at the expense of neighboring counties. Henry, Barkley, and Bao (1997) found that the strongest predictor for rural population and employment growth was proximity to high-growth urban cores; highways had no significant impact on rate or location of growth. Singletary, Henry, Brooks, and London (1995) focused on durable and nondurable manufacturing employment growth in South Carolina and found highway access to be generally significant for durable manufacturing but generally not significant for nondurable manufacturing.

TABLE 9.2. Summary of Selected Regional Highway Impact Studies

Author	Location	Description	Data and Method	Results
Funderburg, Nixon, Boarnet, and Ferguson (2010)	California	Study of population and employment change to test impact of highway investment.	1980–2000 population and employment; census tract and TAZ level. Difference in difference estimates, and lagged adjustment model. Observed changes compared with no-build counterfactuals.	+ Impact on employment in urban counties, - for rural.
Boarnet (1998)	California	Study of economic output to test role of highway capital.	1969–1988 county-level annual data. Production function model, allowing for intercounty spillover effects.	Economic output + for within-county highway capital, – for adjacent county highway capital.
Henry, Barkley, and Bao (1997)	South Carolina, Georgia, North Carolina mixed urban–rural region	Study of urban and rural population and employment growth to determine linkages between rural and urban growth.	1980, 1990 census tract population and employment data. Simultaneous model of regional development.	Coefficient for highway access not significant.
Singletary, Henry, Brooks, and London (1995)	South Carolina	Study of new manufacturing employment to test impact of highway investment.	1980–1989 new manufacturing employment, durable and nondurable; ZIP code level. Regression, accounting for spatial autocorrelation and heterogeneity.	For durable manufacturing, coefficient for highway access generally +. For nondurable manufacturing, coefficient for highway access generally not significant.

Funderburg, Nixon, Boarnet, and Ferguson (2010) examined land use impacts of highway projects constructed in three California counties in the mid-1990s. Results were consistent with general economic conditions: the project had a positive impact on population and employment growth in the high-growth county, no effect in the slow-growing county, and a slight negative effect in the declining county.

These studies suggest that under the right circumstances highways often do influence patterns of economic growth, but fundamental regional economic factors are more important.

Table 9.3 summarizes four intraregional studies selected on the same criteria. Both the Boarnet and Chalermpong (2001) and Voith (1993) studies show consistently significant and positive impacts of the highways on residential property values (a negative coefficient on a distance variable means prices decline with distance from a ramp). Bollinger and Ihlanfeldt (2003) found that in Atlanta, expenditure on highway improvements was associated positively with employment growth, but not with population growth.

TABLE 9.3. Summary of Selected Intraregional Highway Impact Studies

Author	Location	Description	Data and Method	Results
Giuliano et al. (2012)	Los Angeles CMSA	Study of association between employment center growth and transport network access	1990 and 2000 census tract-level employment and population data. ordinary least square (OLS) model; employment center growth as a function of airport access, highway and arterial network access, and labor force accessibility, plus control variables	Coefficients positive and significant for highway and arterial network access and labor force accessibility; airport access not significant
Bollinger and Ihlandfeldt (2003)	Atlanta metropolitan area	Study of change in intrametropolitan employment distribution to test for the impacts of local tax incentive programs and transportation infrastructure improvements.	1985–1997 census tract-level data on employment. OLS model; share of metropolitan private employment inside each tract as a function of three tax incentive measures, spending on highway improvements (> $8 million), and percentage of MARTA station impact area (1/4 mile radius around each station) in the tract.	Coefficient for highway expenditure significant and + for employment share growth; insignificant effect of a MARTA station.
Boarnet and Chalermpong (2001)	Orange County, CA	Study of residential sales to test impact of Orange County toll roads.	1988–1999 residential sales data for two corridors plus control corridor. Hedonic regressions, residential sales, before/after corridor open.	Coefficient for distance from ramp – in after period, both corridors, no difference before/after in control corridor.
Voith (1993)	Philadelphia metropolitan area	Study of residential sales to test significance of highway and commuter rail access to CBD.	1970–1988 residential sales data for Montgomery County. Hedonic regressions, residential sales (individual sales, census tract averages).	Coefficient for presence of station + and increasing in value over period. Coefficient for highway access to CBD – and increasing in value over period.

Giuliano, Redfearn, Agarwal, and He (2012) examined the association between employment center growth and transport network access in the Los Angeles Consolidated Metropolitan Area from 1990 to 2000. They found that access to the region's transport network positively influences employment center growth, primarily because it facilitates labor force accessibility. These intraregional studies suggest that highway access continues to influence land use, although magnitude of the impacts depends on the local context.

Conclusions on Highway Impacts

Several decades of empirical research lead to the following conclusions:

1. Highway investments increase accessibility. The greatest benefits are generated at the site of the new investment. Relative accessibility changes throughout the network.

2. Highway impacts are context-specific. Where there is growing population and employment, and land is available for development or redevelopment, impacts are likely to be significant. When these conditions do not exist, impacts are not likely to be significant.

3. There is little evidence that highway investment is an effective means for promoting *net* economic development. Even when economic growth is documented, it may be a result of redistributing activities from other locations.

What about the future? Might we expect different results from future investments? The strongest argument against future highway impacts having more than highly localized impacts is the incremental nature of most investments. However, it is possible that trends in transportation infrastructure finance (e.g., more local funding, borrowed funding, and user fees) will make future investment more demand-driven. If so, such investments are likely to have more consistent impacts on both property values and economic development patterns.

Transit Impacts

National policy began to focus on public transit in the mid-1960s. Public transit systems in several major cities were in financial trouble, and the federal government was called upon to provide capital for rebuilding the nation's transit industry. The financial needs of big city transit systems, together with a growing public awareness of the adverse environmental and societal impacts of highway building, created a broad consensus for federal support of public transit. The federal transit capital grants program eventually led to the construction of the "new generation" mass transit systems in the 1970s.

These rail systems proved to be very costly public projects. In contrast to highway projects, whose direct costs are funded largely by user revenues earmarked for highway use, transit capital projects were (and are) funded from general revenues (income and other general taxes). Justifying these projects on the basis of indirect benefits therefore became very important.

The new-generation rail systems—BART in San Francisco, METRO in Washington, D.C., and MARTA in Atlanta—failed to meet expectations. With the possible exception of the Washington METRO, these systems did not achieve predicted ridership levels and failed to attract auto users to transit in significant numbers (Kain, 1999). However, public support for mass transit investments continued, and new rail systems (mostly light rail) proliferated. Despite these investments, the transit mode share has not materially increased, but public support for mass transit continues to grow. Transit—particularly rail transit—is still viewed as a means for improving air quality, increasing energy efficiency, redeveloping central cities, providing mobility for the transportation-disadvantaged, and creating more sustainable cities. Indeed, transit advocates argue that more investment in transit is warranted: the only way to compete with the private auto is to provide extensive, high-quality, low-fare service. More recently, high-speed rail (HSR) advocates have made similar arguments to generate support for HSR investments (see Box 9.2).[8]

Theoretical Expectations of Rail Transit Impacts

The construction of a rail transit system should improve accessibility along the rail line corridors and increase the relative advantage of rail corridors compared to areas

---

BOX 9.2

High-Speed Rail

High-speed rail (HSR) refers to intercity passenger railway service with high cruise speeds (150 mph or more). The primary advantages of HSR include potentially shorter travel time between specific locations, shorter loading and unloading times, greater safety, and potentially greater energy efficiency as compared to the competing modes (car or airplane). The disadvantages include high fixed costs and local environmental impacts (noise, interference with surface traffic). HSR systems have been operating in Japan and Europe for decades, and China is rapidly developing an extensive HSR network. There is now growing interest in the United States to invest in HSR. HSR is proposed as a means to address air system capacity constraints; to promote more sustainable intercity transport; to provide a mode that better fits in dense urban environments; and to promote central city revitalization.

Would a new HSR system generate economic and land use impacts in the United States? This depends on the same factors we have discussed for rail transit: (1) the extent to which HSR affects accessibility, and hence attracts passengers; (2) local economic conditions and land/development opportunities; and (3) supportive zoning, parking, and network connectivity policies. Since the large metropolitan areas to be connected by the proposed HSR network in the United States already possess high intercity accessibility (via air and car travel), the marginal increases in overall intercity accessibility from HSR may not be significant. Absence of a well-developed feeder transit system to complement the proposed HSR system, and the small share of potential destinations represented by U.S. city centers, are additional potential limitations to the HSR market.

that are not served by the rail system. All other things being equal, then, activity location should shift toward the rail corridors, and this shift should be reflected in increased land values. In addition, since rail systems are focused on the CBD, the position of the CBD as the most accessible point in the area should be enhanced, leading to an increase in activities and land values in the CBD.

Two additional points follow. First, the potential effect of rail transit is defined by the extent to which accessibility is changed. Because rail service often replaces existing bus service, its effect on accessibility can actually be quite small. Moreover, the transit system accounts for a very small portion of the entire transportation network. Therefore, we should not expect any regional effects, as is the case for highways. Second, a transit improvement is another form of transportation system improvement, and therefore should have the same *decentralizing effect*. To the extent that the transit improvement reduces transport costs, people will use some of the reduced costs to consume more travel. Thus the difference between a highway improvement and a transit improvement should be one of degree: the land value gradient should be steeper for rail than for highway, but any reduction in transport cost will cause the gradient to become flatter, as we saw in Figure 9.4.

Empirical Evidence

The extensive and growing literature on the land use impacts of rail transit is far too large to summarize in one part of one chapter.[9] Here we present results from the most robust studies using reliable data sets. We will discuss the results of heavy rail and light rail separately.[10] Heavy rail has substantially larger passenger carrying capacity and higher average speed than light rail. Correspondingly, potential accessibility benefits of heavy rail are greater than those of light rail. Therefore, heavy rail is expected to produce more land use impacts than light rail.

HEAVY RAIL

Studies of heavy rail conducted during the 1970s showed few significant impacts on land values or development patterns (Allen & Boyce, 1974; Knight & Trygg, 1977; Puskarev & Zupan, 1977; Weber, 1976). These studies suggested: (1) a growing local economy is a necessary but not a sufficient condition for land use impacts, and (2) supportive zoning and development policies must be in place.

It has now been more than 40 years since the opening of first-generation heavy rail systems such as BART and MARTA. These systems have been studied extensively for their impacts since their opening. Results are summarized below.

The San Francisco Bay Area Rapid Transit (BART) System. BART was the first of the "new generation" rail systems to be constructed. The first line began operation in 1972 with 28 miles of rail. Presently, the system consists of 104 miles of rail and 44 stations. Cervero and Landis (1997) conducted a "BART @ 20 Update Study" to examine the land use and other impacts of BART 20 years after its opening. Table 9.4 provides select results. The overall conclusion is that BART impacts have varied dramatically across the system. BART had the greatest impact on downtown San

TABLE 9.4. Selected Results of BART@20 Impact Study

Comparison	Results
County business patterns data, 1981–1990. Shift-share analysis of differences in job growth, ZIP codes with BART station vs. ZIP codes without BART station.	Job growth in ZIP codes with BART station greater: 57% of total job growth (in all ZIP codes) was inside BART ZIP codes. Result due to job growth in downtown San Francisco; all other BART station ZIP codes had less job growth than non-BART ZIP codes.
Employment density, census-tract level, 1980, 1990. Data from Census Transportation Planning Package.	Employment density increases in San Francisco CBD, Oakland CBD, Concord Line corridor, Fremont Line corridor, but also North Bay, San Mateo SR-101 corridor, Silicon Valley.
Office space in various locations, before BART (1962), 1963–1974, 1975–1992.	Downtown SF: about 2/3 new San Francisco office space 1963–1974 within ¼ mile of BART station; about ¾ 1975–1992. Downtown Oakland: less than 10% of new East Bay office space, entire period; most new office space away from BART stations.
Matched-pair comparisons of nine BART stations and similar freeway interchanges.	More multifamily housing construction near BART stations, slightly more nonresidential construction near BART stations, 1979–1993.

Francisco, greatly increasing labor force access and hence allowing an already vibrant downtown to grow even larger. Favorable zoning and properties available for redevelopment were also factors. Between 1975 and 1992, 40 million square feet of new office space was built in San Francisco, about one-half of which was located within one to two blocks of the Embarcadero station.

Across the Bay in downtown Oakland, despite supportive public policies, most of the development that did occur was either public or heavily subsidized. It bears noting that downtown Oakland is the most accessible node in the BART system, so theoretically it should have benefited most. Local conditions, such as strong support or opposition from the municipality or local residents, have mediated BART impacts, with certain stations on the same line attracting significant commercial and residential development, while other stations have not.

Atlanta's MARTA. In contrast to the San Francisco Bay Area, where the physical geography constrains the spatial extent of development, the Atlanta metropolitan area has been able to expand in all directions. Many critics identify Atlanta as one of the most sprawled metro areas in the United States. Population density of the urbanized area is 1,707 persons per square mile, compared to 6,266 for San Francisco–Oakland.[11] The metro area has experienced double-digit growth every decade since 1950; between 2000 and 2010 the population increased by nearly 29%, to 4.5 million. As with the other major rail systems, the intent of MARTA was to increase transit use, reduce private vehicle use, revitalize the downtown, and promote growth within the rail corridors. However, MARTA has not been able to increase transit use commensurate with the region's population growth and has not had the desired land use impacts (Bollinger & Ilhanfeldt, 1997, 2003; Bowes & Ilhanfeldt, 2001). Indeed, between 1995 and 2003, linked trips per capita on MARTA declined at 3% per year (Brown & Thomson, 2008).

Bollinger and Ihlandfeldt (1997) developed models of population and employment growth, using census tract-level data from 1980 and 1990 to measure MARTA impacts. They found that MARTA had neither a positive nor a negative impact on population and employment growth, but it did affect the composition of employment in some station areas, increasing the share of public employment. Findings are attributed to (1) MARTA's insignificant impact on accessibility, (2) the absence of a significant increase in transit use, and (3) modest supportive public policy efforts.

LIGHT RAIL

Light rail transit (LRT) typically replaces existing bus service and operates at grade, so impacts on accessibility are small. The expansion of LRT to most large cities around the United States has spawned numerous studies of their impacts. The literature is quite varied in terms of context (urban, suburban, exurban), extent of study area (within 1/4 mile to 1 mile of rail station/line), and method (e.g., hedonic price model, matched-pair analysis). Most have used changes in property values as the measure for impacts of transit investments and are limited to the transit corridor of interest. Table 9.5 presents a sample of studies that represent the mixed results in this literature.

These studies illustrate the difficulty of controlling for all confounding factors. Hess and Almeida (2007) did not control for highway access; hence the positive premium could well be due to both highway and LRT access. In other studies that found a positive premium for proximity to LRT, the effect may be due more to

TABLE 9.5. Select Empirical Studies of Light Rail Transit Impact on Property Values

Author/System	Dependent Variable	Time Context	Premium Effect
Knapp et al. (2001) Hillsboro Light Rail extension, Portland, Oregon	Sales price per acre of land; residential parcels only	Preservice; analysis for before and after the station locations were announced	Large positive premium within 1/2 mile of stations; positive but smaller premium 1/2–1 mile of stations (after the station locations were announced)
McMillen and McDonald (2004) Midway Rapid Transit Line, Chicago	Residential sales price between 1983 and 1999	Postconstruction > 5 years	Positive premium within 1.5 miles of transit line
Hess and Almeida (2007) Buffalo LRT	Assessed property value of residential properties in 2002	Post-construction: > 15 years	Positive premium within 1/2 mile of a LRT station.
Redfearn (2009) Metro Red Line (heavy rail), and Metro Gold Line (light rail), Los Angeles	1997 and 2002 single-family home sales price for the Red Line; 2002 and 2004 single-family home sales price for the Gold Line	Before/after service started	Results either inconclusive or do not reveal any premium effect
Duncan (2011) San Diego LRT	Sales price of single-family homes between 1997 and 2001	Post-construction: > 17 years	Positive premium for a property located within 1/3 miles of a LRT station

accompanying land use policies than the transit service itself. In the Portland study, much of the proposed rail corridor passed through then undeveloped greenfields; the observed premium could have been in anticipation of development incentives typically offered near new transit stations. More recent studies have acknowledged the effect of land use policies. For example, Duncan (2011) found that while proximity to a LRT station in San Diego has a positive premium effect on residential property values, the effect is conditional upon permissive zoning regulations.

Conclusions on Rail Transit Impacts

Forty years of research on the land use impacts of rail transit provides a rich base from which to draw conclusions. The extent to which research results are consistent over such a long period of time is striking. We conclude the following:

1. A significant impact on accessibility is a necessary but not sufficient condition for land use impacts. Compared to the road system (or even the expressway system), rail lines make up a very small proportion of the transportation network. Even a large rail investment has little effect on regional accessibility, and new rail lines typically replace existing bus lines, further reducing any influence on accessibility.
2. If follows that impacts are highly localized. Impacts are most likely to occur in fast-growing, heavily congested core areas, where demand for commercial or residential development is strong.
3. Impacts depend on complementary zoning, parking, and traffic policy, and especially on development subsidies.
4. Rail investment is not sufficient to promote economic development in declining areas.

Conclusions on Land Use Impacts

Empirical evidence on the land use impacts of both highways and transit indicates that transportation investments do not have consistent or predictable impacts on land use. Land use change does not necessarily follow transportation investments, even when the dollar value of these investments is large. Nevertheless, land use impacts continue to be a major policy issue in transportation planning. Metropolitan areas around the United States expect to solve congestion and environmental problems, revitalize central cities, reduce urban sprawl, and increase livability through investment in rail transit. Highway expansion is resisted on the grounds that it promotes sprawl and auto dependence. Given the empirical evidence, the reader might ask why. We conclude with some possible explanations.

First, urban planners and local officials are heavily vested in the preservation and revitalization of central cities. Many decades of decentralization have left the central city with a decreasing share of economic activity, but an increasing share of the disadvantaged population. Rail transit, like sports stadiums and festival marketplaces, is seen as a tool for attracting the middle class back to the city by making commuting to central-city jobs easier. Rail transit has the added advantage of federal and state subsidies—"free" money to a local jurisdiction.

Second, rail transit is often promoted for its economic development potential.[12] Cities are usually willing to subsidize local station area development by assembling and preparing land for development, offering density offsets or reduced parking requirements, low-cost loans, or outright subsidies. Many urban planners believe that there is significant demand for transit-oriented city living among young professionals and older baby boomers and hence see such efforts as a way of attracting more market-rate development to the city. Some cities (e.g., Portland and San Diego) seem to be successful in such efforts whereas others (e.g., Oakland or Buffalo) are not. In all cases, outcomes depend more on the underlying conditions of the city (population characteristics, economic potential, aesthetics, competition, etc.) than on the presence or absence of a rail system. Austin has a thriving downtown without rail transit, while many downtowns with rail transit continue to struggle.

Third, many urban planners believe that public transit investment is one of the few feasible alternatives available for solving transportation and sustainability problems. It is argued that rail transit, with its superior line-haul travel time and higher-quality ride, has more potential than bus transit for attracting discretionary riders out of their cars. It is further argued that public transit is an appropriate "second-best" policy. Although charging auto users full marginal costs is the most efficient solution to transportation externalities (it would both reduce auto use and increase transit use), it has proved politically difficult to implement.

Fourth, some planners and policymakers are convinced that if transportation capacity increases are restricted to transit, eventually travelers will shift to transit as congestion becomes intolerable. Increased transit mode share will then lead to accessibility-related changes favoring higher density, more compact development. This logic holds only in a closed system; if people and firms had no choice to move to another city or region, or to shift travel to other times and places, rising congestion would indeed shift some travelers to transit as the travel time advantage of the auto erodes.

Finally, it has been argued that rail transit systems are expanding because their construction is politically acceptable (Altshuler & Luberoff, 2003). Rail systems impose a much smaller footprint on urban areas. They require less land; noise and vibration are highly localized. In the dense parts of central cities, rail transit is obviously more appropriate to the urban fabric and to high land values and dense development.

Metropolitan areas throughout the United States have highly developed transportation systems. The extent and redundancy of U.S. transportation systems makes them flexible; system users can adapt to existing capacity in many ways. Even in heavily congested metropolitan areas like Los Angeles or Washington, D.C., developable land with reasonable access remains available. Metropolitan growth is a function of economic conditions—the general state of the economy, labor force quality, the regulatory environment—and increasingly of the preferences of affluent households.

Broader economic and technological trends suggest that the relationship between transportation and urban form will continue to be uncertain. The shift to an information-based economy is changing the nature of accessibility and increasing the space–time flexibility of activities. Whether it ultimately is a force for concentration or deconcentration remains to be seen, although we have no evidence to date that dense clusters of activities are disappearing. A world of flexible work changes the calculus of households, and perhaps makes access to job opportunities more important in location choice than access to any given job. Given the flexibility of the urban transportation system and the many factors that affect the location and travel choices of individuals and firms, it is understandable that the link between transportation and land use in contemporary metropolitan areas is complex and difficult to identify.

ACKNOWLEDGMENTS

Valuable and greatly appreciated assistance in writing this chapter was provided by graduate students in the Sol Price School of Public Policy at the University of Southern California, and the School of Urban and Regional Planning at Queen's University in Kingston, Ontario, Canada. Data and maps for the Los Angeles region were supported by a research grant from the California Department of Transportation. All errors and omissions are the responsibility of the authors.

NOTES

1. For formal presentation of the standard theory and explanation of these relationships, see Mills and Hamilton (1993).

2. The value of time is the monetary value an individual places on time savings. For example, you have the choice of taking a shuttle van or a taxi to the airport. The shuttle van fare is lower, but the trip is longer compared to the taxi. Research has shown that the value of time is related to the individual's wage rate. Therefore, we observe that college students are more likely to take the shuttle, while business executives will take the taxi.

3. For a recent summary, see Giuliano and colleagues (2009).

4. Source: computed by the authors from 2007–2011 American Community Survey data.

5. Source: *www.census.gov/const/C25Ann/sftotalmedavgsqft.pdf*.

6. The I-105 Century Freeway provides an extreme example. The freeway was adopted as part of the State Highway Plan in 1959. Planning and environmental studies began shortly thereafter. The final environmental impact statement was approved in 1978; subsequent litigation delayed the start of construction until 1982 (Hestermann, DiMento, van Hengel, & Nordenstam, 1993). The facility opened in 1993.

7. Over the past decade, some cities have eliminated freeways (e.g., San Francisco). We are not aware of studies of land use impacts of removing links from a network.

8. The nine presently proposed regional HSR hubs in the United States are Los Angeles, Atlanta, New York, Chicago, Dallas, Phoenix, Seattle, and Denver.

9. For a comprehensive review of rail system impacts, see Giuliano and Agarwal (2010).

10. Heavy Rail Transit (HRT) is a rapid transit service that operates on completely grade-separated rights of way. Light Rail Transit (LRT) is a rapid transit service that operates primarily at grade, typically over exclusive rights of way. LRT systems require less and simpler infrastructure and are normally less expensive to build than HRT systems. A light rail line can carry more passengers through a given corridor than buses, but less than a HRT line.

11. Source: *www.demographia.com/db-uza2000.htm*. The Los Angeles–Long Beach urbanized area has the highest population density in the country, at 6,999 persons per square mile.

12. The economic impact from construction of any public project is a transfer from general taxpayers to the local jurisdiction. It is also temporary.

REFERENCES

Allen, W., & Boyce, D. (1974). Impact of a high speed rapid transit facility on residential property values. *High Speed Ground Transportation Journal, 8*(2), 53–60.

Alonso, W. (1964). *Location and land use.* Cambridge, MA: Harvard University Press.

Altshuler, A., & Luberoff, D. (2003). *Mega-projects: The changing politics of urban public investment.* Washington, DC: Brookings Institution.

Anas, A., Arnott, R., & Small, K. (1998). Urban spatial structure. *Journal of Economic Literature, 36,* 1426–1464.

Boarnet, M. (1997). Highways and economic productivity: Interpreting recent evidence. *Journal of Planning Literature, 11*(4), 476–486.

Boarnet, M. (1998). Spillovers and locational effect of public infrastructure. *Journal of Regional Science, 38*(3), 381–400.

Boarnet, M., & Chalermpong, S. (2001). New highways, house prices, and urban development: A case study of toll roads in Orange County, CA. *Housing Policy Debate, 12*(3), 575–605.

Bollinger, C., & Ihlandfeldt, K. (1997). The impact of rapid rail transit on economic development: The case of Atlanta's MARTA. *Journal of Urban Economics, 42,* 179–204.

Bollinger, C., & Ihlandfeldt, K. (2003). The intraurban spatial distribution of employment: Which government interventions make a difference? *Journal of Urban Economics, 53,* 396–412.

Bowes D., &. Ilhanfeldt, K. (2001). Identifying the impacts of rail transit stations on residential property values. *Journal of Urban Economics, 50,* 1–25.

Brown, J., &. Thomson, G. (2008). The relationship between transit ridership and urban decentralisation: Insights from Atlanta. *Urban Studies, 45*(5–6), 1119–1139.

Brueckner, J. (1978). Urban general equilibrium models with non-central production. *Journal of Regional Science, 18,* 203–215.

Calthorpe, P. (1993). *The next American metropolis: Ecology, community, and the American dream* Princeton, NJ: Princeton Architectural Press.

Cervero, R., & Landis, J. (1997). Twenty years of the Bay Area Rapid Transit System: Land use and development impacts. *Transportation Research A, 31A*(4), 309–333.

Crane, R. (1996). The influence of uncertain job location on urban form and the journey to work. *Journal of Urban Economics, 39*(3), 342–358.

Duncan, M. (2011). The synergistic influence of light rail stations and zoning on home prices. *Environment and Planning A, 43,* 2125–2142.

Funderburg, R., Nixon, H., Boarnet, M., & Ferguson, G. (2010). New highways and land use change: Results from a quasi-experimental research design. *Transportation Research Part A, 44,* 76–98.

Garrison, W., Berry, B., Marble, D., Nystuen, J., & Morrill, R. (1959). *Studies of highway development and geographic change.* Seattle: University of Washington Press.

Giuliano, G., & Agarwal, A. (2010). Public transit as a metropolitan growth and development strategy. In H. Wolman, N. Pindus, & H. Wial (Eds.), *Urban and regional policy and its effects* (Vol. 3, pp. 205–252). Washington, DC: Brookings Institution.

Giuliano, G., Agarwal, A., & Redfearn, C. (2009), *Metropolitan spatial trends in employment and housing: Literature review.* Commissioned paper for the Committee for the Study on the Relationships among Development Patterns, Vehicle Miles Traveled, and Energy Consumption, Transportation Research Board and Board on Energy and Environmental Systems, National Research Council. Washington DC: Transportation Research Board.

Giuliano, G., Redfearn, C., Agarwal, A., & He, S. (2012). Network accessibility and employment centers. *Urban Studies, 49*(1), 77–95.

Giuliano, G., & Small, K. (1993). Is the journey to work explained by urban structure? *Urban Studies, 30*(9), 1485–1500.

Glaeser, E., Kahn, M., & Jordan, R. (2008). Why do the poor live in cities?: The role of public transportation. *Journal of Urban Economics, 63,* 1–24.

Glaeser, E., & Kohlhase, J. (2004). Cities, regions and the decline of transport costs. *Papers in Regional Science, 83*(1), 197–228.

Henry, M., Barkley, D., & Bao, S. (1997). The hinterland's stake in metropolitan growth: Evidence from selected southern regions. *Journal of Regional Science, 37*(3), 479–501.

Hershberg, T., Light, D., Cox, H., & Greenfield, R. (1981). The journey to work: An empirical investigation of work, residence and transportation, Philadelphia, 1950 and 1880. In T. Hershman (Ed.), *Philadelphia: Work, space and group experience in the nineteenth century* (pp. 128–173). New York: Oxford University Press.

Hess, D., & Almeida, T. (2007). Impact of proximity to light-rail rapid transit on station-area property values in Buffalo, New York. *Urban Studies, 44*(5–6), 1041–1068.

Hestermann, D., DiMento, J., van Hengel, D., & Nordenstam, B. (1993). Impacts of a consent decree on "the last urban freeway": I-105 in Los Angeles County. *Transportation Research A, 27A*(4), 299–313.

Ihlandfeldt, K. T., & Sjoquist, D. L. (1998) The spatial mismatch hypothesis: A review of recent studies and their implications for welfare reform. *Housing Policy Debate, 9*(4), 849–892.

Kain, J. (1999). The urban transportation problem: A reexamination and update. In J. Gomez-Ibañez, W. Tye, & C. Winston (Eds.), *Essays in transportation economics and policy* (pp. 359–401). Washington, DC: Brookings Institution.

Kim, S.. (2007). Changes in the nature of urban spatial structure in the United States, 1890–2000. *Journal of Regional Science, 47*(2), 273–287.

Knight, R., & Trygg, L. (1977). *Land use impacts of rapid transit: Implications of recent experiences* (Final Report No. DOT-TPI-10-77-29, U.S. Department of Transportation). San Francisco: De Leuw Cather and Co.

Ko, K., & Cao, J. (2013). The impact of Hiawatha Light Rail on commercial and industrial property values in Minneapolis. *Journal of Public Transportation, 16*(1), 47–66.

Koster, H., & Rouwendal, J. (2013). Agglomeration, commuting costs, and the internal structure of cities. *Regional Science and Urban Economics, 43*(2), 352–366.

Ma, K.-R., & Banister, D. (2006). Excess commuting: A critical review. *Transport Reviews, 26*(6), 749–767.

McMillen, D., & MacDonald, J. (2004). Reaction of house prices to a new rapid transit line: Chicago's Midway Transit Line, 1983–99. *Real Estate Economics, 32*(3), 463–486.

Mieskowski, P., & Mills, E. S. (1993). The causes of metropolitan suburbanization. *Journal of Economic Perspectives, 7*(3), 135–147.

Mills, E. S. (1972). *Studies in the structure of the urban economy.* Baltimore: Johns Hopkins University Press.

Mills, E. S., & Hamilton, B. (1993). *Urban economics* (5th ed.). New York: HarperCollins.

Muth, R. (1969). *Cities and housing.* Chicago: University of Chicago Press.

Newman, P., & Kenworthy, J. (1998). *Sustainability and cities: Overcoming automobile dependence.* Washington, DC: Island Press.

Payne-Maxie Consultants. (1980). *The land use and urban development impacts of beltways* (Final Report No. DOT-OS-90079, U.S. Department of Transportation and Department of Housing and Urban Development). Washington, DC: Author.

Pushkarev, B., & Zupan, J. (1977). *Public transportation and land use policy.* Bloomington: Indiana University Press.

Quigley, J., & Raphael, S. (2004). Is housing unaffordable?: Why isn't it more affordable? *Journal of Economic Perspectives, 18*(1), 191–214.

Redfearn, C. (2009). How informative are average effects?: Hedonic regression and amenity capitalization in complex urban housing markets. *Regional Science and Urban Economics, 39*(3), 297–306.

Roberto, E. (2008). *Commuting to opportunity: The working poor and commuting in the United States.* Washington, DC: Brookings Institution.

Singletary, L., Henry, M., Brooks, K., & London, J. (1995). The impact of highway investment on new manufacturing employment in South Carolina: A small region spatial analysis. *Review of Regional Studies, 25,* 37–55.

Voith, R. (1993). Changing capitalization of CBD-oriented transportation systems: Evidence from Philadelphia, 1970–1988. *Journal of Urban Economics, 33,* 361–376.

Weber, M. (1976). The BART experience—What have we learned? *Public Interest, 4,* 79–108.

CHAPTER 10

The Geography of Urban Transportation Finance

BRIAN D. TAYLOR

"Follow the money."
—Deep Throat in Bernstein and Woodward's
All the President's Men (1974)

The geography of urban transportation in U.S. cities is profoundly shaped by the geography of urban transportation finance. Geopolitical struggles over the spatial distribution of public expenditures on transportation have significantly shaped the geography of transportation systems, spatial patterns of travel, and urban form. From where taxes are collected and to where they are distributed is part-and-parcel of political life but has often been ignored by transportation analysts. In addition to the spatial *distribution* of public resources, the appropriate *level* of public spending on transportation has long been hotly debated and varies significantly among both places and transportation modes. Furthermore, urban and transportation economists have long argued that patterns of urban development, economic trade, and travel are strongly influenced by the price of travel, which is, in turn, governed to a large degree by public policies. Put simply, money matters.

As we saw in Chapter 3, around the turn of the last century, American cities expanded with the deployment of privately financed electric streetcars. Investors were attracted to streetcars because of the near-monopoly they held on urban travel prior to the widespread use of automobiles and because new lines and extensions were tied to remunerative suburban land development schemes. A half-century later, plans for a spectacular interstate system of freeways between—and within—metropolitan areas lay largely dormant for a dozen years until a funding package was adopted in 1956, unleashing a torrent of highway construction that would transform cities developed by streetcars into metropolitan areas defined by cars and freeways. And today, efforts to tame the auto in cities often require convincing local voters to raise their taxes in order to pay for lists of popular local public transit, cycling, and pedestrian projects.

So to understand the geography of urban transportation, one must first understand the geography of urban transportation finance.

This chapter begins with an overview of the economic geography of transportation finance; I explore economic rationales for public investments in transportation systems, as well as the central role of political geography in determining transportation investments. I then turn to an overview of current surface transportation finance in the United States, with an emphasis on metropolitan areas. Here I examine broad trends in revenues for transportation and patterns of expenditures by both geography and mode. So to appreciate how and why U.S. cities and their transportation systems developed as they have, we heed Deep Throat's advice and "follow the money."

THE PUBLIC ROLE IN FINANCING URBAN TRANSPORTATION SYSTEMS

Why should the public sector get involved in financing transportation systems? Why not leave it to the private sector? Large portions of urban transportation systems are financed privately, why not all? People pay for their own shoes, bicycles, taxi fares, and automobiles; firms pay to move their goods from factory to warehouse to store; retailers pay for parking lots so that shoppers can park for free; advertisers pay transit systems for display space on buses and transit shelters. But in addition to all of this private investment in transportation, governments—local, regional, state, and national—finance large parts of transportation systems. But just how the public sector should collect funds for transportation, what systems or modes should receive funding priority, and where the funds should be expended is subject to considerable and ongoing political debate. For example:

- Should transportation revenues be collected from user fees (like bridge tolls, transit fares, or fuel taxes), or more general instruments of taxation (like income, sales, and property taxes)?
- Should people pay for transportation systems based on ability to pay? Benefits received? Costs imposed? Should, for example, income-regressive sales tax revenues be used to pay for commuter rail lines that primarily serve higher-income suburban commuters?
- Should streets and highways receive funding priority because they are so heavily used, or should public transit and bicycling receive priority to create more environmentally friendly alternatives to private vehicle use?
- Should transportation taxes and fees collected in one jurisdiction be spent in other places? If so, on what basis should the funds be geographically redistributed?
- Should transportation projects be distributed equally among states and jurisdictions, or should the most-needed projects be funded first, regardless of location?

Such questions about public versus private roles, taxes versus user fees, and redistribution of funding among both individuals and places are at the heart of urban transportation finance.

The Political Economy of Transportation Finance

Public finance economists distinguish *public goods* from *private goods* in deter-mining the appropriate level of public involvement. Private goods—like basketball shoes and smart phones—are allocated by markets. Public goods, on the other hand, are typically not sold, so markets cannot allocate them. National defense and the National Weather Service are examples of public goods. So-called pure public goods share three characteristics: (1) no one can be excluded from consuming them, (2) consumption by one person does not affect consumption by others, and (3) they can only be provided through collective action.

But transportation systems lie on a continuum between private and public goods. At the private end of the spectrum, most public transit systems began as private enterprises funded entirely by passenger fares and the sale of advertising space. People can be excluded from transit vehicles (e.g., for lack of the fare), and consumption of transit capacity does affect others. But except for some specialty services like airport shuttles, the costs of operating comprehensive transit networks when the demand for service is increasingly dispersed spatially and concentrated temporally means that public transit service today cannot be financed without collective action. At the other end of the spectrum are sidewalks. Outside of enclosed shopping malls and amuse-ment parks, it is very difficult to exclude people from sidewalks. Sidewalks are only rarely congested, so use of sidewalks does not normally affect others. And without public regulation, interconnected networks of sidewalks would not exist.

Some observers contend that the public sector has come to play too large a role in the finance and operation of transportation systems (O'Toole, 2010; Poole, 2001). They argue that increasing reliance on private markets to allocate transportation resources would greatly increase economic efficiency. Such arguments led to efforts in the 1980s to increase public contracting with private firms to provide public transit service and in the 2000s to privately finance new toll roads. Political debates over the appropriate scope and scale of public involvement in transportation systems are often drawn along ideological lines, especially regarding taxation: conservatives typically favor lower taxes and a larger private-sector role, whereas liberals more often favor increased collection of revenues for transportation and substantial public control of transportation systems.

Benefits of Public-Sector Transportation Investment

Regardless of ideology, however, transportation public works projects are usually popular with voters and the people whom they elect. Elected officials frequently jus-tify their support for large transportation projects in terms of the jobs created by such projects (Holtz-Eakin & Wachs, 2011; Richmond, 2005), though usually without regard to whether the new jobs are added as a result of the transportation improve-ment or whether the jobs were simply shifted from some other location (Boarnet, 1997). As legislators have increasingly advocated for projects in their home states and districts, the proportion of federal transportation funds "earmarked" for spe-cific highway projects increased dramatically with each successive federal surface transportation legislation—from the 11 earmarks for "demonstration," "priority," or "high-priority" projects in the *Surface Transportation Assistance Act* of 1982,

TABLE 10.1. Earmarked Projects in Federal-Aid Highway Acts

Act	Amount ($ billions)	No. of projects	Percent of authorization for demonstration, high-priority, or other earmarked projects
STAA, 1982	0.6	11	1.4
STURAA, 1987	1.0	152	1.6
ISTEA, 1991	6.2	539	6.0
TEA-21, 1998	9.4	1,850	6.3
SAFETEA-LU, 2005	14.8	5,700	10.6

Note. Earmarked projects are those designated in the Acts as "Demonstration," "Priority," or "High-Priority" authorizations. STAA, Surface Transportation Assistance Act; STURAA, Surface Transportation and Uniform Relocation Assistance Act; ISTEA, Intermodal Surface Transportation Equity Act; TEA-21, Transportation Equity Act for the 21st Century.

Source: Transportation Research Board (2006, Table 2.8).

to the 5,700 earmarks in the *Safe, Accountable, Flexible, Efficient Transportation Equity Act: A Legacy for Users* (SAFETEA-LU) of 2005 (Table 10.1) (Transportation Research Board, 2006). New rail transit projects have been especially popular with elected officials since the 1970s, when public ambivalence toward the disruptions caused by urban freeway projects began to mount.

Early in the 21st century earmarks in the reauthorizations of the federal surface transportation legislation became so widely viewed as having spiraled out of control that Alaska's infamous "Bridge to Nowhere" became a central debate in the 2008 presidential election. Alaska's congressional delegation fought hard to earmark federal funds in 2005 to help build a $400 million bridge roughly the size of the Golden Gate Bridge that would connect Ketchikan, Alaska (population 8,050), to the Ketchikan Airport on Gravina Island (population 50). Bipartisan embarrassment over widespread perceptions of wasteful earmarking of projects led Congress to eschew all legislative earmarks, including in 2012's reauthorization bill (dubbed Moving Ahead for Progress in the 21st Century, or MAP-21), though some observers have argued that the absence of earmarks helps to explain why this reauthorization was years overdue, covered only a 2-year period, and contained no meaningful revenue enhancements.

EXPENDITURE EFFECTS

When announcing federal funding of new transportation projects to constituents, Congress members often cite the economic benefits such projects will bring to the affected districts. Over the past two decades, for example, rail transit projects in Buffalo, Los Angeles, Milwaukee, and elsewhere have been touted by elected officials as tools for job creation and economic revitalization in depressed areas (Metropolitan Council, 2012; Richmond, 2005; Valley Metro, 2012). While being debated in Congress, the MAP-21 legislation was repeatedly presented by officials first and foremost as a jobs bill. When the bill was passed by the Senate and sent to the House of Representatives, Democratic senators sent a letter to all members of the House of

Representatives emphasizing estimates that the bill would create or sustain 2.9 million jobs. The letter included a chart indicating the number of jobs that would be created or sustained in each state (United States Senate Democrats, 2012).

But transportation economists have for decades cautioned policymakers not to confuse the economic effects of expenditures of public funds on transportation facilities and operations with the economic benefits of an improved transportation system brought about by such expenditures. The former—the *expenditure effects*—concern balancing the diminution of economic activity caused by taxation, on the one hand, with stimulation and redistribution of economic activity of tax revenues, on the other. From this perspective, the expenditure effects of a new highway maintenance facility do not directly concern transportation; rather the expenditure effects of such a project should properly be compared with other possible public expenditures (for schools, health care, etc.) or with not taxing to collect the funds in the first place.

Expenditure effects result directly from the outlay of money, such as when subsidies pay for transit workers' salaries or when public funds pay a construction firm to build a new suburban freeway. These direct payments are then respent in the economy, generating additional economic activity. This expanded effect, known as the multiplier process, is the reason for dividing the expenditure effects into direct, indirect, and induced effects in Figure 10.1.

So while local economic activity can be generated by transportation expenditure effects, much—if not most—of this activity may simply be redistributed from other people (taxpayers) and other places (that lost out in the geographic competition for subsidy dollars). To transportation economists, policymakers simply miss the point when they focus on the local expenditure effects of transportation expenditure decisions (Holtz-Eakin & Wachs, 2011; Lewis, 1991).

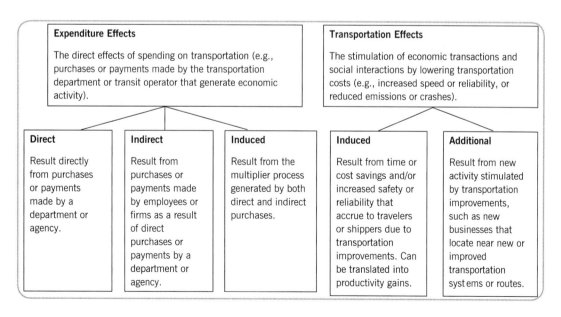

FIGURE 10.1. Economic benefits of transportation expenditures.

TRANSPORTATION EFFECTS

In contrast, most transportation economists would argue that public expenditures on transportation should properly be judged on their *transportation effects*. In other words, the real economic benefits of public investments in transportation come from improvements to the transportation system. Decreased travel time, increased reliability, reduced emissions, increased safety, and the like all benefit the economy and society by lowering transportation costs both to users of the transportation system and to society at large. Lower transportation costs facilitate economic transactions and social interactions; they make it cheaper to produce current goods and services and they make new forms of goods and services possible. Thus, transportation improvements "help us do things differently and help us do different things" (Gillen, 1997).

In the case of a highway improvement, travelers who accrue time or cost savings from reduced congestion delay realize the transportation effects of highway expenditures. A second source of transportation effects comes from the benefits accruing to others (termed "additional effects" in Figure 10.1). For example, a new high-occupancy vehicle lane that is successful in motivating some travelers to carpool may reduce congestion on adjacent unrestricted lanes, thus benefiting all roadway users, not just carpoolers. A new public transit investment may generate new economic activity by attracting new businesses to locate near transit stations. The ability of firms to restructure their logistics and distribution networks to reduce production and distribution costs is another source of economic benefit. As with the expenditure effects, however, the transportation effects of transportation expenditures must be netted against the economic costs of taxation (Lewis, 1991; Small & Verhoef, 2007).

But to many policymakers, the transportation effects of public investments can seem rather abstract, arcane, and arbitrary. To a member of the Oregon congressional delegation, a study showing that rail transit investments in New York City yield far greater transportation benefit than those in Portland is not terribly relevant to debates over the equitable geographic distribution of federal transit subsidies. In contrast, the local expenditure effects of transportation facility construction projects in individual congressional districts are clear and unambiguous. A new highway funded with substantial federal funding is a dramatic and highly visible public investment that, during construction, clearly and directly generates employment and economic activity in particular jurisdictions. That much of this economic activity is simply shifted from taxpayers in other jurisdictions is almost beside the political point. In a democracy, like the United States, composed primarily of geographically based representatives at the federal, state, and local levels, a stable and popular transportation investment program emerges from a never-ending geopolitical struggle over the collection and distribution of funding (Garrett & Taylor, 1999; Taylor, 1995).

THE GEOPOLITICAL EQUITY IMPERATIVE

This overriding concern with the geographic equity of transportation funding among states, districts, and jurisdictions ensures a political focus on the expenditure effects of transportation investments, which makes it all but impossible to consider the transportation effects in making transportation funding decisions. Thus, from the perspective of most policymakers, it's the transportation economists who simply miss the point by focusing all of their attention on the transportation effects of funding decisions.

To understand the public finance of urban transportation systems, we must evaluate both the geopolitics of the transportation finance *program* and the effects of this finance program on the deployment and use of the transportation *system*. Table 10.2 offers an overview of how we might simultaneously evaluate the performance of a transportation finance program in each of these realms.

Program performance criteria evaluate how well a finance mechanism meets tests of political acceptability and administrative ease. These questions tend to be prominent in policy debates, and most of them will be familiar even to those new to transportation issues.

System performance criteria, on the other hand, address how finance mechanisms influence the use and performance of the transportation system itself. These criteria acknowledge that finance policies are not just about raising and spending money. Finance instruments also profoundly affect the way transportation services are provided and the way citizens use them. The fares, fees, tolls, and taxes paid by travelers affect their decisions on where to live and work; where, when, and how to travel; and even whether to travel. Use of the transportation system in turn greatly influences land use as well as the maintenance and new capacity "needs" of the transportation system, which in turn affects the finance system. Thus, the transportation finance system, the performance of the transportation system, and urban development systems are mutually reinforcing.

How both the supply of and demand for transportation are influenced by user costs is neither abstract nor trivial. The issue of truck-weight fees provides an example of how the transportation finance system affects user decisions. The pavement damage caused by heavy trucks increases significantly with the weight per axle. Many people are surprised to learn that a relatively small share of trucks with heavy axle loads does most of the damage to roads (Small, Winston, & Evans, 1989; Straus &

TABLE 10.2. Program Performance and System Performance Criteria

	System performance	Program performance
Effectiveness	• Optimizes utilization of existing capacity. • Lowers transportation costs and promotes economic development.	• Is politically feasible: has stable political support, is popular with voters, and has little opposition from powerful stakeholders. • Revenues generated meet needs and are stable and predictable.
Efficiency	• Optimizes provision of transportation service for a given level of expenditure.	• Has low administrative and overhead costs relative to the revenue collected.
Equity	• Provides all users with transportation access, regardless of circumstances (age, income, disability, etc.). • Is progressive based on ability to pay. • Charges users in proportion to the costs they impose on the system and society.	• Is perceived as treating places and jurisdictions fairly. • Major stakeholders and interest groups perceive they are treated fairly.

Source: Brown et al. (1999).

Semmens, 2006). Yet for decades many states levied truck weight fees based on the weight of *empty* trucks; and tollways frequently set rates based on the number of axles per vehicle. Both policies encourage truckers to load heavy weights onto as few axles as possible, thereby *maximizing* damage to roadways. Such truck fee systems increase maintenance and rehabilitation costs in comparison to jurisdictions where fees are assessed in ways to encourage truckers to reduce axle weights. Thus, changing the way that fees are levied on trucks can change truckers' behavior, and, in turn, substantially lower maintenance costs without necessarily increasing taxes or revenues.

That transportation finance programs are guided first and foremost by concerns over equity may seem a puzzling, even counterintuitive, assertion. But the way that public officials think of equity in transportation finance is far different from the way that most social scientists, students, or transportation analysts would define the term. Thus, "equity" gets defined quite differently by different interests at different times. To paraphrase former Supreme Court Justice Potter Stewart on the question of pornography, most of us can't precisely define equity or inequity in transportation finance, but we think that we know it when we see it.

Much of the confusion and debate over equity in transportation finance arises from the competing and contradictory ways that equity is both framed and evaluated (Table 10.3). Debates over equity in transportation finance can arise over the type of equity to be considered—market equity, opportunity equity, or outcome equity—or the appropriate unit of analysis with which to frame the debate—geographic, group, or individual—or both. While "unit of analysis" may seem an abstract concept, it is perhaps the most important consideration in transportation equity (Taylor & Norton, 2009).

TABLE 10.3. Confounding Notions of Equity in Transportation Finance

Unit of analysis	Type of equity		
	Market equity	Opportunity equity	Outcome equity
Geographic			
States, counties, legislative districts, etc.	Transportation spending in each jurisdiction matches revenue collections in that jurisdiction.	Transportation spending is proportionally equal across jurisdictions.	Spending in each jurisdiction produces equal levels of transportation capacity/service.
Group			
Modal interests, racial/ethnic groups, etc.	Each group receives transportation spending/benefits in proportion to taxes paid.	Each group receives a proportionally equal share of transportation resources.	Transportation spending produces equal levels of access or mobility across groups.
Individual			
Residents, voters, travelers, etc.	The prices/taxes paid by individuals for transportation should be proportional to the costs imposed on society.	Transportation spending per person is equal.	Transportation spending equalizes individual levels of access or mobility.

Source: Data from Taylor and Norton (2009).

In general, public finance scholars tend to focus on *individual equity,* advocates and activists are more likely to focus on *group equity,* while elected officials are concerned most with *geographic equity.* This is because representation in the United States is organized geographically into a hierarchy of jurisdictions. And because it is elected officials who oversee the collection and distribution of transportation funds, most debates in transportation finance center first and foremost on questions of *geographic equity.*

Geographic equity questions arise frequently in the context of federal transportation policy. For example, the more populous, urbanized states have historically tended to generate more in federal motor fuels tax revenues than they receive in fuel-tax-funded federal expenditures, whereas less populous, rural states have tended to receive more in federal transportation funding than their motorists generate in federal fuel taxes. This redistribution of federal fuel tax revenues from "donor" states to "donee" states has been hotly debated in Washington for decades, has several times delayed the passage of federal surface transportation legislation, and has gradually declined over time as political support for geographic redistribution of transportation funds across state lines has withered.

Supporters of redistribution argue that it enables wealthier states to cross-subsidize poorer states, it allows us to have an interconnected national highway system, and that this redistribution justifies federal involvement in transportation finance. Critics of the redistribution of federal fuel taxes have countered that the redistribution reflects a rural bias in the federal transportation program (especially highways), and research has shown that it has actually redistributed funds from poorer states (those with less fiscal capacity) to richer states (with more fiscal capacity) (Lem, 1997). Redistribution critics contend further that the national highway system is largely in place and the most significant transportation investment needs are in congested urban areas. Beginning with the 2005 federal surface transportation legislation, SAFETEA-LU, an "equity bonus" program has guaranteed that at least 92.5% of each state's federal fuel tax contribution would be returned (Schank & Rudnick-Thorpe, 2011). Critics of the equity bonus program argue that this allocates transportation funding to states without regard for the states' funding needs and undermines the redistributive logic of federal transportation programs (Cooper & Griffith, 2012). If all federal fuel tax funds were simply returned to states (or state fuel tax revenues to counties) exactly proportional to their collection, there would be no rationale for a federal (or state) fuel tax; it could be eliminated and states (or localities) would then be free to collect as much as they needed from higher fuel taxes. Thus, some have argued that federal transportation tax collections should be dropped and that each state should be left to make do on its own (Roth, 1996).

Given the overriding political concern with geographic equity, distortions arise when transportation use or demand does not vary somewhat equally across jurisdictions. Public transit is perhaps the most striking example of this. Transit ridership is concentrated spatially in the largest, most densely developed cities. Better than one-third of all transit trips in the United States are in the New York metropolitan area. The 10 largest U.S. transit systems carry over 60% of all trips; the hundreds of other, smaller systems carry less than 40% (Federal Transit Administration, 2012). In the *realpolitik* of public transit finance, however, debates center on how resources are doled out to jurisdictions and the *suppliers* of transit service, with little regard for the enormous spatial variation in the *consumers* of transit service.

For example, the New York Metropolitan Transit Authority (NY MTA) alone carries about 30% of the nation's transit trips each year (Federal Transit Administration, 2012). During the 10 years between 2000 and 2009, federal capital and operating subsidies combined averaged $0.32 per unlinked passenger trip on NY MTA. In contrast, riders on Chapel Hill Transit in North Carolina, which carries five ten-thousandths (0.05%) of the nation's transit riders, enjoyed federal transit subsidies almost three times higher ($0.89 per trip) (Federal Transit Administration, 2012). Such geographic disparities are not confined to federal transportation finance. In California, the San Francisco Municipal Railway carries nearly half (45%) of all Bay Area transit riders, but receives just 13% of the subsidies allocated through the state Transportation Development Act (TDA). On the other hand, Santa Clara Valley Transit Authority, in the San Jose area, carries less than one-tenth of all Bay Area transit riders yet receives nearly a third (29%) of the region's TDA transit subsidies (Federal Transit Administration, 2012; Metropolitan Transportation Commission, 2012).

The reason for these disparities is quite straightforward: representation in the U.S. House of Representatives and most state legislatures match the geographic distribution of voters, and not urban transit patrons. Geographic equity, therefore, allocates public transit funding "equally" among jurisdictions, regardless of how they are utilized. The centrality of the geographic equity imperative in transportation policy and planning can hardly be overemphasized. It explains why Texas received $4.2 billion *less* in federal fuel tax revenues between 1956 and 2010 than motorists in Texas paid in federal fuel taxes, while Hawaii has received $3.7 billion *more* than motorists there paid in federal fuel taxes; for every $1.00 in federal fuel tax generated in Hawaii, the state has received $2.97 in fuel-tax funded appropriations (U.S. Department of Transportation Federal Highway Administration [USDOT FHWA], 2011). It also explains why new rail transit systems were built in Atlanta, Miami, and many other sprawling Sunbelt cities over the last quarter century, while plans for the Second Avenue subway in transit-oriented Manhattan date to 1929, but it did not carry its first paying passenger until 2017.

Given this overview of the theories and rationales for the finance of metropolitan transportation systems, we now turn to the structure of urban transportation finance.

WHO PAYS FOR WHAT?: THE STRUCTURE OF URBAN TRANSPORTATION FINANCE

Transportation is big business. The movement of people, goods, and information is central to both human culture and economic activity. The U.S. Department of Transportation estimates that over 10% of the U.S. economy relates in some way to transportation (U.S. Department of Transportation, Bureau of Transportation Statistics [USDOT BTS], 2013). Transportation systems are financed from a labyrinth of public and private sources that defy simple characterization; the structure of transportation finance varies greatly over space, from state to state, and even from county to county. Table 10.4 presents a schematic overview of how five principal metropolitan transportation systems are financed.

TABLE 10.4. Schematic Overview of Metropolitan Transportation Finance in the United States

Transport system	Share of metropolitan person trips in 2009	Payments by direct beneficiaries		Payments by others	
		Paid by travelers	Paid by others	Direct subsidies	Indirect subsidies
Pedestrian systems	11.2%	Time, shoes, socks, energy	Sidewalks usually initially funded by developers, and later by property owners via property taxes.	Few	Few
Bicycle systems	1.1%	Time, bicycles, energy	Local streets often paid for initially by developers, later by property owners via property taxes.	Bikeways, bike paths, bike lanes financed from fuel taxes and other tax revenues for transportation.	Roads and highways financed from fuel taxes and other tax revenues for transportation.
Public transit systems	4.3%	Time, fares	Local streets on which buses operate often paid for initially by developers, later by property owners via property taxes and transportation impact fees.	Transit capital and operations paid from fuel and general taxes. Highways and freeways financed with fuel taxes, license fees.	Minor delay costs borne by other travelers; pollution, noise, and resource-depletion costs borne by all.
Local streets/ roads	83.8%	Time, vehicles, fuel, insurance. Some fuel taxes, license fees, etc. for operations, maintenance.	Rights-of-way provided and initial construction often paid by developers. Local streets financed by property owners via property taxes.	Few	Delay costs borne by other travelers; pollution, noise, and resource-depletion costs borne by all.
Metropolitan highways and freeways	83.8%	Time, vehicles, fuel, insurance. Fuel taxes, tolls, license fees, etc. for rights of way, construction, operations, maintenance.	Minimal	Some sales taxes and other general taxes for rights-of way, construction, maintenance.	Significant delay costs borne by other travelers; pollution, noise, and resource-depletion costs borne by all.

Mode share source: U.S. Department of Transportation, Federal Highway Administration. (2009).

Private Investment

Individuals and firms pay for private vehicles, insurance, fuel, and fares. Shippers pay fees for the construction and operation of seaports, and passengers pay fees for the development and expansion of airports. Pedestrians pay for shoes, while property owners typically pay for sidewalks.

Parking is a significant, but often overlooked, private investment in transportation—one that profoundly shapes both local land uses and metropolitan form. One reason it is overlooked is that those who benefit most from it rarely pay for it directly.

While drivers park free for 99% of all vehicle trips, parking is not free to provide (Shoup, 2011). Free parking at a grocery store, for example, is paid indirectly by shoppers, whose groceries are priced to include the cost of construction and maintenance of parking spaces that typically occupy more land than the store. Shoppers who walk to and from a grocery store thus subsidize those who drive and park for free.

Free parking at work is essentially a matching grant for commuting by private vehicle—employers pay the cost of parking at work only if commuters are willing to drive to work. Commuters who do not drive to work do not receive a subsidy. While many commuters would drive even if they had to pay for parking at work, some would choose to carpool, ride public transit, walk, or bike to work instead; such commuters, therefore, drive to work because they park for free. Shoup (2005) has shown that employer-paid parking increases the number of cars driven to work by about 33% when compared with employment sites with driver-paid parking.

Recently, some employers have begun to offer workers the option to "cash out" free parking by taking the cash equivalent of any parking subsidy offered. Offering commuters the choice between a parking subsidy or its cash equivalent shows commuters that free parking has an opportunity cost, in this case foregone cash. At such work sites, commuters can continue to park free, but the option to receive cash in lieu of free parking also rewards commuters who choose to carpool, ride public transit, walk, or bike to work instead. When given this option, a surprising proportion of employees choose to pocket the cash and commute to work by means other than driving alone (Shoup, 2005).

Finally, public regulations strongly encourage the provision of unpriced parking for drivers. Substantial portions of street and road systems are devoted to curbside parking instead of traffic flow. Most curb parking is free, and even where curb spaces are metered they are typically priced below the market-clearing price (i.e., the price that would ensure at least a few available spots at most times of the day to serve the needs of those willing to pay for parking). And because the demand for free or cheap on-street parking often exceeds available curb space, local governments typically require all developments (grocery stores, apartment buildings, medical offices, etc.) to provide enough off-street parking to satisfy the peak level of demand for free parking at that land use. Such minimum parking requirements subsidize private vehicle use, increase the share of urban space devoted to parking, and substantially increase development costs—which are borne by everyone, not just parkers. Shoup (2011) has estimated that providing four parking spaces per 1,000 square feet of floor area increases the cost of constructing an office building by between $40 and $100 per square foot of office space. Thus, the high cost of free parking is an important, and often overlooked, part of both transportation finance and urban development patterns.

Public Investment

In addition to the many large private investments in urban transportation systems, public investment in U.S. transportation systems is immense. In 2009, total public (federal, state, and local) expenditures on surface transportation systems exceeded $200 billion. Table 10.5 breaks these expenditures down by mode and expenditure

TABLE 10.5. Government Expenditures on Surface Transportation, 2009 (in Millions of Dollars)

	Capital funding	Share	Maintenance/ operations funding	Share	Total	Share	Funding per passenger mile (transit) or vehicle mile (streets/highways)
Public transit[1]	$17,919[1]	15%	$37,245[2]	44%	$55,164	27%	$1.00[3]
Local streets/county roads[4]	31,419	27%	26,290	31%	57,709	29%	$0.03[5]
State/federal Highways[6]	66,612	57%	21,122	25%	87,734	44%	$0.09[5]
Total	$115,950	100%	$84,657	100%	$200,607	100%	

[1]Source: APTA 2012 Factbook, Table 44.   [4]Source: Highway Statistics 2010, Table HF 2.
[2]Source: APTA 2012 Factbook, Table 49.   [5]Calculated from Highway Statistics 2010, Table VM 202.
[3]Calculated from APTA 2012 Factbook, Table 3.   [6]Calculated from Highway Statistics 2010, Table HF 2.

type. It shows that public transit subsidies exceeded $55 billion, local streets and county roads expenditures were over $58 billion, and state and federal highways expenditures were nearly $88 billion (USDOT FHWA, 2002). This amounted to $1.00 for every transit passenger mile in the United States, and about $0.09 for every vehicle mile of highway travel in the United States.

Sources of Public Funds for Transportation

The revenue raised for transportation comes from a variety of transportation and nontransportation sources, ranging from vehicle registration and weight fees to property and sales taxes. When transportation revenue sources are used to fund nontransportation purposes (like parks or schools), or when revenues from general instruments of taxation (like sales or income taxes) are used for transportation purposes, transportation is part of larger programs of public finance. But when revenues from transportation sources are used for transportation purposes, the finance system can be viewed as one of indirect user fees, rather than taxes (see Table 10.6). These distinctions are important because highway and, later, public transit interests for decades

TABLE 10.6. User Fees and General Taxes in Transportation Finance

	Expenditures	
Revenues	Transportation purposes	Nontransportation purposes
Transportation sources	*Transportation user fees* • Motor fuel taxes for highways and transit service • Transit fares • Bridge tolls to retire bridge bonds	*Transportation taxes for general purposes* • Fuel taxes for "deficit reduction" • Parking meter revenue to fund libraries
Nontransportation sources	*General taxes for transportation* • Sales taxes dedicated to transportation • General obligation bonds for transportation	*General taxes for general purposes* • Income taxes for education, health care, and national defense

successfully lobbied Congress and most states to create special accounts, or trust funds, into which transportation revenues are deposited and out of which transportation programs and projects are funded. These separate accounts brought considerable stability to the finance of highways and public transit and for decades reflected a user fee approach to transportation finance. In recent years, however, the philosophical commitment to user fee finance and trust funds has eroded considerably.

User Fees for Transportation Purposes

Two types of user fees are common in transportation finance. Motor fuel taxes are by far the most common, while direct road pricing in the form of tolls is common in some places and unknown in others. New forms of electronic roadway tolling are favored by many economists, but have to date proven politically challenging.

FUEL TAXES

Since the 1920s, the principal user fee by which revenues have been raised for the construction and maintenance of U.S. highways (and later public transit systems) has been the motor fuels tax. The motor fuels tax is unique in many respects. Because drivers of motor vehicles impose costs on the transportation network that, to a certain extent, are proportional to their use of fuel, the motor fuels tax has been considered a transportation user fee since its inception in the 1920s. Those who drive more pay more, and those who drive large, fuel-guzzling vehicles pay more as well. During the Great Depression in the 1930s, and then again beginning in the 1990s, the federal government used federal fuel tax revenues for nontransportation purposes, but for the most part fuel tax revenues finance transportation.

As an instrument of taxation, the motor fuels tax has much to recommend it fiscally, politically, and administratively. First, as motor fuel consumption soared during the 20th century, so did tax proceeds. The motor fuels tax can be a phenomenal revenue producer, yielding $34.8 billion in the United States in fiscal year 2011 (USDOT FHWA, 2012). Second, the tax is paid in relatively small increments; the current average levy of $0.391 per gallon ($0.184 federal levy plus an average state levy of $0.207) is relatively hidden in the price of motor fuel (USDOT FHWA, 2011). This particular feature of the tax has tended to minimize organized public opposition to it, though other taxes collected in even smaller increments (like sales taxes) are eclipsing the fuel tax in popularity. Finally, the tax is, both from the taxpayer's and the government's point of view, easy to administer and collect. The gasoline tax is collected from gasoline distributors rather than directly from retailers or consumers, which serves to minimize the opportunities for gas tax evasion and to reduce the cost of collection to an historical average of only 0.5% of tax proceeds (National Surface Transportation Infrastructure Financing Commission, 2009).

As population, personal travel, and, especially, vehicle use have increased dramatically in recent years, the motor fuels tax has gradually faltered. Four factors have combined to make it difficult for fuels taxes to keep up with expanding needs: (1) increasing vehicle fuel efficiency, (2) increasing transportation program commitments, (3) the fact that per-gallon fuel tax revenues do not increase with inflation, and (4) a waxing perception that it is not a fee but a tax, and an unpopular one at that (Taylor, 1995; Transportation Research Board, 2006).

First, automobile fuel efficiency has increased significantly over the past 40 years. Table 10.7 shows that for each gallon of fuel consumed, motor vehicles traveled 46% farther in 2010 than they did in 1970. And since most fuel taxes are levied on a per-gallon basis, fuel tax revenues cannot keep pace with the growth in travel without regular increases in the per gallon levy (USDOT BTS, 2002). While improvements in vehicle fuel efficiency have undoubtedly benefited the environment, they have also substantially reduced fuel tax revenues per mile driven. The gradual conversion of the automobile fleet to hybrids, alternative fuel vehicles, and electric power (discussed in Chapter 12) further threaten fuel tax revenues. Alternative fuel and electric powered vehicles use roadways to the same extent as traditional gasoline- and diesel-powered vehicles, but they do not produce fuel tax revenues (Rufolo & Bertini, 2003; Transportation Research Board, 2006; Wachs, 2003).

Second, the demands on faltering motor fuel tax revenues have increased as increasing safety standards and environmental safeguards have been added to highway projects, and new types of projects (public transit, bicycle, etc.) have been made eligible for fuel tax funding. This so-called program creep has helped to increase the demands on faltering motor fuel tax revenues (Brown et al., 1999).

Third, inflation has diminished the purchasing power of the motor fuels tax. Because the cost of materials used in transportation projects and the cost of land for transportation facilities have tended to rise faster than the general rate of inflation, the buying power of fuel tax revenues has eroded even faster than the inflation rate would suggest. Many taxes, such as sales, property, and income taxes, are able to maintain their productivity in the face of inflation because the tax base rises with inflation. Motor fuel taxes, however, are generally levied on a per-gallon basis, and thus their proceeds do not increase in response to inflation (Sorensen, Ecola, & Wachs, 2012; Transportation Research Board, 2006).

TABLE 10.7. Trend in Average U.S. Vehicle Fleet Fuel Economy since 1970

	1970	1980	1990	2000	2010
Million vehicle miles traveled[1]	1,109,724[2]	1,527,295	2,144,362	2,746,925	2,966,506
Million gallons fuel consumed[3]	92,967	115,538	132,918	162,595	170,776
Average fuel economy (mpg)	11.9	13.2	16.1	16.9	17.4
10-year fuel increase (mpg)	—	1.3	2.9	0.8	0.5
10-year change (%)	—	11%	22%	5%	3%
Increase since 1970 (mpg)	—	1.3	4.2	5.0	5.4
Change since 1970 (mpg)	—	11%	35%	42%	46%

[1]Source: Highway Statistics 2010, Table VM 202
[2]Source: Highway Statistics 2008, Table VM 203
[3]Source: Highway Statistics 2010, Table MF 202

Fourth, to keep pace with rising costs, gas and diesel fuel taxes must be increased periodically by act of a legislature and approval of a governor or the president. Despite public concern over congestion and, to a lesser extent, deteriorating transportation infrastructure, achieving the political consensus necessary to raise fuel taxes has become increasingly difficult. User fee or not, changes to fuel tax levies have been central to many partisan debates over tax increases since the so-called tax revolts of the 1970s. As such, legislators have become increasingly wary of potential voter hostility toward tax increases of any sort, and have been reluctant to accept regular increases in the motor fuels tax levies to keep pace with increasing fuel efficiency and inflation, especially given voter discontent with rising and volatile petroleum prices.

The result of these four factors is a widening gap since the 1970s between transportation finance revenues and transportation construction and maintenance needs. Legislators have responded by enacting periodic stopgap revenue enhancement measures, such as the transportation projects in the federal stimulus bill in 2009, but no meaningful structural reforms have yet emerged to stabilize the transportation finance system. Combined state and federal fuel taxes have increased somewhat since the early 1980s, but overall the tax increases have failed to keep pace with the combined effects of inflation, increasing vehicle fuel efficiency, and new program responsibilities. While some states have occasionally increased their fuel tax rates, the federal fuel tax was last increased in 1993 to $0.184 per gallon. Figure 10.2 shows that to return the buying power of the 2010 fuel tax to its 1961 level, the combined state and federal gas tax would have to more than double to nearly $1.00 per gallon.

Some fixes for the eroding buying power of fuel taxes have been proposed and have been tried in a few states. One is to index the per-gallon fuel tax to inflation,

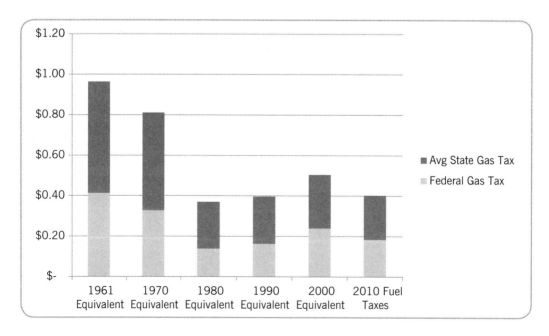

FIGURE 10.2. Changes in per-gallon fuel taxes required in 2010 to restore inflation-adjusted revenues per vehicle mile of travel to level of prior decades.

highway construction and maintenance costs, or even programmatic needs. Thus, during inflationary times, the per- gallon rate would need to be adjusted up regularly to keep pace with increasing costs. Indexing has been tried in some states, but because the tax rate tends to increase at precisely the times of high fuel prices, indexing has often proven politically unpopular and some states have subsequently dropped it.

In the late 2000s, three separate bipartisan commissions conducted studies of transportation finance in the United States, and concluded with calls for fuel tax increases (Bipartisan Policy Center, 2009; National Surface Transportation Infrastructure Financing Commission, 2009; National Surface Transportation Policy and Revenue Study Commission, 2007). But this chorus of calls for increases in fuel taxes to rebuild crumbling bridges and highways and increase road and public transit capacity has largely fallen on deaf ears. Elected officials, on both sides of the aisle, are ever more loathe to increase fuel taxes, and have looked to other finance mechanisms, including tolls, privatization, and increasingly general taxes for transportation. Beginning in 2008, fuel tax revenues no longer covered federal surface transportation program obligations, and $8 billion was transferred from the general fund to cover the shortfall. This was, at the time, argued to be part of federal deficit "stimulus" spending in response to the Great Recession. But the general fund transfers became necessary again in 2009 ($7 billion), 2010 ($13.6 billion), and 2011 ($6.2 billion), further eroding the user fee/trust fund logic of federal transportation finance (USDOT FHWA, 2016).

Despite their erosion, state and federal motor fuels taxes still account for 64% of all transportation-related revenue sources (including motor fuel taxes, registration fees, weight fees, tolls, and driver's license fees) (Transportation Research Board, 2006). Taken as a whole, transportation-related revenue sources accounted for 78% of U.S. revenues for highways in 2004; in 1961, when inflation-adjusted revenues per vehicle mile were much higher, the figure was 84% (Transportation Research Board, 2006).

TOLLING

Fees imposed on users in proportion to the costs users impose on society are typically the finance mechanisms that will help optimize transportation system performance. User fees raise people's awareness of the costs of travel (in the form of wear and tear on the system, delay imposed on others, environmental damage, etc.). Such information encourages drivers to shift low-priority trips to less socially costly times of day, routes, modes, or destinations.

Most transportation economists agree that transportation finance programs should, as much as possible, charge users the *marginal social cost* of travel (Small et al., 1989; Small & Verhoef, 2007). The term *marginal* refers to the cost of providing for one additional trip, given that others are already using the system at the same time. For example, when a car gets on the freeway, it takes up space that other motor vehicles can no longer occupy, it imposes some delay on vehicles upstream, and it also causes some amount of pavement damage. If there are very few vehicles already on the freeway, then the cost of providing for that one additional car is very small. On the other hand, if the freeway is crowded, one additional vehicle can slow other cars upstream and increase congestion to a surprising degree. In such cases, the marginal cost of accommodating an additional car is large. The term *social* refers to the costs that society pays for accommodating that one additional vehicle. These social costs result mostly from congestion, pollution, noise, and road wear and tear from a trip.

A large body of research shows that the current transportation finance programs do not make users pay the marginal social cost of vehicle use (Bipartisan Policy Center, 2009; National Surface Transportation Infrastructure Financing Commission, 2009; National Surface Transportation Policy and Revenue Study Commission, 2007; Transportation Research Board, 2006). Yet as the role of the motor fuel tax has declined relative to non-transportation-related finance instruments like sales taxes and bonds, we are actually moving further away from marginal social cost pricing of transportation (Goldman & Wachs, 2003).

The current transportation finance program generally charges users according to the "average," and not the marginal, cost of using the transportation system. That is, when a vehicle gets on the highway and contributes to congestion or an overloaded truck damages a roadway, the "cost" of that delay or road damage is paid by *everyone* experiencing the delay or rough road. The cost is averaged out so that individual users are not made aware of the total costs they are individually imposing on everyone else, and thus have little incentive to alter their travel choices (Pozdena, 1995).

A commonly posed argument against marginal social cost pricing of transportation is that poor people will simply be priced off roads and transit vehicles, leaving uncrowded, free-flowing systems for the wealthy. Such social equity concerns are indeed important, but they ignore the social inequities of our current transportation finance system. The two most common equity principles in public finance are the benefit principle and the ability-to-pay principle (Musgrave, 1959). Under the benefit principle, equitable taxes are those levied on each individual in proportion to the benefits received by that individual. Within this rubric, charging users according to the incremental social costs they impose on society when using the transportation system is equitable. On the other hand, the ability-to-pay principle suggests that a method of finance based solely on benefits received may disproportionately burden the poor. From this perspective, an equitable finance program will treat fairly people who have different abilities to pay, with ability measured primarily by income. Current transportation user fees, like the motor fuels tax and the driver's license fee, fare well under the benefit principle but poorly under the ability-to-pay principle (Schweitzer, 2009; Transportation Research Board, 2011).

Sales taxes on consumer items are essentially unrelated to use of transportation systems. Yet in recent years many regions around the United States have turned to sales tax increases to pay for transportation improvements (Goldman & Wachs, 2003). Unfortunately, such transportation sales taxes fare poorly under *both* the benefit and ability-to-pay principles.

Thus, in comparison with our current system of transportation finance, a user fee system based on the principles of marginal cost pricing would clearly increase equity based on the benefit principle and may well increase overall equity under the ability-to-pay principle as well. Use of the highway system in congested conditions is positively correlated with income; that is, higher-income travelers tend to spend a larger share of their travel time in traffic congestion than do lower-income travelers.

Thus, a shift to a transportation finance system that charges drivers more on congested routes and less elsewhere would fare well under the ability-to-pay principle in comparison to our current finance system.

Given the many potential advantages of a marginal social cost system of transportation pricing, there have been some recent, promising experiments with road pricing in the United States, and more ambitious plans for mileage-based user fees on the drawing boards. In 1995, a four-lane facility opened in the median of State Route 91 freeway through Santa Ana Canyon in metropolitan Los Angeles (see Figure 10.3). The facility, called the "91 Express Lanes" was entirely privately financed, and is funded through electronically collected tolls. Use of the 10-mile facility (which doubled its length in 2017) allows toll-paying users to by-pass congestion on the adjacent free lanes and save up to 30 minutes per trip. The tolls vary from a low of $1.35 when the adjacent free lanes are free-flowing, to a high (in 2013) of $9.55 during times when adjacent lanes are most congested. Evaluations of the facility have found that it has increased the throughput of traffic and has decreased congestion in both the toll facility *and* in the adjacent free lanes (Mastako, Rillet, & Sullivan, 1998; Sullivan & El Harake, 1998). Over time, however, increased demand has increased congestion in the adjacent free lanes, which has meant that the peak toll has had to be increased from time to time to keep the express lanes' traffic free-flowing.

An equity analysis of the 91 Express Lanes found that, in comparison to sales tax finance, the congestion tolls shifted the finance burden off of the lowest-income residents (in relative terms) and highest-income residents (in absolute terms) and on to middle-income drivers who use the lanes and voluntarily pay the tolls (Schweitzer & Taylor, 2008)

In 1998, underutilized high-occupancy vehicle (carpool) lanes in San Diego, California, were converted to allow single-occupant vehicles to pay a toll to use the

FIGURE 10.3. State Route 91 freeway through Santa Ana Canyon in metropolitan Los Angeles. Photo by Maya Blumenberg-Taylor.

carpool lanes and by-pass congestion on the adjacent unrestricted lanes. The San Diego facility is unique because the tolls are set in "real time": they change every few minutes from $0.50 to $8.00 based on congestion levels. Following the success of these facilities, variable price projects have since opened in southeast Florida, northern Virginia, the Puget Sound region, the Twin Cities, and Houston (USDOT FHWA, 2015). The trend toward managed toll lanes continues to accelerate, with new tolling projects currently planned or underway in Atlanta, Chicago, Dallas, Denver, Indianapolis, Los Angeles, Orlando, San Francisco, and Seattle (Poole, 2012).

Collectively, these congestion-priced facilities have worked well in practice. Traffic flow is increased, congestion delay is reduced, and the people who voluntarily pay tolls to by-pass congestion report high levels of satisfaction. Even travelers in congested, adjacent facilities are generally satisfied with the congestion-priced facilities, because they tend to experience improved traffic flow in the free lanes and have the option to buy out of congestion if they wish (Mastako et al., 1998). Despite the success and local popularity of these congestion-pricing and related mileage-based user fee experiments, however, many elected officials remain wary of political backlashes to congestion pricing by drivers angry about paying for something that was formerly "free" (from the driver's perspective). So while congestion pricing has perhaps the greatest promise to reduce traffic congestion significantly in metropolitan areas, its deployment is likely to continue to be slow and gradual (Taylor & Kalauskas, 2010).

NONUSER FEES FOR TRANSPORTATION PURPOSES

Property Taxes. Except for a few private communities that own and maintain their streets and sidewalks, the vast majority of local streets and sidewalks are policed, cleaned, and maintained by local governments (whether directly or by contract). These local governments sometimes build new streets and sidewalks but often depend on private landowners to do this.

When cities and counties approve property owners' requests to rezone and subdivide property for sale, they often require that the owner make improvements to the property to suit the proposed new uses. This describes the fiscal model employed to create much of suburban America. For example, if a farm on the edge of a city is divided into residential lots on which new houses are built and sold, the city might require the developer to build a local street network to their defined standards, construct schools and parks, and dedicate some land for public uses like houses of worship. These new streets, sidewalks, and schools are then turned over to the city, which is then responsible for operating and maintaining them.

Because the principal purpose of these local street networks is private property access and not serving through traffic, the benefits of property-serving streets have long been thought to accrue more to property owners than the drivers, delivery trucks, and emergency vehicles that use them (McShane, 1995). Thus, while property taxes levied to maintain local streets are not a transportation user fee, there is a benefit–principle spatial nexus between property taxes and street maintenance.

Problems arise when streets need reconstruction and cities have not budgeted prudently in advance. This is particularly an issue during times of economic recession when tax revenues falter. Because street maintenance is rightly viewed as less pressing than, say, police and fire services, street maintenance budgets are often the first to be cut. Unfortunately, once street pavements have deteriorated to the point where water infiltrates asphalt cracks causing potholes, it is usually too late to reseal the street, and complete reconstruction is required to stop potholes from recurring. But reconstructing poorly maintained streets is far more expensive in the long run than simply sealing and maintaining them properly, so cutting street maintenance budgets can prove a very expensive way to save money in the long run.

One recent innovation to better link the cost of street operations and maintenance to the costs imposed by various land uses is the transportation utility fee. Dozens of cities around the United States (mostly in the Northwest, and particularly in Oregon) have shifted from property tax finance to transportation utility fees in recent years. Property taxes assess landowners based on the estimated value of their property, regardless of how much a given parcel generates vehicle trips. So an expensive home that generates very few vehicle trips might be assessed more in property taxes than a drive-through fast-food restaurant that generates many such trips. Instead of assessed valuation, transportation utility fees assess property owners based on the number of vehicle trips land uses of that type and size typically generate. The rationale for such fees is to increase equity by better linking the fees paid to the use of streets (Junge & Levinson, 2012).

Sales Taxes. Because it has proven so politically difficult to regularly adjust the per-gallon fuel tax levy to reflect inflation and increasing vehicle fuel efficiency, many places have turned to sales taxes to fund transportation systems. Sales tax revenues for transportation are subsidies and not user fees, because they are collected independently of transportation system use. Subsidies can come from state or local general fund monies, but the most popular trend in transportation finance is toward the imposition of local option taxes (LOTs), particularly on sales. LOTs are usually collected, not at the state level, but by localities, usually counties. They are typically approved directly by voters through ballot initiatives. LOTs generally hike the local sales tax, with the extra revenue earmarked for transportation. These incremental tax increases are small in percentage rate terms, but can raise considerable amounts of revenue since they are levied on nearly every retail transaction. Because they are paid in small increments over a very large number of transactions, they are often less visible to people than other types of taxes, making them relatively popular with voters. People typically pay property taxes twice per year in large lump sums and reconcile their income taxes once per year, but sales taxes are paid a few cents at a time over hundreds or even thousands of transactions per year, meaning that most people are unaware of their annual sales tax bill.

LOT revenues are usually dedicated to specific, local projects by law. Voters are typically presented with a package of road and/or transit spending proposals specified in the ballot measure. While LOT proposals are regularly rejected by voters, they have succeeded more often than not and are an important and growing source of transportation revenues. In most cases, LOTs "sunset," meaning they need to be periodically renewed by the voters (Goldman & Wachs, 2003).

As noted earlier, while LOTs have proven politically effective and have helped to shift the geography of transportation finance from the federal and state levels to the regions and communities that levy them. They also disconnect transportation revenues from transportation system use, thereby weakening the price signals that fuel taxes and tolls send to travelers.

Debt Finance. Governments have long borrowed money to pay for "big-ticket" items, and public borrowing for transportation projects has grown substantially in recent years. A common way for governments to borrow money is through the issuance of bonds. Bonds are essentially IOUs in which the bond issuer promises to pay the bond holder back with interest at some later date. In recent years such forms of government debt financing for transportation have been referred to as "innovative finance," but the process of government borrowing to fund pressing present needs is in fact an ancient one. In many cases debt finance has been highly successful; in others, the results have not justified expense. Common examples of bond-financed projects are schools, dams, and sewage treatment plants, which are often financed by bonds because they require large lump-sum payments up front and generate a steady stream of benefits over many generations. Typically, the heaviest reliance on bonds comes in wartime, when the present benefits of victory are deemed so worthwhile and lasting that part of the cost is billed to future generations.

While bonds have an important and long-established role in public finance, bonds are not a revenue *source,* they are a finance *technique* because, in the end, bonds must be repaid with interest from some revenue source. Because bonds raise funds for public expenditure from private capital markets, they require certain assurances. Typically, government issuers must guarantee that the bonds will be repaid from a particular source. When bonds are backed by a dedicated revenue stream, such as from fuel tax revenues, truck weight fees, or bridge toll revenues, they are called revenue bonds; when they are to be repaid from a government's general tax revenues, they are called general obligation bonds (Taylor et al., 2006).

While there can be strong justifications for borrowing money via bonds to pay for large up-front expenditures on projects that will provide many years of benefits, it is harder to justify borrowing money to pay for ongoing operations or maintenance expenditures, or to adopt excessive payback periods for capital projects. Doing so simply puts off the uncomfortable task of raising revenues or cutting expenditures by saddling future generations (who will have to pay for their own operating and maintenance costs) with debt service unconnected to any future stream of benefits. While enjoying transportation benefits now and worrying about how to pay for it later is politically tempting and increasingly common, doing so is a costly way to put off difficult fiscal decisions.

CONCLUSION: LOOKING TO THE FUTURE

Given this overview of the geography of urban transportation finance, what trends do we see as we look to the future? First, and perhaps foremost, the geopolitics of transportation finance are firmly established at all levels of government and are likely to remain so for the foreseeable future: public investment in transportation systems will continue to be determined more by political struggles over the spatial distribution of resources than by other factors such as transportation system use, costs imposed, or ability to pay.

The emphasis on transportation capital investments over spending on operations and maintenance also shows few signs of abating. While the overcapitalization of transportation systems is clearly economically inefficient, it is politically popular and is likely to remain so. Transportation investments will probably continue to emphasize projects over programs, lines and modes over integrated systems, and expenditures over savings.

Finally, the current trend away from user fees (like the motor fuels tax and distance-based transit fares) is also likely to continue, at least in the short term. Funding for transportation is likely to come increasingly from general public finance instruments like sales taxes and general obligation bonds. This trend is popular with elected officials and voters, but it is likely to reduce both the economic efficiency and social equity of transportation systems. Economic efficiency is diminished because sales taxes and general obligation bonds separate the prices travelers pay from the costs they impose on society. And social equity is reduced because the relative burden of transportation finance is shifted toward lower-income people (who pay proportionally more in sales taxes) and less extensive users of transportation systems.

Over the longer term, however, it is possible that both the introduction of new propulsion technologies and electronic payment media will begin to turn the tide back toward user fees. New propulsion technologies, like gas-electric hybrid engines, fully electric motors, natural gas engines, and fuel cells, will further erode the ability of motor fuel taxes to keep pace with the growth in vehicle travel. And smart card and smart phone payment media that remotely deduct payments as travelers pass through fare gates or under payment gantries on roads are becoming increasingly widespread. We are witnessing a rapid expansion of electronic payment mechanisms across sectors, and these developments could usher in a new era of easy-to-pay and inexpensive-to-collect transportation user charges to replace the fuel, sales, and other taxes we currently use to finance transportation. Such payment systems could allow marginal cost pricing of roads, transit services, and so on to reflect the costs individual travelers impose on transportation systems. This development, in turn, would encourage travelers to use transportation systems in places and at times when excess capacity exists and would signal travelers to be more judicious in their consumption of roads and transit seats in places and at times of peak demand. Such changes would both increase economic efficiency and social equity of transportation finance, but widespread implementation of such user charge systems remains in developmental stages.

In this chapter we have "followed the money" to argue that the political geography of transportation finance has enormous influence on the deployment of urban transportation systems, travel in cities, and patterns of metropolitan development to show that the geography of urban transportation is to a large degree determined by the geography of urban transportation finance.

ACKNOWLEDGMENT

UCLA Urban Planning PhD student Carole Turley Voulgaris assisted with the preparation of this chapter.

REFERENCES

Bernstein, C., & Woodward, B. (1974). *All the president's men.* New York: Simon & Schuster.

Bipartisan Policy Center. (2009). *Performance driven: A new vision for U.S. transportation policy.* Washington, DC: Bipartisan Policy Center National Transportation Policy Project.

Boarnet, M. G. (1997). Highways and economic productivity: Interpreting recent evidence. *Journal of Planning Literature, 11*(4), 476–486.

Brown, J., DiFrancia, M., Hill, M. C., Law, P., Olson, J., Taylor, B. D., et al. (1999). *The future of California highway finance.* Berkeley: California Policy Research Center.

Cooper, D., & Griffith, J. (2012). *Highway robbery: How Congress put politics before need in federal highway and transit funding.* Washington, DC: Center for American Progress.

Federal Transit Administration. (2012). National transit database. Retrieved from *www.ntdprogram.gov/ntdprogram/data.htm.*

Garrett, M., & Taylor, B. D. (1999). Reconsidering social equity in public transit. *Berkeley Planning Journal, 13,* 6–27.

Gillen, D. (1997, December). *Evaluating economic benefits of the transportation system.* Paper presented at the meeting of Transportation and the Economy: The Transportation/Land Use/Air Quality Connection, Lake Arrowhead, CA.

Goldman, T., & Wachs, M. (2003). A quiet revolution in transportation finance: The rise of local option transportation taxes. *Transportation Quarterly, 57*(1), 19–32.

Holtz-Eakin, D., & Wachs, M. (2011). *Strengthening connections between transportation investments and economic growth.* Washington, DC: Bipartisan Policy Center National Transportation Policy Project.

Junge, J. R., & Levinson, D. (2012). Prospects for transportation utility fees. *Journal of Transport and Land Use, 5*(1), 33–47.

Lem, L. L. (1997). Dividing the federal pie. *Access, 10,* 10–14.

Lewis, D. (1991). *Primer on transportation, productivity, and economic development* (National Cooperative Highway Research Program Report 342). Washington, DC: Transportation Research Board.

Mastako, K. A., Rillet, L. R., & Sullivan, E. C. (1998). Commuter behavior on California State Route 91 after introducing variable-toll express lanes. *Transportation Research Record: Journal of the Transportation Research Board,* No. 1649, 47–54.

McShane, C. (1995). *Down the asphalt path: The automobile and the American city.* New York: Columbia University Press.

Metropolitan Council. (2012). Central corridor light rail transit: Project profile and facts. Retrieved from *https://metrocouncil.org/getattachment/28c5ddf9-b959-45f7-b3e3-f379ceef7d44/CC-ProjectFacts-pdf.aspx.*

Metropolitan Transportation Commission. (2012). 2011 annual report. Retrieved from *http://files.mtc.ca.gov/library/pub/4943_2011.pdf.*

Musgrave, R. A. (1959). *The theory of public finance: A study in public economy.* New York: McGraw-Hill.

National Surface Transportation Infrastructure Financing Commission. (2009). *Paying our way: A new framework for transportation finance.* Washington, DC: National Surface Transportation Infrastructure Financing Commission.

National Surface Transportation Policy and Revenue Study Commission. (2007). *Transportation for tomorrow.* Washington, DC: National Surface Transportation Policy and Revenue Study Commission.

O'Toole, R. (2010). *Fixing transit: The case for privatization* (Policy Analysis No. 670). Washington, DC: Cato Institute.

Poole, Jr., R. W. (2001). *Commercializing highways: A "road-utility" paradigm for the 21st century.* Los Angeles: Reason Foundation.

Poole, Jr., R. W. (2012). Surface transportation. In L. Gilroy & H. Kenny (Eds.), *Annual privatization report 2011.* Los Angeles: Reason Foundation.

Pozdena, R. J. (1995). *Where the rubber meets the road: Reforming California's roadway system.* Los Angeles: Reason Foundation.

Richmond, J. (2005). *Transport of delight: The mythical conception of rail transit in Los Angeles.* Akron, OH: University of Akron Press.

Roth, G. (1996). *Roads in a market economy.* London: Avebury Technical.

Rufolo, A. M., & Bertini, R. L. (2003). Designing alternatives to state motor fuel taxes. *Transportation Quarterly, 57*(1), 33–46.

Schank, J., & Rudnick-Thorpe, N. (2011, January). *End of the highway trust fund?: Long-term options for funding federal surface transportation.* Paper presented at the 90th annual meeting of the Transportation Research Board, Washington, DC.

Schweitzer, L. (2009). The empirical research on the social equity of gas taxes, emissions fees, and congestion charges. In *Special Report 303: Equity of evolving transportation finance mechanisms.* Washington, DC: Transportation Research Board.

Schweitzer, L., & Taylor, B. D. (2008). Just pricing: The distributional effects of congestion pricing and sales taxes. *Transportation, 35*(6), 797–812.

Shoup, D. (2005). *Parking cash out.* Chicago: Planning Advisory Service.

Shoup, D. (2011). *The high cost of free parking.* Chicago: Planners Press.

Small, K. A., & Verhoef, E. (2007). *The economics of urban transportation.* New York: Routledge.

Small, K. A., Winston, C., & Evans, C. (1989). *Road work: A new highway pricing and investment policy.* Washington, DC: Brookings Institution Press.

Sorensen, P., Ecola, L., & Wachs, M. (2012). *Mileage-based user fees for transportation funding: A primer for state and local decisionmakers.* Santa Monica, CA: RAND Corporation.

Straus, S. H., & Semmens, J. (2006). *Estimating the cost of overweight vehicle travel on Arizona highways.* Phoenix: Arizona Department of Transportation.

Sullivan, E. C., & El Harake, J. (1998). California Route 91 toll lanes—Impacts and other observations. *Transportation Research Record: Journal of the Transportation Research Board,* No. 1649, 55–62.

Taylor, B. D. (1995). Program performance versus transit performance: Explanation for the ineffectiveness of performance-based transit subsidy programs. *Transportation Research Record: Journal of the Transportation Research Board,* No. 1496, 43–51.

Taylor, B. D., & Kalauskas, R. (2010). Addressing equity in political debates over road pricing: Lessons from recent projects. *Transportation Research Record: Journal of the Transportation Research Board,* No. 2187, 44–52.

Taylor, B. D., & Norton, A. T. (2009). Paying for transportation: What's a fair price? *Journal of Planning Literature, 24*(1), 22–36.

Taylor, B. D., Wachs, M., McCullough, B., Morris, E., Evans, A., Smirti, M., et al. (2006). *Transportation pricing and finance options for California.* Los Angeles: California Department of Transportation.

Transportation Research Board. (2006). *Special report 285: The fuel tax and alternatives for transportation funding.* Washington, DC: Transportation Research Board.

Transportation Research Board. (2011). *Special report 303: Equity of evolving transportation finance mechanisms.* Washington, DC: Transportation Research Board.

U.S. Department of Transportation, Bureau of Transportation Statistics. (2002). *National Transportation Statistics: 2001.* Washington, DC: U.S. Government Printing Office.

U.S. Department of Transportation, Bureau of Transportation Statistics. (2013). *Pocket guide to transportation: 2013.* Washington, DC: U.S. Department of Transportation Research and Innovative Technology Administration.

U.S. Department of Transportation, Federal Highway Administration. (2002). Highway statistics 2001: Total disbursements for highways, all units of government. Retrieved from *www.fhwa. dot.gov/ohim/hs01/hf2.htm.*

U.S. Department of Transportation, Federal Highway Administration. (2009). 2009 National Household Travel Survey. Retrieved from *http://nhts.ornl.gov.*

U.S. Department of Transportation, Federal Highway Administration. (2011). Highway statistics 2010. Retrieved from *www.fhwa.dot.gov/policyinformation/statistics/2010.*

U.S. Department of Transportation, Federal Highway Administration. (2012). Highway statistics 2011. Retrieved from *www.fhwa.dot.gov/policyinformation/statistics/2011.*

U.S. Department of Transportation, Federal Highway Administration. (2015). *Congestion pricing—A primer: Evolution of second generation pricing projects* (FHWA-HOP-15-036). Washington, DC: Author. Retrieved from *http://ops.fhwa.dot.gov/publications/fhwa-hop15036/index.htm#toc.*

U.S. Department of Transportation, Federal Highway Administration. (2016). Status of the Highway Trust Fund, Table FE-1. Retrieved from *www.fhwa.dot.gov/highwaytrustfund.*

United States Senate Democrats. (2012). Senators Boxer and Johnson call on House to pass Bipartisan Senate Transportation Jobs Bill. Retrieved from *http://democrats.senate.gov/2012/03/16/senators-boxer-and-johnson-call-on-house-to-pass-bipartisan-senate-transportation-jobs-bill.*

Valley Metro. (2012). U.S. Transportation Secretary LaHood announces $75 million to help fund light rail extension. Retrieved from *www.valleymetro.org/pressreleases/detail/u.s._secretary_lahood_announces_75_million_will_go_to_help_fund_3.1_mile_li.*

Wachs, M. (2003). Commentary: A dozen reasons to raise the gas tax. *Public Works Management and Policy, 7*(4), 235–242.

CHAPTER 11

Transportation and Environmental Impacts and Policy

SCOTT LE VINE
MARTIN LEE-GOSSELIN

Movements of people and freight interact with the environment in a myriad of complex ways. It is a two-way relationship, with transportation causing external impacts, and environmental conditions affecting how transportation systems are conceived, built, and used.

The planning, design, and operations of transportation infrastructure must be sensitive to the physical environment, which varies greatly from place to place and season to season. Transport networks simply cannot function properly without addressing the requirements that the local environment (geological, topographical, biological, human, meteorological, etc.) impose on construction or operations. But beyond these inherent physical realities, transportation planning must also follow national, regional, and local environmental regulations. In high-income countries, this regulatory framework has taken shape since the 1960s through a combination of legislation, rule making by executive agencies, court decisions, and standard professional practices, and it continues to evolve today. Environmental regulation has led to major changes in the transportation sector. While maintaining environmental quality undoubtedly entails costs, it also yields great benefits. As we shall see, much of the environment today is cleaner than in decades past.

MULTIPLE POINTS OF INTERSECTION
BETWEEN TRANSPORTATION AND THE ENVIRONMENT

The infrastructure-project development process is only one way in which transportation and the environment interact. Households and businesses decide where to locate in part on the basis of accessibility. Their locational decisions structure the type and amount of day-to-day travel required, which in turn determines the level of environmental and public health impacts from vehicle use. But, while accessibility is attractive to people and businesses, being directly next to major transport infrastructure often is not, for a variety of reasons that will be discussed in this chapter.

The discussion in this chapter focuses principally on road transportation, which is the dominant form of travel in the United States and many other countries. We will review in lesser detail the environmental issues and mitigation strategies associated with aviation, mass transit, and seaports.

ENVIRONMENTAL CHALLENGES AT A RANGE OF SPATIAL SCALES

The direct impacts of the construction, maintenance, and use of transport infrastructure—including roads, parking facilities, railways, airports, canals, and maritime ports—are inescapably enormous. Indeed, road networks comprise the largest human artifact on the planet (Forman et al., 2003). There are 4 million miles of public roads in the United States alone, enough for eight round-trips to the moon.

Environmental impacts start with the construction phase of transportation facilities, in which the immediately surrounding landscape is thoroughly transformed: habitats are fragmented, noise levels from machinery and blasting are extreme, watercourses and wetlands are altered permanently, new opportunities are created for invasive species, air quality is impaired (especially from particulate matter), and sudden changes in artificial illumination may disrupt the natural cycle of light and dark in the vicinity of many transportation facilities.

Transport infrastructure also has indirect environmental consequences. For instance, road improvements may increase market pressures to allow residential and commercial sprawl (decentralized suburban development), with potentially major impacts through the loss of green space.

When considering environmental issues, policymakers must bear in mind impacts at a variety of spatial scales, which can be in tension with one another. For example, road infrastructure investments to ease stop-and-go traffic conditions may improve local air quality conditions, but at the same time induce traffic, thereby increasing greenhouse gas emissions. As we shall see later in this chapter, many of the available measures to mitigate specific environmental problems have benefits and drawbacks, and their effectiveness can be context-dependent.

EMERGENCE OF THE ENVIRONMENT AS A PUBLIC ISSUE

Humans have shaped their external environment for millennia, and the undesirable environmental effects of human activity have long been recognized. The stench and smokiness of the air in Rome was noted two millennia ago. In the year 1272, King Edward I prohibited the burning of certain types of coal in England. Some 500 years later, in the early 19th century, "noxious industry" was banned from the financial district of central London. An early transportation-related example is the hot ash that coal-powered steam trains were known to deposit on neighboring properties, occasionally starting fires. By the early 20th century environmental conditions led cities to physically separate land uses deemed to be incompatible, such as housing and heavy industry, which has had major impacts on urban geography.

Many major transportation facilities that date from the mid-20th century, urban freeways in particular, are today regarded as insufficiently sensitive to their surroundings. The first generation of freeway builders genuinely believed, however, they were improving the local environment by removing heavy traffic flows from the surface street network and by facilitating access to the green spaces of the urban hinterlands.

By the end of the 1960s—as the first of the baby boomers were coming of age—environmental quality had become a mainstream concern in developed countries, and the effects of transportation were front and center. Freeway revolts occurred as residents objected to new highways that were planned through dense urban areas, with neighborhood destruction and environmental degradation common themes of the antifreeway activists.

A reported 20 million Americans took part in the first Earth Day events in 1970. In that year, 53% of Americans said pollution should be among the "top three problems the government should devote most of its attention to in the next year or so," compared to 17% five years earlier (Norton, 1991). The first major international summit on environmental quality took place in Stockholm, Sweden, in 1972.

But during the late 1970s, environmental protection lost its preeminent position on the list of the public's priorities. By 1980, the year conservative Republican Ronald Reagan was first elected president, only 24% of Americans saw the environment as a top-three issue for government.

From the 1980s through the end of the 20th century, a majority of Americans prioritized environmental protection over economic growth, though this fell sharply in the early 2000s. The post-2007 period of economic recession and slow recovery has also affected the public's views, as can be seen in Figure 11.1. The latest data show that Americans once again tend to say that environmental protection should be prioritized over economic growth. Despite these shifting views, majority opinion in the United States (62% in 2010) remains that the environmental movement has done more good than harm overall (Gallup, 2013).

Sensitivity to environmental issues varies among different groups, with better-educated people, younger people, women, and people living in urban areas more likely to support environmental protection. On the political spectrum, liberals (in the American sense of those to the left of center) tend to be more environmentally sensitive than conservatives, and environmental groups are frequently on the opposite side of policy debates from the "business lobby," such as chambers of commerce and industry associations. Not everyone agrees, however, that inherent trade-offs exist between economic development and environmental protection. Environmental damage from transportation can impose costs (monetary or otherwise) that are not priced in markets. In many cases regulation can be more economically efficient than not taking action, though this outcome is context-dependent. Some proponents of strong environmental policies also believe that clear and consistent regulations provide firms with greater certainty about future business conditions and can nurture the emergence of technologically advanced "green" economic sectors.

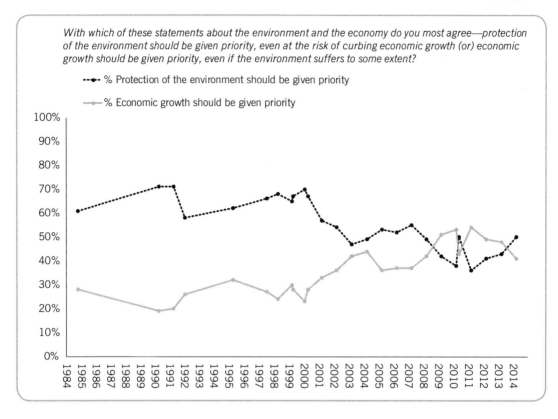

With which of these statements about the environment and the economy do you most agree—protection of the environment should be given priority, even at the risk of curbing economic growth (or) economic growth should be given priority, even if the environment suffers to some extent?

--●-- % Protection of the environment should be given priority

——— % Economic growth should be given priority

FIGURE 11.1. Preferences for environmental protection versus economic growth. Copyright © 2013 Gallup Inc. All rights reserved. The content is used with permission; however, Gallup retains all rights of republication.

Caution must be exercised when studying trends in public awareness, as environmental issues themselves have evolved over time. As we shall see, much progress has been made in addressing the first generation of environmental problems linked with transportation, such as urban air pollution, while new issues have emerged that present very different public policy challenges. Climate change, for instance, has increasingly become a focus of concern in the 21st century (see Chapter 12).

The reasons for the downward trend in environmental sensitivity since the early 2000s are not clear. One interpretation is that the success of earlier policies at addressing high-visibility environmental problems (e.g., urban smog) may have led people to prioritize other political issues. The emergence of climate change as the main environmental issue on the public agenda may also explain part of this recent trend, as policies to address it would arguably require more substantial changes to the lifestyles of individual people (North Americans in particular) than those necessitated by earlier environmental problems such as depletion of the high-level layer of ozone in the atmosphere.

It may also be that as stewardship of environmental quality has become institutionalized, it has receded as a frontline political issue in the public's mind. Today many public and private organizations, especially large institutions, are incorporating sustainability principles into their day-to-day working practices and their strategic planning. Sustainability is based on the notion that future generations have the same rights to health, prosperity, and a clean environment as people alive today. The U.S. Environmental Protection Agency (2011) formally defines it as follows:

> Sustainability is based on a simple principle: Everything that we need for our survival and well-being depends, either directly or indirectly, on our natural environment. Sustainability creates and maintains the conditions under which humans and nature can exist in productive harmony, permitting fulfilling the social, economic, and other requirements of present and future generations.

Participating publicly in the sustainability agenda means that people and institutions are committed to taking actions to minimize environmental harm, though the degree to which such commitments affect decision making is unclear (Millard-Ball, 2013).

As we shall see in the next section, the peak in environmental awareness in the early 1970s left an important legacy; the cornerstones of the environmental-protection system in place today were laid nearly a half-century ago, and the changes since then have been mostly incremental.

REGULATORY FRAMEWORK

Environmental laws vary from country to country. Many countries, particularly developed ones, have environmental review processes that are based on the principle that actions with potentially significant environmental impacts must be subject to official review. In India (as well as other countries) environmental protection is enshrined in the nation's constitution. The discussion in the remainder of this section focuses on the regulatory framework in the United States, which is relatively well developed by international standards. We begin with the landmark federal environmental legislation that took shape in the 1960s and 1970s, alongside the development of public concern about the environment.

National Environmental Policy Act

The National Environmental Policy Act (NEPA; 1970) is the centerpiece of environmental legislation in the United States—it is a "statute of productive dimensions because it adjusts the relationships of the body politic" (Yost, 2003, p. ii). The legislation set the groundwork for environmental review processes by directing federal agencies to follow standardized procedures for all actions they perform, fund, or actively enable, even if ultimately performed by others. In general, NEPA requirements apply to discretionary actions, meaning that operations and maintenance (including major rebuilding projects) do *not* trigger NEPA processes. A rebuilding project in which vehicle or passenger capacity is significantly expanded, however, would in principle require a NEPA evaluation.

NEPA requires federal agencies to follow a detailed process when making decisions that affect the environment. It imposes rigor on public-sector decision making by requiring actions to follow from clearly defined goals and transparent selection of the preferred alternative from the set of feasible options. The formal requirements of the NEPA process are shown in Figure 11.2. Potentially significant environmental impacts (damage) must be identified early on and then thoroughly considered in the review process. If "significant" impacts are found in the course of the review, the agency must show evidence that maximum feasible measures to mitigate the impacts are taken. A NEPA process must include a rigorous alternatives-analysis stage, which documents that alternative courses of action to achieve the agency's objectives were studied and considered in good faith. Crucially, NEPA requires engagement with outside stakeholders and interested members of the public throughout the process, including the opportunity to provide formal feedback in the form of comments that become part of the public record of the NEPA process.

Funding from the federal Department of Transportation (USDOT) triggers NEPA review for most urban transportation projects in the United States. In other circumstances the role of USDOT is operational: for instance, the USDOT's Federal Aviation Administration (FAA) prepared a formal plan in 2011 documenting how it will comply with NEPA's requirements for its NextGen project to upgrade the nation's air traffic management system. In 2011, the most recent data available, USDOT filed 237 Environmental Assessment and 20 Environmental Impact Statement documents.

Beyond USDOT, a range of federal agencies may take actions that affect the transportation network and its relationship with the natural and built environment. Almost 10% of the United States' public roads are overseen by the U.S. Forest Service (Forman et al., 2003), and the U.S. Army Corps of Engineers regulates transportation projects that affect wetlands or watercourses.

An agency decision is not the final word, however, as NEPA processes are frequently the subject of lawsuits. Disagreement with the *outcome* of a NEPA process is typically not a strong legal argument, if the decision-making agency is able to show that the steps mandated by NEPA were faithfully followed. Rather, environmental review processes instead are typically challenged in court for failing to adhere to NEPA's procedural requirements (see Box 11.1).

The distinction between actions that involve the federal government and those that do not is fundamental to understanding how NEPA works. NEPA does not apply *at all* to projects without a link to a federal action; such projects are those carried out by state and local levels of government or the private sector that do not require money or specific permission from a federal agency. Many states, however, have laws inspired by NEPA that apply to actions by state and local government agencies.

A key feature of NEPA is that it requires specific processes to be followed when a NEPA process is triggered, but following the NEPA processes does not in itself prohibit environmental damage. The project sponsor must show that the mandated processes have been followed. But if the appropriate process has been performed, the sponsoring agency may still ultimately choose to take actions that harm the environment.

NEPA is a relatively brief piece of legislation; it consists of broad guidance that was later codified into a set of regulations. Because NEPA covers federal actions only (but see Box 11.2), it does not at all address many important actions in the transportation sector. These include the activities of private firms and ways that consumers use the transport system. For instance, the rapidly increasing popularity of sport utility

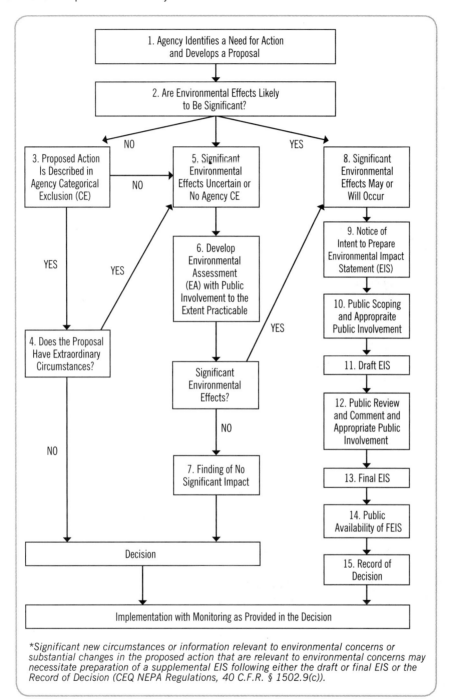

FIGURE 11.2. Overview of NEPA-mandated environmental review process for actions involving federal agencies.

BOX 11.1

NEPA, the Courts, and Westway

Westway was an interstate highway planned in the 1970s on the west side of Manhattan, New York City, along the then-derelict Hudson River piers. Pursuant to NEPA, an environmental impact statement (EIS) was prepared by the New York State Department of Transportation and USDOT. The EIS considered alternatives and concluded in favor of the proposed highway facility.

Westway had local and federal political support. It received a required landfill permit from the Army Corps of Engineers and was allocated funds by USDOT (construction was estimated to cost $1.2 billion in 1977 dollars).

Interstate highway construction in urban areas had by this time become contentious. Beginning with the publication of the Draft EIS (in 1974) Westway was challenged in court in several ways. Initial challenges to the validity of the traffic and air quality analyses resulted in modifications in the Final EIS, which was published in 1977.

In 1982, however, the U.S. District Court found that the EIS was, despite being adequate in other respects, deficient in failing to disclose impacts Westway would have on the breeding grounds for the striped bass, a commercially fished species.

On the basis of the failure to disclose this significant impact, the EIS was rejected; had the EIS process acknowledged the impact on striped bass breeding areas, the claim would have been dismissed. Though there was no legal barrier to the project sponsors undertaking the NEPA process again, funding priorities changed and this did not happen. Westway was not built as an interstate highway, though the right of way was eventually used for a medium-capacity urban boulevard that exists today.

BOX 11.2

State- and Local-Level Environmental Review

Sixteen U.S. states (and some municipalities, including New York City) have "Little NEPA" laws (and others have requirements enacted through executive order rather than legislation). California has consistently been a leader in enacting such laws—the California Environmental Quality Act (CEQA) became law months after the federal NEPA in 1970. CEQA employs slightly different definitions from NEPA, but shares similar aims and codified processes. Indeed, it is possible to satisfy the requirements of both NEPA and CEQA through joint review processes such as an EIS/EIR (a single process that produces a NEPA Environmental Impact Statement and a CEQA Environmental Impact Report)—this was done for segments of California's planned high-speed rail system. In New York City, a project requiring discretionary actions by federal, state, and city agencies is subject to environmental review processes incorporating requirements from all three levels of government.

As with NEPA itself, state-level environmental regulation that would affect transport projects is the subject of contemporary political debate (IPANM, 2013).

vehicles in the 1990s, which had major implications on energy consumption, was never reviewable under NEPA, nor are any of the emerging technologically enabled transportation services such as carsharing, ridesharing, and private-hire taxi apps.

The NEPA review process tends to lengthen the time it takes to deliver transportation infrastructure. In addition to the stages for formal public input, opponents can challenge the way the process has been implemented, which can delay a project and in extreme cases lead to its cancellation. Supporters of a transportation project may therefore attempt to downplay its likely impacts in order to advance it through environmental review. Opponents, by contrast, may then claim that the project will have impacts that the review process did not properly consider, and that the process is therefore faulty. Even the threat of a lawsuit may be enough to incentivize supporters to compromise with opponents by making modifications to the project, such as reducing the planned number of lanes of a new road.

Environmental impact statements (EISs) have become much more complex and lengthy. Over time the average EIS document for highway projects has grown from 22 pages to 1,000 pages, and the duration of the process has lengthened from 2 to 8 years. The current complexity can discourage innovation: an agency may choose to maintain a transportation facility as-is rather than replace it with a more up-to-date piece of infrastructure because the latter would trigger NEPA processes. The increasing complexity of environmental documents is in part a response to court rulings that have led to lengthy EIS documents by penalizing agencies for failing to mention potential impacts.

Many experts recognize that the practice of environmental review in the 21st century leads to undue delays and other undesirable outcomes (see Box 11.3), which was not the intent of NEPA's drafters. The Moving Ahead for Progress in the 21st Century (MAP-21) transportation funding law in 2012 made several efforts at streamlining the NEPA process, by, among other changes, exempting projects that receive less than $5 million in federal funds. MAP-21 also set overall time limits for environmental review and required agencies to act swiftly after an EIS is completed and accepted. Other streamlining initiatives in recent years allow individual U.S. states to take greater responsibility for NEPA processes and to consolidate parts of the transportation planning and NEPA processes that have typically been performed separately.

One long-standing simplification strategy is "tiering," whereby the environmental review process for very large-scale projects is separated into two steps. A Tier 1 analysis assesses the project's macroscale issues, and when a decision at that level is made, a number of Tier 2 processes are initiated that focus on impacts specific to different sites affected by the project. The efficiency arises from the ability of multiple Tier 2 processes to proceed independently of each other because the large-scale environmental challenges were addressed in Tier 1. Examples of recent projects advanced under tiered environmental review include California's planned high-speed rail network and the Interstate 66 corridor in northern Virginia.

Although NEPA is the centerpiece of the United States' environmental legislation, other laws also relate directly to environmental issues. The most prominent of these are discussed in the next two sections.

BOX 11.3

Smoothing the NEPA Process

While some suggest that the NEPA process is in need of updating via new legislation, others point to delays caused by outdated and sometimes conflicting practices within federal agencies (e.g., how agency staff interpret the legislation) or wider issues related to governmental operations, such as the procurement processes through which public-sector agencies hire consultants that prepare many EISs.

Eccleston (2000, p. 4) writes that "NEPA is often pursued more as a permitting requirement for documenting decisions already made than as a true planning process," and lists commonly observed problems in NEPA processes. Typical problems Eccleston identifies include:

1. Fragmented or uncoordinated plans
2. Failure to integrate NEPA with early project schedules
3. Lack of impartiality in analyzing a proposed action versus reasonable alternatives
4. Miscommunications

In 2011, the Regional Plan Association convened an expert roundtable seminar on Re-Thinking NEPA: Leveraging Lessons Learned to Expedite Project Delivery. The subsequent study report found that "It is the misguided implementation of the law that has significantly affected project delivery time" (2012, p. 3). Rather than proposing legislated changes, the following recommendations were put forward:

1. **Integrate planning and environmental reviews.** Establish broad agreement among agencies and stakeholders on project goals and carry them forward into the environmental process to help prevent controversies from arising later on.

2. **Front-load agreements among agencies.** Spend more time at the beginning of the process establishing memoranda of understanding among participating agencies on timelines, procedures, language, and environmental outcomes.

3. **Establish stronger federal leadership on major projects.** Strengthen federal leadership on major employment-generating projects and reduce federal involvement in minor projects. Allocate sufficient funding and staff capacity to federal agencies to take on stronger leadership roles.

4. **Train the next generation of environmental practitioners.** Educate environmental professionals to adopt and share best practices, such as streamlined stewardship and environmental performance commitments, for more efficient and effective reviews.

5. **Increase transparency and accountability.** Environmental practitioners should focus on producing a more thorough administrative record, as opposed to excessive analysis of unlikely impacts. Greater transparency and accountability can also be achieved by posting the deadlines for key decision points online.

6. **Update procedures for the 21st century.** Take advantage of information technologies, such as digital submission and transmission of environmental documents, and web-based interactive stakeholder involvement tools to improve the efficiency of the NEPA process.

Sources: Eccleston (2000); Regional Plan Association (2012).

Air Quality Legislation

The 1970 amendments to the Clean Air Act directed the EPA to develop air quality standards for certain "criteria pollutants"; these amendments brought major consequences for the transport sector. The 1970 amendments targeted a 90% reduction in emissions per mile of the criteria pollutants of new cars by 1975, an ambitious and contentious goal (and the subject of a major lawsuit between carmakers and the EPA) that ultimately was nearly met. In large part because of these regulations, the United States has experienced sustained improvements in air quality since the 1970s although Americans currently drive many more miles than they did in the 1970s.

All U.S. states must prepare and keep an up-to-date state implementation plans (SIPs) for maintaining or improving air quality, which the EPA must approve. When a violation of the National Ambient Air Quality Standards (NAAQS) occurs (see Figure 11.3) the region is deemed to be in nonattainment of the NAAQS for the relevant pollutant(s). No transportation infrastructure project planned for the region can receive federal funding unless it is shown that it would not worsen air quality. States with nonattainment regions must also provide a SIP outlining their plan to improve air quality.

Other Environmental Laws

The Clean Water Act

Wetlands are ecologically important for a variety of reasons, and therefore the United States and many other countries have legislation to protect them. The Clean Water Act directs the U.S. Army Corps of Engineers to provide a wetlands-disturbance permit to a project (such as a new road) only if (1) efforts have been made to avoid disturbance, (2) unavoidable impacts are minimized, and (3) any unavoidable impacts are compensated by creating or protecting other areas of wetlands of similar ecological value.

Many states also have laws that go beyond federal legislation in protecting wetlands, which include restrictions on building within defined buffer zones adjacent to a wetland, in addition to the wetland itself. While wetland protections preserve wetlands' ecological function, building around wetlands can lead to poorly connected street networks and neighborhoods, which can discourage pedestrian activity and thereby lead to more car travel. A strategy to prevent such disruption to urban form is to use "wetland banking" mechanisms that enhance or preserve wetlands off-site to compensate for a project's impact on wetlands.

The Endangered Species Act

Together with complementary local laws in all U.S. states, the Endangered Species Act (1973) provides protection for animal and plant species that are threatened with extinction. The main impact on transportation projects is that actions of federal agencies (including, as ever, funding) are prohibited from interfering with the habitat of species that have been determined to be endangered or threatened (as of 2013 these two categories collectively include approximately 1,500 species).

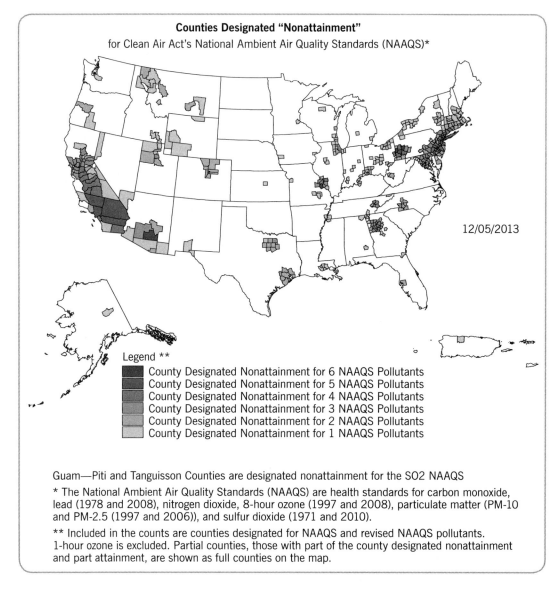

Counties Designated "Nonattainment"
for Clean Air Act's National Ambient Air Quality Standards (NAAQS)*

12/05/2013

Legend **
- County Designated Nonattainment for 6 NAAQS Pollutants
- County Designated Nonattainment for 5 NAAQS Pollutants
- County Designated Nonattainment for 4 NAAQS Pollutants
- County Designated Nonattainment for 3 NAAQS Pollutants
- County Designated Nonattainment for 2 NAAQS Pollutants
- County Designated Nonattainment for 1 NAAQS Pollutants

Guam—Piti and Tanguisson Counties are designated nonattainment for the SO2 NAAQS

* The National Ambient Air Quality Standards (NAAQS) are health standards for carbon monoxide, lead (1978 and 2008), nitrogen dioxide, 8-hour ozone (1997 and 2008), particulate matter (PM-10 and PM-2.5 (1997 and 2006)), and sulfur dioxide (1971 and 2010).

** Included in the counts are counties designated for NAAQS and revised NAAQS pollutants. 1-hour ozone is excluded. Partial counties, those with part of the county designated nonattainment and part attainment, are shown as full counties on the map.

FIGURE 11.3. U.S. counties that are designated "nonattainment status" for the Clean Air Act's National Ambient Air Quality Standards (NAAQS).

Special-Purpose Districts

At the state and local level, special-purpose districts are an increasingly common approach to managing environmental issues that are localized but cross municipal boundaries, such as regional air quality. Special-purpose districts may include transportation as part of their remit, and possibly also related issues such as land use planning. One noteworthy special-purpose district aims to protect water quality and other environmental resources in the Highlands region of New Jersey, located on the outer suburban fringe of the greater New York City metropolitan area (see Box 11.4).

> ## BOX 11.4
> ## New Jersey's Highlands Council
>
> Rapid suburban and exurban development in areas that serve as the watershed for more than half of New Jersey's population led to the 2004 Highlands Water Protection and Planning Act, with the aim of sustainable stewardship of water quality and other environmental resources in the Highlands. The law set up the Highlands Council and required it to prepare a regional master plan with a transportation element.
>
> The council reviews all transportation projects and local land use plans in the Highlands for consistency with the master plan. The master plan is based on the principle of "carrying capacity," and explicitly recognizes that the transportation network is a factor that limits the amount of development that the area can support. The council's stated policy when reviewing transportation projects for conformance with the master plan is to consider "potential growth-inducing effects such as substantial new land use, new residents, or new employment that could occur as a result of road improvements for increased motorized vehicle traffic capacity." Transportation projects that are likely to be growth-inducing for areas without the capacity for additional land development cannot be approved.

Anthropogenic climate change is of increasing relevance to transportation policymakers, as the use of fossil fuels for transport is a major source of greenhouse gas (GHG) emissions (see Chapter 12). In the United States, federal efforts have sought to legislate reductions in GHG emissions, but none have become law. There is much activity at the state level, however. California, for instance, has mandated that by 2020 GHG emissions in the state be reduced to no more than the level in 1990. Local governments are required to prepare transportation and land use plans documenting how the GHG reductions will be achieved.

ENVIRONMENTAL CHALLENGES AND STRATEGIES FOR MITIGATION

To examine the range of impacts of transport systems on the environment, it is useful to distinguish between human ecosystems and plant/wildlife ecosystems. Although some of the effects are similar, for example, the disruption of sleep by traffic noise, the mechanisms for identifying problems, designing potential solutions, and evaluating their costs, benefits, and disbenefits are quite different. Moreover, "plants and wildlife don't vote."[1]

The following subsections provide a brief discussion of the main domains of impact of transport on human ecosystems and the broad possibilities for their mitigation, followed by a similar treatment of impacts on plant and wildlife ecosystems. Several examples of interventions addressing one or more ecosystem vulnerability are given in boxes. We then briefly examine the reverse—situations where transport systems are affected by the natural environment.

In general, the order of the domains of impact shown in the text and in Tables 11.1 and 11.2 reflects their relative importance (from higher to lower) to the transport–environment debate, and to political action in high-income countries, notably in Europe and North America. The review that follows is necessarily limited by the constraints of space; ecosystems are connected to other systems in many ways, and a complete mapping of those connections is not possible within the scope of this chapter.

Effects of Transportation on Human Ecosystems

Table 11.1 lists transportation's principal impacts on human ecosystems, as defined in the following subsections, and for each impact the main types of mitigation that are in use.

Local and Regional Air Quality

Air quality regulations in many countries focus primarily on what in North America are known as "criteria air contaminants" (CAC). Those associated with motor

---

TABLE 11.1. Domains of Impacts of Transportation on Human Ecosystems and Major Examples of Mitigation, in Approximate Order of Importance (as Evidenced by Political Action)

Local and Regional Air Quality
- (Chronic air quality situations) Broad government actions to limit emissions, such as the imposition of pollutant emissions performance standards on new and used motor vehicles, enforced at point of sale, or through periodic compliance testing
- (Chronic air quality situations) Low-emission vehicles given exclusive or preferential access to zones, or to express lanes in major corridors
- Individual strategic actions such as moving house (in response to chronic problems of air quality)
- (Acute air pollution episodes) Temporary restrictions on the use of motor vehicles and the encouragement of alternative modes (see Box 11.5)

Global Atmospheric Effects/Climate Change*
- Performance standards for new and used motor vehicles to reduce GHG emissions, notably by energy efficiency
- Integrated community energy systems
- Noise performance standards guiding interventions such as:
  o Roadside walls to deflect or absorb noise
  o Retrofitted soundproofing of buildings
  o Noise level criteria for siting new transport infrastructure
  o Airport operations curfews

Blighting of Human Habitat by Transport Infrastructure
- Pedestrian- and bicycle-friendly routes across road and rail corridors
- Greening of arterial streets as activity destinations
- Radical removal or rerouting of traffic corridors

Impervious Surface Effects (Heat Islands)
Replace heat-absorbing surfaces of pavements and buildings with reflective surfaces
Plant trees to increase shade
Implementation of recent developments in LED outdoor lighting that limit "escaping" light

*Also has important impacts on plant/wildlife ecosystems.

vehicles include: nitrogen oxides (NOx) consisting of nitrogen oxide (NO), which oxidizes to form nitrogen dioxide (NO_2); carbon monoxide (CO); volatile organic compounds (VOCs); particulate matter (PM10); and fine (PM2.5) or ultrafine (PM1) particulate matter. Ground-level ozone (O_3), a major component of smog, is formed when NOx and VOCs react with sunlight, and smog is worsened by PM. Motor vehicle tailpipe emissions are typically responsible for about one-third of VOCs and more than half of NOx, and in addition are an important source of PM. These pollutants diffuse within built-up areas, and there is also a growing concern about unhealthy air quality inside moving vehicles (Müller, Klingelhöfer, Uibel, & Groneberg, 2011).

NO_2, O_3, smog, and PM have been associated with lung disease. NO_2 and O_3 may be causes of decreased immune function, CO reduces the oxygen capacity of the bloodstream, and certain VOCs and PM may increase the risk of cancer. The effects of exposure to $PM_{2.5}$ and PM_1, on their own or in combination with other pollutants, are as yet poorly understood.

Of key importance is the effect on pollutant loads of the growth in road vehicle use (see Chapter 1, Figure 1.10).

Additional challenges to urban air quality come from commercial vehicle operations (National Research Council, 2014) and ships. Furthermore, some point facilities, such as logistics centers, ports, and airports, present major challenges. Airports and seaports generate vast numbers of vehicle movements, both commercial and private. Seaports can become pollution hotspots because of the presence of ocean-going ships; very few are registered in the United States, and therefore most large ships are not subject to U.S. emissions regulations.

As noted in Table 11.1, interventions designed to reduce transport's impact on local and regional air quality—and therefore on public health—vary according to whether the effects are "chronic," such as may be evidenced by a growing number of diagnoses of asthma or other respiratory diseases, or "acute," such as when hospital admissions for respiratory distress peak in a serious way.

Many chronic situations can be addressed only by drastic actions by the individuals affected, such as moving to a region with better air quality, or by broad government actions such as imposing mandatory emissions performance standards on new and used motor vehicles. The latter approach has been the centerpiece of government policy to reduce CACs in many areas of the world, including the European Union and North America. On the other hand, governments may offer incentives or disincentives for people and companies to use low-emission vehicles.

Despite the success of efforts to address chronic air quality problems, acute situations affect some urban areas when key pollution indicators, notably ground-level ozone, exceed predetermined thresholds. Initial stages may involve appeals by public authorities for voluntary reductions in vehicle use[2] and advisories to vulnerable population groups, notably the elderly and the infirm, to stay indoors. But at higher pollution levels, an air quality emergency may be declared, triggering mandatory measures such as restrictions on car use. An example from Paris is described in Box 11.5. This was notable for including recent innovations such as carsharing and bikesharing in its complementary measures.

BOX 11.5

Short-Term Interventions during an Air Pollution Emergency: Paris, France, March 2014

Paris, with a metropolitan population of 10 million, experienced unusually mild weather in March 2014. Calm winds, warm days, and cool nights prevented the normal mixing of air layers, which led to dangerously elevated concentrations of airborne particulates (more than double the maximum alert level). To mitigate the pollution emergency, public authorities implemented short-term measures that greatly affected the transportation sector for 4 days.

The short-term measures included both incentives to use mass transit and disincentives to use private vehicles. Public transportation fares were eliminated, at a reported cost to public agencies of €4 million ($5.6 million) per day—even the bikesharing (*Vélib'*) and battery-electric-vehicle carsharing (*Autolib'*) systems were made available for the public to use free of charge. On the other hand, private-car use was restricted; speed limits were reduced; and an alternating "odd-even" system (based on the last digit of a car's license plate) was introduced in which drivers could only use their cars on alternate days, in theory reducing the size of the private car fleet by one-half.

Odd/even car-use systems have been previously implemented in other cities. Mexico City, for instance, has had similar restrictions since the 1980s. The system's impact in reducing traffic levels tends to lessen over time; people may, for instance, buy second cars (which are likely to be older and more polluting) to circumvent the restriction. In Paris, the penalty for not complying with the restrictions was not high: at €22 ($31), the fine for noncompliance was less than double the standard daily congestion charge in London.

Noise and Human Settlements

The effects of traffic noise on urban residents, and specifically on those living close to major road arteries, rail corridors, or airports, were among the earliest environmental concerns of policymakers (Long, 2000). Several factors influence the extent to which audible noise harms people. Most prominent is intensity, which depends on distance, traffic volume, and speed. However, frequency and variability in noise levels are also important, as are perceptions of disturbance. Noise thresholds are sometimes incorporated into criteria for requiring mitigation actions, such as walls and berms to deflect or absorb noise, retrofitting soundproofing onto buildings, and imposing curfews on airport operations.

Aircraft noise is a fact of life for many urban areas, and once flight paths have been set, alternatives for mitigation are limited. London's Heathrow airport has a number of mitigation programs; about 8,500 homes around Heathrow are inside the affected zones.

Blighting of Human Habitat by Transport Infrastructure

The visual and functional intrusion of road and rail infrastructure can affect both urban and rural environments. A harsh reality of major transportation facilities, especially in cities, is that once built, they are extremely costly and politically difficult to change or remove. The displacement and deterioration of homes, businesses, and green spaces may occur over a wider band than just the path of the transport infrastructure, and communities can be fragmented by the loss of pedestrian-friendly environments and the degradation of personal security and road safety (Buchanan, 1963; Institution of Highways and Transportation Management, 1990).

Nevertheless, mitigating actions are sometimes possible. Connectivity across road and rail corridors can be improved for pedestrians and bicycles (sometimes only at substantial cost), although without careful design and ongoing maintenance such connections can be uninviting. The barrier effects of urban arterial streets can be reduced in some situations by making them attractive as destinations for commercial activity, coupled with greater priority for pedestrians (at the expense of vehicle speeds). In extreme cases, a radical reversion of cityscape is made possible by placing urban freeways in tunnels (e.g., Boston) or removing them completely (e.g., San Francisco, Seoul). For other proposed removals, see Congress for the New Urbanism (2012).

Impervious Surface Effects (Heat Islands)

Transport infrastructure is a major contributor to the "hardening" of surfaces in urban areas; in large cities the proportion of surface area that has been made impervious to water infiltration can reach 60% or more. Such surfaces are in part responsible for *heat islands* of elevated temperature in which the health of vulnerable residents can be seriously compromised during heat waves. The additional heat also leads to increased peak electricity demand for air conditioning, with further environmental effects that vary according to the energy source used for electricity generation. Akbari (2005) estimates that 5–10% of urban peak electricity demand can be attributed to heat island effects.

In addition, impervious surfaces in cities result in concentrations of storm water that move faster and are much warmer than water in the natural watercourses into which they are discharged, often with serious effects on aquatic life and amenities.

In recent years, the main thrust of mitigation has been to introduce, to the extent feasible, reflective surfaces to replace dark, heat-absorbent surfaces of pavements and buildings, and to plant trees to provide shade. Some parts of the transportation network (e.g., low-traffic parking areas) can also be made less impervious to water infiltration through careful selection of surface materials.

Light Pollution

Human diurnal rhythms are regulated, in part, by exposure to light. While it is possible to regulate the amount of light inside a dwelling, increasingly we have become a world of city dwellers, where the amount of artificial light, notably from street lights, is beyond the control of the individual.

Some recent developments in outdoor LED lighting technologies may mitigate these effects. Proponents claim that it is possible to reduce polluting losses of light from about 50% for conventional sodium or mercury lighting to a few percent, and to reduce energy consumption by 40–60% at the same time. But some dispute these advantages and raise concerns about unintended health effects, notably the disruption of sleep by the blue-rich color temperature of LED lights commonly installed in street lighting and some other applications.

Effects of Transport on Plant and Wildlife Ecosystems

Table 11.2 lists the principal domains of transportation's impacts on plant and wildlife ecosystems and the main types of mitigation that are in use.

As in the discussion above regarding human ecosystems, the effects on plant and wildlife ecosystems come primarily from road traffic. Interested readers will find in Forman and colleagues' (2003) seminal work on road ecology an unprecedented assembly of relevant research findings.[3]

Habitat Fragmentation

The fragmentation of natural habitats by road networks presents a twofold challenge to plant and wildlife ecosystems. First, increasing road densities lead to ever smaller pieces of the landscape between roads, which may be incompatible with the spatial requirements of species found in the area. Second, roads limit connectivity between

---

TABLE 11.2. Domains of Impacts of Transportation on Plant/Wildlife Ecosystems and Major Examples of Mitigation, in Approximate Order of Importance (as Evidenced by Political Action)

Habitat Fragmentation
- Improved connectivity for animals: over-/underpasses that are integrated into the natural landscape
- Road and roadside design in residential neighborhoods (see Box 11.6)

Permanent Loss of Crucial Wetlands
- Improved water flow within watersheds, using culverts with profiles that also enable the mobility of wildlife
- Restoration or reconstruction of wetlands
- Integrated approaches to wetland complexes and their relationship to drainage basins and watersheds
- Compensatory intervention to conserve and improve wetlands in a substitute location or bank

Pollution of Watercourses and Wetlands
- Retention basins to hold and/or treat contaminated water
- Reduced release of contaminants at source through improved management of roads, roadsides, bridges, and vehicle fleets

The Effects of Traffic Disturbance on Fauna: Noise, Vibration, and Vehicle Lights
- Measures to reduce congregation of wildlife near roads
- Traffic reduction and management (see Box 11.7)

Collisions between Vehicles and Wildlife
- Animal barriers (such as exclusion fences) in combination with designated crossings, some of which are specially built animal over-/underpasses

Invasive Species along Road and Water Transport Corridors
- Legal ban on introduction of invasive plants
- Guidelines to favor native species
- Barriers against invasive species (e.g., Asian carp)

fragments that might be satisfactory collectively but are not individually. This problem is particularly acute in the case of wetlands, discussed in the next section. Natural crossings of roads, such as overhanging trees that meet in the middle of a road, are relatively rare and unusable by many animals. Roads can be crossed on the surface but not without risk of collision with a motor vehicle.

The effects of fragmentation on habitats are not just a problem in open country. A striking example of the cumulative effects in urban areas can be found in Figure 11.4 (Balej & Lee-Gosselin, 2002). Approximate effect distances for woodland birds in the suburban areas northwest of the center of Quebec City, Canada, were mapped for 1960 and 1994, taking into account road classification. In 1960, before the construction of freeways and most of the neighborhoods, about 60% of the landscape was largely untouched by the effects of roads. By 1996, only about 20% was untouched, and the remaining landscape fragments were much smaller.

Mitigating habitat fragmentation is difficult and inevitably involves compromise. In rural areas, connectivity can be improved with overpasses that are part of the natural landscape. As shown in Figure 11.5, with careful planning these facilities can provide continuity of habitat. However, the need to channel animal movements to such facilities to obtain safety benefits requires the use of roadside barriers that inevitably reinforce fragmentation.

In residential neighborhoods, environmentally friendly road and roadside designs can help reduce the fragmentation and degradation of natural habitats and provide support for native species (see Box 11.6).

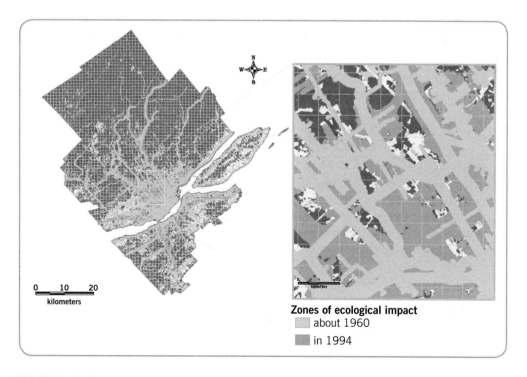

FIGURE 11.4. The evolving ecological footprint of roads, Quebec City, Canada, 1960 to 1994.

FIGURE 11.5. Wildlife crossing over a section of Interstate 78 in suburban Berkeley Heights, New Jersey. Source: Google Maps.

BOX 11.6
Roads and Roadside Design in Residential Neighborhoods

In many cases sensitive design of road rights of way can greatly lessen environmental impacts with little or no negative impact on the utility of the road infrastructure. "Headwater" residential streets that primarily accommodate access rather than through movements typically take up half or more of the total surface area of urban road networks. They can be designed with narrower widths than is the case in many suburban areas, providing cost savings while also reducing traffic speeds and minimizing land consumption. Indeed, the average width of new residential streets increased by half in the decades after World War II. Such treatments require compromise between road planners and fire and waste-management agencies, which traditionally have expected to be able to freely maneuver large trucks on even the narrowest residential streets.

Permeable pavements can also minimize the environmental footprint by allowing a greater proportion of precipitation to infiltrate into the soil (rather than entering the engineered drainage system). They work best in low-speed traffic environments where heavy vehicles do not have to be accommodated, such as driveways, parking lots, and sidewalks.

Roadside vegetation can be an important tool for managing environmental impacts. Low-cut grassy surfaces directly adjacent to roads protect drivers' sight lines. But in many instances this is not necessary across the entire right-of-way cross-section. Areas further away from traffic lanes can be mowed higher and less frequently than the roadside nearest to moving traffic. In parts of rights of way, larger bushes and trees can provide habitat connectivity. These strategies can reduce maintenance costs while providing a greater diversity of ecological habitat within a given physical right of way.

Permanent Loss of Crucial Wetlands

Wetlands (areas that are permanently or temporarily covered with surface water) act as buffer zones to reduce flooding, filter contaminants upstream from watercourses, and also contribute to biodiversity. Historically, agriculture has been responsible for the majority of wetland losses in the United States (about half of the natural extent of wetlands has been destroyed), but draining or filling wetlands for road construction now plays an increasingly important role. As discussed above, federal and local regulations to protect wetlands exist, but the protection is not absolute.

Wetlands consist mostly of grassy marshes, which are inundated for long periods, and swamps, which contain trees and shrubs, and are inundated intermittently. Much disruption results from the barrier effect of roads, leading to radical changes in water flow. In addition, as discussed below, chemical contamination can occur. The combined effects of roads on plant and wildlife habitats in adjacent wetlands can be detected up to 2 kilometers from the nearest paved road (Forman et al., 2003).

Diagnosing wetland problems and selecting mitigation measures are difficult not only because of the large number of distinct animal and plant species involved, but also because the effects can lag the disruption by decades. Interventions may involve managing water flow through engineering mechanisms and the use of specially designed culverts to enable animal movements. However, regulatory responses and interventions may be limited to isolated locations where the effects of roads are most conspicuous, whereas research suggests that an integrated, multilevel approach at the scale of wetland complexes (and their relationship to drainage basins and watersheds) would be more effective.

When road construction affects wetlands, compensatory mitigation measures are frequently required. Mitigation can be in the form of expanded or reengineered wetlands directly adjacent to the road itself, or it can be part of a wetlands complex in another location that is an offset against the road's impacts. The criteria for compensatory wetlands are the subject of much debate; hydrological experts typically argue that a balance between size and functional attributes should be sought (Bartoldus, Garbisch, & Kraus, 1994).

Pollution of Watercourses and Wetlands

The release of chemical contaminants into water by roads and road use is a complex problem because of the number of different classes of contaminants, the variety of their sources, and the diversity of their effects on plants and wildlife. Insofar as the problem reaches downstream to sources and storage of potable water, this contamination is a problem that also affects the human ecosystem. The contaminants include substantial quantities of metals; organic compounds (mostly from petroleum products); solid matter from sediment, pavement surfaces, and tires; and chemicals used for the maintenance of roadside land or de-icing. Small quantities of toxic metals and of other chemicals from vehicles and road structures are added to the mix. The most important source of all these contaminants is the motor vehicle, through the leakage of oils, greases, and hydraulic fluids, the wear of mechanical components, brake linings and tires, corrosion, and exhaust emissions. However, road maintenance activities are also major contributors.

The main mechanisms of pollution are rain runoff, snow melt, and wind. Water contaminants concentrate principally in roadside ditches, which can constitute extensive linear wetlands. Mitigation is problematic because effects can be spatially diffused, crossing jurisdictional boundaries. Targeting a particular contaminant may also involve dealing with multiple sources, which can be expensive to the point of being impractical. As a consequence, two main approaches are taken. The first is to build engineered structures to retain and/or treat contaminated water. The second is to reduce the release of pollutants, primarily through changing management and maintenance practices for vehicles, roads, roadsides, bridges, and nearby agriculture. A recent example is the reduction of the use of road de-icing chemicals in regions with severe winter weather through real-time optimization: this involves locating mobile equipment using GPS, and assigning them to distribute the lowest possible quantities of chemicals at the most appropriate time for de-icing purposes.

The Effects of Traffic Disturbance on Fauna: Noise, Vibration, and Vehicle Lights

While it is difficult to assess the sensitivity of animals to different types of sound, the frequency ranges that are audible to various types of animals are known. In general, mammals can hear the widest range of frequencies, birds a much more limited range, and reptiles a range that is narrower still. Reptiles, however, do best at detecting vibrations that indicate the presence of predators, but little is known about how wildlife respond to vibrations associated with traffic. Noise can attract some wildlife, and animals can eventually become tolerant of it, encouraging them into the danger of close proximity to fast-moving vehicles. Other types of wildlife are averse to traffic noise, leading to "avoidance zones" that vary by species. Effects may also differ between busy and low-volume roads and whether the noise is constant or varies over time. Similar observations pertain to how wildlife reacts to vehicle headlights and taillights, although this is not well understood.

Remedies for traffic disturbance are limited, and in areas of exceptional environmental importance, restricting the use of vehicles is sometimes the only option. National and state parks that are within easy reach of large urban populations have implemented such restrictions, including permanently closing some roads or not allowing any new road construction. Parks may also manage visitor motor traffic through limited opening hours or car parking spaces, improved public transportation of various kinds (especially on a park-and-ride basis), and ultimately by limits on the number of vehicles allowed to enter a park. A consensus on solutions has proven elusive given the role that major parks play in the tourist economy and their status as publicly owned amenities. In the case of Yosemite National Park (see Box 11.7), efforts to manage the impacts of motorized tourism go back more than 60 years.

Collisions between Vehicles and Wildlife

For many people, collisions between vehicles and large animals symbolize the ecological problems of roads. However, the problem also extends to small wildlife such as birds, reptiles, and amphibians, whose numbers make up the vast majority of "road

BOX 11.7

Traffic Management in Yosemite National Park

The popularity of California's Yosemite National Park has led to increasingly acute stewardship challenges. In 2011 visitors spent $380 million in the region, supporting an estimated 5,000 jobs. However, over 80% of park visitors reported feeling some degree of crowding during their visit (Whittaker & Shelby, 2012).

Traffic congestion at Yosemite is worst during weekends, especially in summer. The sparse access-road network provides few options for routing or parking, but higher-capacity road infrastructure would be deeply controversial. Indeed, in other parts of the Sierras the *removal* of some roads in sensitive areas is considered an option (United States Geological Survey, 1997).

Yosemite's natural systems are highly sensitive to the heavy traffic levels. The park's 1980 General Management Plan identified environmental damage from cars as the "single greatest threat to enjoyment of the natural and scenic qualities of Yosemite," and set a goal to remove all car traffic from strategic areas of the park. The plan foresaw that "in the near future automobile congestion will be greatly reduced by restricting people's use of their cars and increasing public transportation," which remains an elusive goal nearly 35 years later.

Today the National Park Service publishes hour-by-hour traffic forecasts. Motorists are encouraged to arrive early and stay late, and to avoid driving from site to site within the park. The $20-per-car entrance fee encourages ridesharing. Shuttle-bus services circulate internally, and connect with transit services to neighboring "gateway" communities. Even setting up the bus service was controversial; one park visitor voiced the views of many opponents that the system was "designed to make it difficult to get to the park (Herremans, 2006, p. 168).

The future of access to Yosemite is unclear. A day-visitor permit system remains under consideration, along with less-contentious strategies to delay or redirect motor traffic, as well as improvements (e.g., pedestrian underpasses) to eliminate pedestrian–vehicle conflicts.

kill" victims. As noted above, vehicle–wildlife collisions are one consequence of the disturbance of fauna and the fragmentation of habitats. They also result from many of the changes that roads bring to their edges, especially the easily reachable road kill that lures predator birds and animals into zones of frequent conflict with vehicles. Ironically, efforts to keep some kinds of wildlife away from roads may reduce the grazing of verges and allow the plants that feed and provide cover for smaller wildlife to flourish, leading to increased populations of such wildlife living in close proximity to moving vehicles.

The main focus of mitigation is to channel wildlife into or over protected crossings, rather than to allow wildlife to cross the roadway at any point along its length. In the best examples, animal road crossings and barriers (notably exclusion fences) are treated together as a system that is well informed by an understanding of animal behavior. Some crossings consist of overpasses built for animals; as discussed above, the most elaborate are wide and planted to provide continuity of the landscape (Figure 11.5). Others are specially built animal underpasses, which may differ considerably in size depending upon the behavior of the targeted wildlife. In addition, small mammals, reptiles, and amphibians often cross roads via existing culverts.

A variety of roadside anticollision technologies have been tested (particularly in the case of large mammals such as deer) including reflectors, lighting, and ultrasonic whistles. These strategies have mostly produced poor results, notably a rapid decay of effectiveness as animals over time begin to lower their response to the devices' operations. Other strategies aim to sensitize motorists or restrict traffic during periods of particular animal vulnerability, such as breeding and migration seasons. In Australia, for instance, night-time vehicle speeds are permanently restricted in certain places that have large marsupial populations.

Invasive Species along Transport Corridors

While roads may create formidable barriers in the natural landscape, their linear corridors can also provide the physical mechanism for invasive species to migrate onto new territory. Displacement of native species by exotic ones such as *Common Reed* can then spread to adjoining habitats and may threaten wildlife that depend on the native species for food and cover (Jodoin et al., 2008). Some invasive species, such as the plant *Multiflora Rose,* have been intentionally introduced along roadsides, while others are accidentally introduced through soiled construction equipment or road vehicles carrying seeds.

In the United States, mitigation includes (since 1999) a ban on the use of federal funds for any action that would introduce known invasive plant species. Interventions also include measures to support or introduce native trees and shrubs, especially those that are resistant to de-icing salts, in order to slow the spread of salt-tolerant invasive species.

Navigable waterways are also vectors for invasive aquatic species, such as zebra mussels and Asian carp. Zebra mussels deplete plankton on which other aquatic species feed, form colonies on top of the shells of native mussels (threatening the natives' ability to filter, feed, and survive), and cause damage to submerged equipment. They have been inadvertently introduced to many of the world's waterways via commercial shipping. The Great Lakes of North America, the largest complex of freshwater lakes on the planet, are particularly affected by zebra mussels and are now also the focus of much concern about the Asian carp. This large fish could potentially devastate native fish species in the Great Lakes, which are connected to the Mississippi River basin (where the Asian carp already dominates) by a commercially important ship canal that passes through Chicago. An electrical barrier system in the ship canal creates an electric field in the water intended to discourage fish from crossing, but its effectiveness is unclear. In an effort to ensure that Asian carp do not enter the Great Lakes, various states abutting the Lakes have unsuccessfully initiated legal action (against Illinois and the federal government) to force the canal to be closed.

Vulnerability of Transport Facilities

While transport networks extensively reshape natural and human ecosystems, they are also themselves vulnerable to damage from external physical forces. These processes occur across a wide range of temporal and geographical scales. Over long periods of time solar radiation degrades roadside infrastructure (such as traffic signs). Many metal components of vehicles and transport infrastructure are subject to oxidation (rusting) that eventually can destroy structural integrity; bridges exposed to road salt or sea water are particularly susceptible. In northern climates freeze–thaw cycles accelerate the rate at which road surfaces crack, and the accumulation of wet leaves in autumn can increase the risk of road crashes and also make it difficult for trains to climb steep grades. Solutions exist, such as engineering roadbeds to a higher standard where repeated freeze–thaw cycles occur, and spraying sand onto rail tracks in autumn to increase friction. Such mitigation measures require additional costs but in general are more efficient than measures to deal with problems of this type after they have already occurred.

Other physical processes affecting transport systems can be quite intense. Seismic events (earthquakes) can sever transport networks, and cause elevated structures to collapse unless they have been specially engineered to withstand major seismic activity. Weather events can occasionally cause flooding or landslides, or can make travel temporarily impossible because of snow and ice buildup (see Box 11.8). High river flows can scour bridge abutments, rendering them structurally unsound. Flooding

---

BOX 11.8

Hurricane Sandy's Impact on the New York Region's Transport Network

After tracking across the Caribbean and the Atlantic Ocean, Hurricane Sandy made landfall in the Mid-Atlantic United States on October 29, 2012. Major road and mass transit facilities, including New York City's subways, buses, and commuter trains, were preemptively shut down ahead of the storm. Mandatory advance evacuations were ordered for low-lying areas.

An estimated $65 billion of damage was caused by Sandy, with over 150 lives lost in the United States alone. The storm surge in New York City occurred at precisely the worst time, coinciding with high tide. It peaked at 14 feet above the mean low water level.

Transport networks sustained extensive damage, with the White House's Hurricane Sandy Rebuilding Task Force declaring it the worst disaster for public transport systems in U.S. history. The federal government authorized $12 billion to repair and protect New York's transportation system.

New York's subways remained completely out of service for 3 days, with eight tunnels flooded with seawater. Critical links in the road network, such as bridges, tunnels, and waterfront highways, were also closed during the storm. Only one road access into Manhattan was kept continuously open.

The region experienced unprecedented congestion following the storm, with typical commute durations doubled or tripled. Damage to the petroleum supply chain led to shortages of gasoline, with lines of many hours (and some scuffles among motorists) reported at service stations. Restrictions on retail purchases of gasoline were imposed, and remained in place until nearly a month after the storm.

It would be impossible to prevent a storm like Sandy from disrupting transport, but the storm highlights lessons to be learned for ensuring that future impacts are kept to a minimum. Subway entrances and air vents can be raised. Back-up generators and fuel for critical infrastructure should also be kept above the expected high-water level. In addition to following good practices in physical design for disaster resiliency, transport agencies should maintain up-to-date emergency management plans with contingencies and clear lines of communication—both internally among staff and externally to the public. An important lesson from Hurricane Sandy was the value of new forms of social media for exchanging information critical to the transport system.

can also be caused by other mechanisms, such as extreme high tides which are exacerbated by the slow (1–3 mm/year) decades-long rise in sea levels. In extreme heat rail tracks can buckle if thermal expansion exceeds the design tolerances; likewise asphalt road surfaces may weaken from extreme heat and subsequently deform from otherwise acceptable traffic loadings. In all of these cases, consideration during the planning and design phases of transport projects can result in lower maintenance and repair costs over time.

FUTURE CHALLENGES AND TRENDS

This chapter has provided an overview of the interactions between transport systems and the environment, with road transport as its main focus. We conclude with a brief look at three challenges and trends that may plausibly alter these interactions in coming years; we then consider possible future directions for environmental regulation.

Impacts of New Technologies

As we have seen, cars have become much less polluting over time as a result of technologies such as catalytic converters, electronic fuel injection, and electric drive trains, which have been spurred by changing public policy toward the environment. Further technological developments in vehicle design will take place in years to come, though they may either exacerbate or lessen the environmental impacts from transportation.

Carmakers are today offering an increasing range of partly automated driving systems, such as adaptive cruise control and lane-keeping assist, and it is expected that more fully automated vehicles will become commercially available. These systems will lead to smoother drive cycles, which will tend to lower emissions of many pollutants on a per-mile-driven basis. However, by making driving easier and less burdensome they may lead to more vehicle miles of travel.

Evidence suggests that people's use of ICT (information and communication technology; see Chapter 4) affects how people organize their daily routines (activity and travel patterns in time and space) (Lee-Gosselin & Miranda-Moreno, 2009). New forms of ICT-enabled lifestyles (e.g., online shopping, telecommuting, telemedicine) are becoming evident. While the net impacts on travel demand are as yet unclear, transportation agencies must guard against overconfidence in estimating future traffic patterns—and related environmental impacts, such as carbon emissions—using only past trends when ICT was much less pervasive.

Technological advances will also enable innovative practices for managing environmental resources. Newly developed materials will help to reduce soil erosion from construction projects. The prices of many types of sensors have fallen sharply in recent years, and through new communication technologies they can provide resource managers with much greater situational awareness to direct mitigation efforts where and when they are most needed. Real-time dispatching and routing of commercial vehicle fleets is also improving, allowing fleet managers to make more effective (and environmentally beneficial) decisions. Another example is the use of mobile (typically vehicle-borne) air-quality sensors to augment the existing network of fixed-location stations.

This will enable finer-grained real-time mapping of air quality hotspots and provide more useful air quality information for those in charge of traffic networks and for the general public, especially particularly vulnerable people (e.g., asthma sufferers).

Future Environmental Regulation

Great strides in environmental quality have taken place in response to sustained action by governments, such as the actions taken by the U.S. federal government discussed in this chapter. While environmental regulation of transportation has imposed new burdens on industry, transportation agencies, and to a lesser degree on individual citizens, such regulation has undoubtedly led to a much cleaner environment with a wide set of benefits for both nature and human welfare. In general, affluent societies demand more environmental protection than lower-income societies. Therefore, as economic activity grows in the future, it is likely that environmental policy will continue to strengthen (particularly in low- and middle-income societies that are developing quickly, such as China and India). In high-income countries with well-developed frameworks to manage urban air quality, delivering further large reductions in criteria pollutant emissions from transportation will likely require higher marginal costs than has previously been the case.

Many informed observers anticipate new regimes of environmental impact management in the future, especially relating to greenhouse gas emissions (which inevitably would include the transport sector). For instance, it is possible that, in the United States, government could take concerted action to reduce the nation's greenhouse gas emissions, either through a direct price mechanism (e.g., a carbon tax or a cap-and-trade system), or via other forms of regulation such as fuel demand restraint, smart-growth neighborhood design standards, improvements in fuel economy standards, or massive investment in public transportation. In the 2000s some U.S. states (notably California) and cities have enacted legislation to reduce greenhouse gas emissions, though such legislation by individual states and municipalities has yet to be proven effective in reducing overall greenhouse gas emissions, where it is the global level of emissions that matters. Some argue that formalizing ambitious emissions-reduction goals into law encourages innovation that can deliver both environmental benefits and economic development, though this view is not universally held. Barring major technological breakthroughs, reducing greenhouse gas emissions will require major lifestyle changes that will be noticeable in people's daily lives. Fundamental change in democratic societies such as the United States hinges on fashioning a majority political coalition that shares a common goal and then maintaining the coalition through multiple successive election cycles, which has not yet occurred in the United States at a national level with respect to greenhouse gas emissions.

It is likely that the environmental review framework put in place by NEPA will remain essentially unchanged for some time to come. Some analysts believe, however, that the legislated standard NEPA process will be modified to reduce the length of the approval process, either with or without new legislation. Despite widespread unease regarding the bureaucratic requirements of NEPA, the Endangered Species Act, and other federal environmental legislation, these regulations have remained essentially intact for decades, through changes in political control of Congress and the presidency. It is also not clear that alternatives to the present framework would deliver more desirable outcomes. As illustrated in Box 11.1, U.S. environmental regulations have in the past been much more fluid at the state level, a pattern that is likely to continue. In California, for instance, a 2013 law changed the focus of state-level environmental review of transportation impacts away from the traditional measure of traffic delay (which could require road widening as a mitigation measure) to vehicle miles traveled (California Office of Planning and Research, 2014).

In the final analysis, the ways in which transport's many and complex impacts on the environment are managed will be subject to deeply political choices that the electorate will make and adjust over time. On occasion important social changes have also been brought about through judicial action, however, and one cannot rule out the possibility of future legal judgments leading to major changes in the way that environmental constraints affect, and are affected by, the transport system.

NOTES

1. However, the distinction between human and plant/wildlife ecosystems is less clear in the case of global climate change. This is briefly covered in Chapter 3, but the primary discussion in this book of the role of transportation in global warming (and how to deal with its contribution to GHG emissions) will be found in Chapter 12.

2. Some experience with designing voluntary and mandatory reductions of private vehicle use are described in the literature from the 1973 and 1979 international oil shortages (e.g., Lee-Gosselin, 2010).

3. Readers are referred, in particular, to Figure 11.6 on page 308 of Forman and colleagues (2003).

REFERENCES

Akbari, H. (2005, May 17–24). *Energy saving potentials and air quality benefits of urban heat island mitigation.* Paper presented at the First International Conference on Passive and Low Energy Cooling for the Built Environment, Athens, Greece. Retrieved February 15, 2015, from *www.osti.gov/scitech/servlets/purl/860475.*

Balej, R., & Lee-Gosselin, M. E. H. (2002). *Route et environnement dans la région de Québec: Pré-diagnostic de l'empreinte écologique des systèmes routiers.* Presentation at the 70th meeting of l'Association francophone pour le savoir (ACFAS), Université Laval, Quebec, Canada.

Bartoldus, C. C., Garbisch, E. W., & Kraus, M. L. (1994). *Evaluation for planned wetlands (EPW).* St. Michaels, MD: Environmental Concern.

Buchanan, C. (1963) *Traffic in towns.* London: Her Majesty's Stationery Office.

California Office of Planning and Research. (2014). Updating transportation impacts analysis in the CEQA guidelines. Retrieved February, 15, 2015, from *www.opr.ca.gov/docs/Final_Preliminary_Discussion_Draft_of_Updates_Implementing_SB_743_080614.pdf.*

Congress for the New Urbanism. (2012). Freeways without futures 2012. Retrieved February 15, 2015, from *www.cnu.org/highways-boulevards/freeways-without-futures/2012.*

Eccleston, C. H. (2000). *Environmental impact statements: A guide to project and strategic planning.* New York: Wiley.

Forman, R. T. T., Sperling, D., Bissonette, J. A., Clevenger, A. P., Cutshall, C. D., Dale, V. H., et al. (2003). *Road ecology: Science and solutions.* Washington, DC: Island Press.

Gallup. (2013). Environment. Retrieved February 15, 2015, from *www.gallup.com/poll/1615/environment.aspx.*

Herremans, I. M. (2006). Yosemite National Park: Parks without private vehicles. In I. M. Herremans (Ed.), *Cases in sustainable tourism: An experiential approach to making decisions* (pp. 167–186). Binghamton, NY: Haworth Press.

Independent Petroleum Association of New Mexico. (2013). Baby NEPA killed. Retrieved February 15, 2015, from *http://ipanm.org/wp-content/uploads/2016/05/Feb-22-NEPA-Killed.pdf.*

Institution of Highways and Transport Management. (1990). *Guidelines for urban safety management.* London: Author.

Jodoin, Y., Lavoie, C., Villeneuve, P., Theriault, M., Beaulieu, J., & Belzile, F. (2008). Highways as corridors and habitats for the invasive common reed *Phragmites australis* in Quebec, Canada. *Journal of Applied Ecology, 45,* 459–466.

Lee-Gosselin, M. E. H. (2010). What can we learn from the demand restraint policies of the 1970s and 1980s and public attitudes to them? *Energy Efficiency, 3*(2), 167–175.

Lee-Gosselin, M. E. H., & Miranda-Moreno, L. (2009). What is different about urban activities of those with access to ICTs?: Some early evidence from Québec, Canada. *Journal of Transport Geography, 17*(2), 104–114.

Long, W. L. (2000): *International environmental issues and the OECD, 1950–2000.* Paris: Organisation for Economic Cooperation and Development.

Millard-Ball, A. (2013). The limits to planning: Causal impacts of city climate action plans. *Journal of Planning Education and Research, 33*(1), 5–19.

Müller, D., Klingelhöfer, D., Uibel, S., & Groneberg, D. A. (2011). Car indoor air pollution—Analysis of potential sources. *Journal of Occupational Medicine and Toxicology, 6*(33), 1–7.

National Research Council. (2014). *Reducing the fuel consumption and greenhouse gas emissions of medium- and heavy-duty vehicles, phase two: First report.* Washington, DC: National Academies Press.

Norton, B. G. (1991). *Towards unity among environmentalists.* Oxford, UK: Oxford University Press.

Regional Plan Association. (2012). Getting infrastructure going: Expediting the environmental review process. Retrieved April 30, 2014, from *http://library.rpa.org/pdf/RPA-Getting-Infrastructure-Going.pdf.*

United States Geological Survey. (1997). *Status of the Sierra Nevada: The Sierra Nevada Ecosystem Project* (Digital Data Series DDS-43). Washington, DC: Author.

Yost, N. J. (2003). *NEPA deskbook* (4th ed.). Washington, DC: Environmental Law Institute.

CHAPTER 12

Transportation and Energy

DAVID L. GREENE

U.S. AND WORLD TRANSPORTATION ENERGY USE

Energy is and always will be essential to transportation.[1] Transportation requires energy to accelerate mass and to overcome forces of friction that oppose motion. Before the Industrial Revolution, humans, animals, and sails provided the power for mobility and the necessary energy came from renewable sources: biomass and wind (Smil, 2010). The Industrial Revolution allowed transportation to tap into fossil fuel resources, transforming the way transportation used energy: first coal for steam engines and later petroleum. The development of internal combustion engines and of petroleum resources in the latter half of the 19th century enabled transportation's second energy transition. The transition to a transportation system dominated by petroleum and internal combustion engines was essentially complete by 1950. More than half a century later, petroleum still provides 95% of the energy for global transport.

The quantity of energy used by modern societies to move people and things is enormous. Preindustrial societies in Europe consumed 10 to 25 gigajoules (GJ; billion [10^9] joules) (see Table 12.1 and Box 12.1) per capita for all purposes (Smil, 2010, p. 61). In 2012 Americans' use of energy for transportation alone exceeded 85 GJ per capita. The U.S. economy uses 27 exajoules (EJ; quintillion [10^{18}] joules) to produce almost 8 trillion passenger kilometers of travel and approximately 6 trillion kilometer-miles of freight shipments (Bureau of Transportation Statistics [BTS], 2014, Tables 1.40 and 1.49). The U.S. transportation system consumes 30% of total global transportation energy use, more than any other nation by a wide margin (Davis, Diegle, & Boundy, 2013).

How the world's transportation systems use energy has important consequences for the environment, the global economy, and the sustainability of human society.

TABLE 12.1. Energy Conversion Factors

	Exajoules	Gigatonnes of oil equivalent	Quadrillion BTU	Million barrels of oil per day
Exajoules	1	2.388×10^{-2}	0.948	0.448
Gigatonnes of oil equivalent	41.868	1	39.68	18.76
Quadrillion BTU (quads)	1.055	2.52×10^{-2}	1	0.472
Million barrels of oil per day	2.233	5.33×10^{-2}	2.117	1

Source: Davis et al. (2013, Table B.6).

Petroleum fuels are hydrocarbons, composed almost entirely of hydrogen and carbon atoms. Combustion of petroleum fuels by internal combustion engines produces a variety of local air pollutants in addition to carbon dioxide (CO_2) and water vapor. Among the more important are fine particulate matter, partially burned hydrocarbons, oxides of nitrogen, and carbon monoxide. Transportation is also responsible for about one-fifth of all anthropogenic greenhouse gas (GHG) emissions in the United States and the world. Transportation's near total dependence on petroleum costs the United States hundreds of billions of dollars each year in direct economic and national security costs (Greene, Lee, & Hopson, 2013).

Transportation's energy use also poses a major challenge for global sustainability. Not only are the fossil fuels on which transportation relies finite, but their combustion at even current global levels is incompatible with avoiding risky levels of global climate change (National Research Council [NRC], 2012). Achieving a sustainable energy basis for the world's transportation system is likely to require major improvements in the efficiency of energy use and a transition from petroleum to renewable and low-carbon energy sources (Global Energy Assessment [GEA], 2012).

BOX 12.1
Measuring Energy

Energy is defined as the ability of a system to do work, and *work* is defined as the application of force over a distance. Since nothing can be moved without applying a force through a distance, transportation is work and cannot be accomplished without the use of energy. Energy is measured in terms work done or heat equivalents: joules in the international system of units or British thermal units (Btu) in the English system. The international system will be used often in this chapter although other more commonly units will be used as well (e.g., barrels of oil, gallons of gasoline). A joule is defined as the work done by applying a force that can accelerate 1 kilogram by 1 meter per second squared over the distance of 1 meter. A Btu is the amount of energy required to raise the temperature of 1 pound of water by 1° F. One Btu equals 1,055 joules. The U.S. transportation system uses such vast amounts of energy that it is convenient to speak in terms of exajoules (1 EJ = 10^{18} joules) or quads (10^{15} Btu). Petroleum fuels are traditionally measured in terms of volume or weight. For example, one gallon of gasoline typically contains 132 megajoules (MJ = million joules) or 125,000 Btu.

The unintended consequences of energy use (GHG emissions, oil dependence, environmental degradation) depend on the level and type of transportation activity, the efficiency of energy use, and the impacts per unit of energy. These, in turn, are determined by the transportation choices we make, the technologies with which we convert energy into transportation services, and the properties of the energy sources we use. For the most part, the unintended consequences of transportation energy use are not taken into account when private individuals and firms make their transportation decisions. As a result, public policies play a critical role in mitigating the problems caused by transportation's use of energy, such as air pollution, GHG emissions, and energy insecurity. Motor vehicle emissions standards, for example, have reduced emissions of most air pollutants from new cars and light trucks to just 1–4% of the levels of 50 years ago (BTS, 2014, Table 4.31). However, effective policies to reduce emissions of GHGs have only begun to be implemented, and how to achieve a transition to sustainable energy remains an unsolved challenge for public policy.

Transportation's 27 exajoules of energy use in 2012 comprised 28% of total U.S. energy use (BTS, 2014, Table 4.6). Among transportation modes, highway vehicles' energy use predominates, accounting for almost four-fifths (78.6%) of the total (Figure 12.1). Light-duty vehicles (passenger cars, SUVs, minivans and pickup trucks) alone use more than half (57.9%) of the total. Combination and single-unit trucks use about one-fifth (19.7%), followed by air travel (9.8%)[2], waterborne transportation (6.5%), rail freight (2.0%), pipelines (2.6%), and finally buses and passenger rail (1.5%). These basic patterns have remained virtually unchanged for decades and have

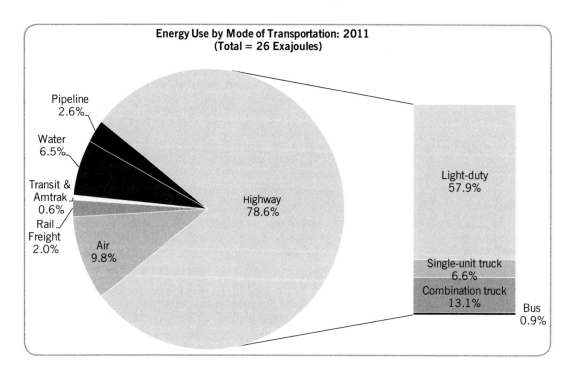

FIGURE 12.1. Energy use by mode of transportation in 2011.

important implications for public policies aimed at reducing petroleum dependence or GHG emissions.

The preeminence of highway vehicles in transportation's energy use is not limited to the United States. The distribution of energy use across modes in the United States is quite similar to the global pattern. (Figure 12.2). Highway vehicles predominate followed by air, water, and rail transport. Some of the apparent differences are due to the fact that as indicated in Figure 12.2, unlike Figure 12.1, U.S. air and water energy use do not include international flights and shipments due to lack of data for 2009, whereas the world totals do.

Fundamentally changing how transportation uses energy will take decades. Motor vehicles typically remain in use for 15 years. Manufacturers need 5–10 years to make major changes in vehicle technology across their entire product lines. Aircraft, railroad locomotives, ships, and the infrastructure to supply fuels all have even longer time constants for change.

The next section of this chapter addresses the transportation sector's dependence on petroleum: why petroleum is such an attractive energy source, the nature of the oil dependence problem, the costs of petroleum dependence, and the challenge of energy security. Section 3 describes transportation's role in climate change: the GHGs produced by transportation energy use and how transportation's GHG emissions are affected by the type of energy used and the way that energy is produced. The critical role of energy efficiency is the subject of section 4: how energy efficiency is measured, the energy efficiencies of alternative modes of transportation, trends in energy efficiency, and special aspects of the market for energy efficiency that have important implications for public policy. Section 5 is concerned with the new technologies and energy sources that offer potential solutions to the unintended consequences of transportation's energy use: biofuels, natural gas, electricity, hybrid and plug-in vehicles, and hydrogen fuel cell vehicles.

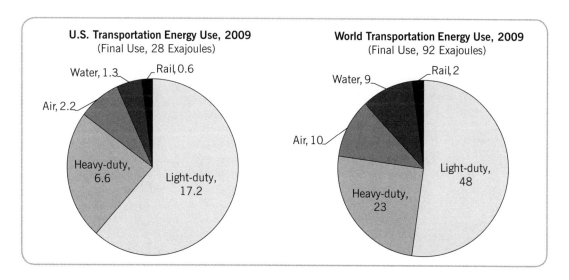

FIGURE 12.2. World and U.S. transportation energy use by mode, 2009.

OIL DEPENDENCE AND ENERGY SECURITY

Today, the U.S. transportation system cannot function without petroleum fuels (see Figure 12.3). Nearly all of the natural gas used in transportation goes to power the pumps of natural gas pipelines. Only about 3% propels natural-gas-fueled buses, trucks, and passenger cars. Electricity is used predominantly for passenger rail transportation and more recently by plug-in electric automobiles, but its share of transportation's total energy use remains minuscule. Only biofuels, mostly ethanol produced from corn and blended with gasoline at up to 10% by volume (gasohol or E10) together with a smaller amount of biodiesel have displaced enough petroleum to be significant, as Figure 12.3 shows.

Petroleum's Dominance as a Fuel

Petroleum has several advantages over alternative transportation fuels that have enabled it to maintain its dominance for more than half a century. Most importantly, the world's petroleum resources are large relative to annual petroleum use. Just over 1 trillion barrels of petroleum have been produced and consumed to date (BP [formerly British Petroleum], 2013); despite uncertainty about the exact quantity, we believe that approximately twice as much conventional petroleum remains to be exploited.

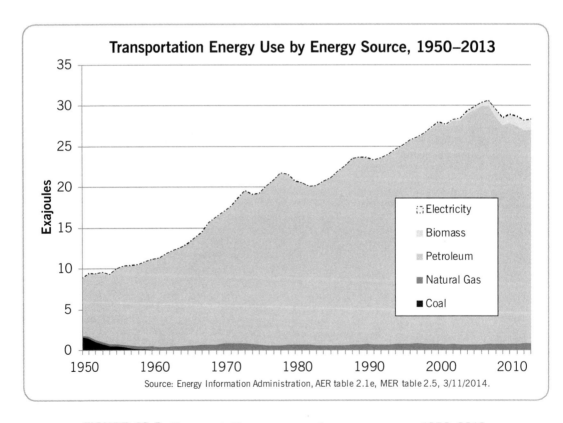

FIGURE 12.3. Transportation energy use by energy source, 1950–2013.

Unconventional petroleum resources from which petroleum fuels like gasoline, diesel, and jet fuel can be made are on the order of 10 trillion barrels.[3] Global oil consumption was 31 billion barrels in 2012. Although all fossil energy resources are ultimately finite, the world is not about to run out of resources from which to make petroleum fuels for transportation.

Petroleum fuels are also relatively easy to transport and store on board transportation vehicles. Alcohols and compressed gases have much lower energy content by volume, and the energy storage capacities of even advanced batteries are an order of magnitude smaller than gasoline (Figure 12.4). Even with oil costing about $100 per barrel, petroleum fuels are competitive with other liquid alternatives.

The Negative Influence of the Organization of Petroleum Exporting Companies

The transportation sector's dependence on petroleum comes with substantial economic and political costs. Oil price shocks have triggered economic recessions (Hamilton, 2009), dependence on imported petroleum has influenced national defense and foreign policies (Deutch, Schlesinger, & Victor, 2006), and higher oil prices brought about by the market power of the Organization of Petroleum Exporting Countries (OPEC) have transferred hundreds of billions of dollars from U.S. consumers to oil-exporting economies in the form of monopoly profits (Greene, Lee, & Hopson, 2013). Although OPEC members supply about 40% of the world's petroleum, they still wield considerable monopoly power in the world oil market owing to the inelasticity of world oil demand and supply. *Elasticity* is defined as the percent change in quantity supplied or demanded for a 1% change in price. World oil demand and

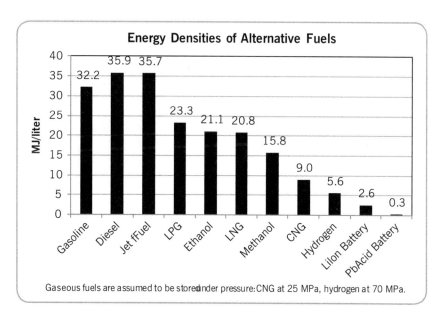

FIGURE 12.4. Energy densities of alternative fuels. Source: Davis, Diegal, and Boundy (2013, Appendix B).

supply are very inelastic. A 10% price increase will cause less than a 1% reduction in demand over the period of 1 year and only about a 5–6% reduction if sustained for 10–20 years. World oil supply outside of OPEC is also inelastic.

The inability of oil supply and demand to respond strongly to oil price changes permits the OPEC cartel to raise prices above what would prevail in a competitive market and creates the potential for major oil price shocks in the event that oil supplies are disrupted. In 1973–1974, world oil supplies were disrupted when the Arab members of OPEC declared a boycott of oil supply to countries that supported Israel during the October War of 1973 (Yergin, 1992). Supplies were again reduced in 1979–1980 by the war between Iran and Iraq, important members of OPEC. Both events caused world oil prices to suddenly increase to two to three times the previous level (Figure 12.5). The Gulf War in 1991 caused a brief run-up in prices. The most recent major increase in 2007–2008 was different. Whereas the earlier price shocks were caused by reductions in the supply of oil, the 2007–2008 shock was caused by a strong increase in demand from developing economies, particularly China, combined with stagnant world production (Hamilton, 2009). When prices rise, competitive businesses increase production. When prices shot up to $100 per barrel in 1980, the OPEC cartel repeatedly reduced its production in an effort to keep prices high. Saudi Arabia, for example, was producing 9.8 million barrels per day in 1979 but by 1985 had cut its output to just over 3.6 million barrels per day. Total OPEC production fell from 28.7 million barrels per day in 1978 to 15.9 in 1985, with a resulting loss of market share from 45% to 28%.

In response to the high prices, electric utilities in the United States and around the world switched from oil to coal and natural gas, and nearly all developed economies adopted fuel economy standards to restrain petroleum use by motor vehicles. At the same time, non-OPEC producers sought and developed new sources of petroleum, especially in northern Alaska and in Europe's North Sea.

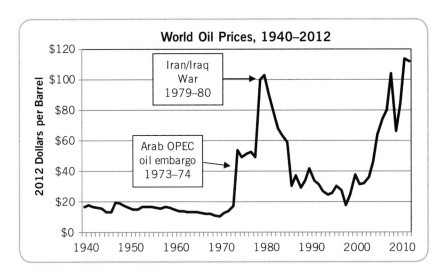

FIGURE 12.5. World oil prices, 1940–2012. Source: *BP Statistical Review of World Energy 2013*, "Crude Oil Prices 1861–2012."

Artificially high prices and sudden price changes have cost the U.S. economy trillions of dollars over the past four decades. Monopolistic prices transfer wealth from the United States to oil-exporting economies. Higher prices, in general, reduce economic output throughout the economy, and price shocks cause dislocations that idle productive resources, causing additional economic losses. The sum of these impacts from 2005 to 2010 alone has been estimated at approximately $2 trillion (Greene, Lee, & Hopson, 2013). Political, strategic, and national defense costs are much more difficult to quantify, yet they may be of a similar magnitude.

The most recent price increase appears to be different in that oil prices near $100 per barrel have been maintained without any reduction in supply by OPEC members. OPEC member states supplied 43% of the world's petroleum in 2006 and retained a 43% market share through 2012 (BP, 2013). This fact is consistent with studies indicating that world oil supply and demand have become even less price-elastic in recent years. Decreased supply-and-demand elasticities amplify the market power of the OPEC cartel. On the demand side, more price-sensitive uses like electricity generation and home heating have switched from oil to other energy sources, leaving most of the remaining oil use in the less price-responsive transportation sector. At the same time, oil use has grown fastest in the world's developing economies whose demand is less responsive to price changes (Dargay & Gately, 2010). On the supply side, production of conventional crude oil (oil that flows in underground reservoirs and can be easily pumped to the surface) appears to have reached a peak outside of the OPEC nations. Increases in production outside of OPEC are coming from unconventional fossil resources like Canadian tar sands and U.S. oil shale formations. Unconventional oil resources are more expensive to produce, tend to have greater local environmental impacts, and produce more carbon dioxide emissions per barrel than conventional petroleum. The world's unconventional oil resources are vast, however, with potentially dire implications for global climate change, as will be discussed below.

Although world oil market supply-and-demand conditions have increased OPEC's market power, the United States has dramatically reduced its dependence on imported petroleum by a combination of increased domestic petroleum production and decreased demand (Figure 12.6). U.S. petroleum use in 2013 was down about 3 million barrels per day from the peak in 2005 owing to a combination of economic recession, higher oil prices, and the early effects of motor vehicle fuel economy standards. At the same time, domestic oil production increased by more than 2 million barrels per day from tapping tight shale formations. As a consequence of these changes plus an increase of about 0.5 million barrels per day of increased blending of ethanol in gasoline, U.S. oil import dependence fell from 60% in 2005 to just 33% in 2013 and 24% in 2015. Oil prices fell to $40–50/barrel in 2015 due to stagnant world oil demand and increased non-OPEC supply (Energy Information Administration [EIA], 2016, Tables 3.1 and 9.1).

In sum, while changes in the world's oil markets have increased the market power of the OPEC cartel, the U.S. has decreased its vulnerability to high oil prices and oil market disruptions by expanding domestic production and reducing oil demand. If current policies to increase motor vehicle fuel economy and reduce GHG emissions remain in effect, U.S. petroleum demand is likely to decline until 2025 and remain

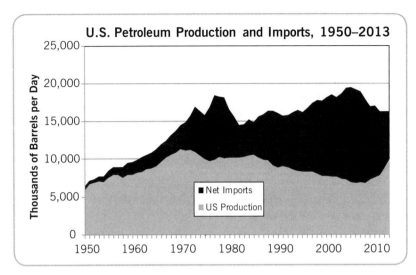

FIGURE 12.6. U.S. petroleum production and imports, 1950–2013. Source: Energy Information Administration (2016).

nearly flat through 2040 (Figure 12.7). Energy use by passenger cars and light trucks is expected to account for most of the decline despite a projected increase of 30% in vehicle miles of travel (EIA, 2013a, Table 7). As a consequence, dependence on imported petroleum is expected to remain at about 30% through 2040. While this is far better than 60% dependence on imports, oil dependence costs will remain a serious economic problem as long as prices remain in the neighborhood of $100 per barrel.

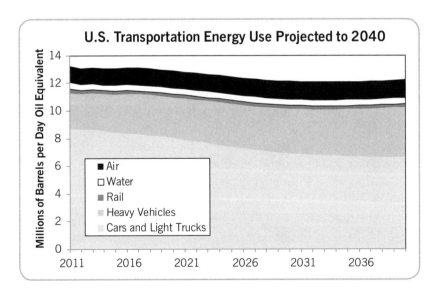

FIGURE 12.7. U.S. transportation energy use projected to 2040. Source: Energy Information Administration (2014, Table 7).

GHG EMISSIONS AND CLIMATE CHANGE

Reducing CO_2 Emissions

The U.S. National Academy of Sciences has concluded that to keep atmospheric CO_2 concentrations stable after 2050 (at about twice preindustrial levels) and avoid increasing climate change impacts, global CO_2 emissions will have to be reduced by about 80% (NRC, 2012).

GHG emissions from transportation are almost entirely comprised of CO_2 from the combustion of petroleum fuels. Transportation vehicles emit 1.8 million metric tons of GHG each year, 25% to 30% of total U.S. GHG emissions and more than any other nation's *total* GHG emissions, with the exceptions of China and Russia (Davis et al., 2013, Table 11.4). Because CO_2 from the combustion of petroleum fuels in internal combustion engines accounts for over 95% of all of transportation's GHG emissions, the distribution of GHG emissions by mode follows the pattern of energy use. Hydrofluorocarbons (HFCs) that leak from vehicle air conditioners are second in importance (Figure 12.8). Although the volume of HFC emissions is very small relative to CO_2, HFCs are among the most potent GHGs known. Nitrous oxide (N_2O) is mainly produced in the catalytic converters of motor vehicles' emission control systems. New emission control regulations are likely to greatly reduce both HFC and N_2O emission over the next decade (Environmental

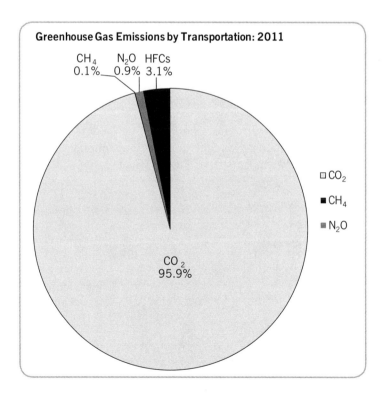

FIGURE 12.8. Greenhouse gas emissions by transportation in the United States. Source: Environmental Protection Agency (2011).

Protection Agency [EPA], 2014). Small quantities of methane (CH_4) are produced by incomplete combustion of petroleum fuels or escape from the relatively small numbers of natural-gas-fueled vehicles during refueling or from onboard storage tanks. Methane is also a potent GHG with approximately 20 times the warming potential per molecule as CO_2.

Motor vehicle use is usually the single largest part of a household's carbon footprint. A typical U.S. passenger car or light truck emits twice its weight in CO_2 each year. Most of the mass of CO_2 is due to the oxygen that a vehicle's engine takes from the surrounding air.[4] Each gallon of gasoline results in 8.89 kilograms of CO_2 emissions, and each gallon of diesel fuel produces 10.18 kilograms because of the higher carbon-to-hydrogen ratio of diesel fuel (EPA, 2011). A typical U.S. passenger car weighing 3,500 pounds (1,591 kilograms) traveling 10,000 miles per year at 25 miles per gallon will use 400 gallons of gasoline and emit 3,556 kilograms (7,823 pounds) of CO_2.

Although emissions directly from transportation vehicles are large, transportation activity induces additional emissions in the extraction and processing of fuels, the manufacturing and disposal of vehicles, and the construction and maintenance of infrastructure (Figure 12.9). Direct emissions from vehicles typically far exceed the other components of full life-cycle emissions, yet the other components can be important. Tailpipe emissions typically comprise 65 to 70% of a passenger car's life-cycle emissions, with upstream emissions from fuel extraction, transportation, and refining making up 15 to 20%. Construction of the vehicle itself (including emissions traceable to the steel, plastic, and other materials from which it is made) accounts for 10 to 15%. The batteries, even for a hybrid vehicle, account for less than 1% of life-cycle energy use.

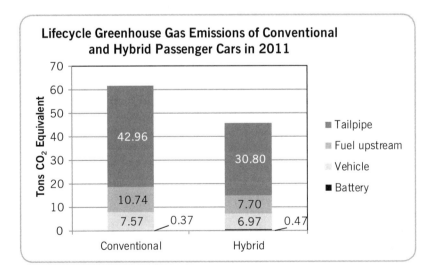

FIGURE 12.9. Lifecycle greenhouse gas emissions of conventional and hybrid passenger cars in 2011. Source: Michalek et al. (2011).

Greenhouse Gases

GHG emissions vary greatly depending on the type of fuel a vehicle uses, from what resource the fuel is produced, and the fuel economy of the vehicle. Well-to-wheel (WTW) GHG estimates include all these factors but exclude the energy used to produce the vehicle itself. The GHG emissions (in grams of CO_2 equivalent per mile) of several alternative power trains and fuel types for passenger cars are compared in Figure 12.10. An average gasoline engine vehicle is responsible for 477 grams per mile, or about 4.8 metric tons of CO_2 equivalent every 10,000 miles. This is higher than the tailpipe emissions cited above because about 15–20% of the emissions occur in the production, refining, and transportation of the gasoline and because the vehicle in Figure 12.10 is assumed to have lower fuel economy. A similar passenger car running on compressed natural gas (methane: CH_4) would be responsible for only 349 grams/mile because burning natural gas produces relatively more H_2O and less CO_2 than burning gasoline. The WTW emissions of methane, however, depend strongly on how much methane leaks in the production, transportation, storage, and dispensing of natural gas. At the time of writing this chapter, there is considerable controversy about the upstream emissions of methane; some investigators argue that upstream emissions are large enough to cancel any advantage over gasoline whereas others consider the numbers shown in Figure 12.10 to be more accurate.

Hybrid vehicles produce fewer GHG emissions because of their greater energy efficiency. Although a hybrid vehicle is partly powered by an electric motor, all its energy is ultimately derived from the gasoline burned in its internal combustion engine. Not

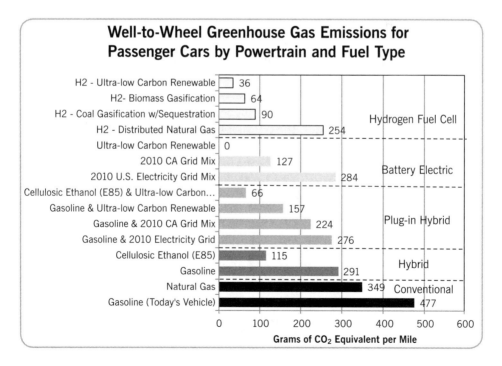

FIGURE 12.10. Well-to-wheel greenhouse gas emissions of passenger cars by powertrain and fuel type. Source: Davis et al. (2013, Figure 11.13).

only does a hybrid burn less gasoline per mile but it is responsible for a smaller share of upstream emissions as a result. A hybrid vehicle running on ethanol made from cellulosic biomass rather than corn, would reduce GHG emissions by about 75% compared with today's gasoline vehicle. Tailpipe emissions would be virtually zero since the CO_2 created in combustion would be offset by an equal amount of CO_2 removed from the atmosphere by the next round of cellulosic crops. Most likely, there would still be substantial emissions from agricultural operations including the use of fertilizer, harvesting, transportation, and processing the feedstock into ethanol.

Plug-in hybrid vehicles have larger batteries and more powerful electric motors than hybrid vehicles and can take electricity from the grid as fuel in addition to gasoline. The plug-in hybrid represented in Figure 12.10 is assumed to have enough onboard electricity storage to travel 10 miles on electricity alone. The additional battery capacity and motor power make plug-in hybrids more energy-efficient. Their GHG emissions, however, also depend on how the electricity they use was generated. In 2013 40% of the electricity in the United States was generated by burning coal (EIA, 2013b, Table 1.1). Because California's electricity generation relies much more on natural gas and renewable energy, upstream emissions for plug-in hybrid and battery-electric vehicles with no tailpipe emissions are much lower in California than in other parts of the United States. If all electricity were produced from renewable energy, battery electric vehicles' WTW GHG emissions could be zero.

Emissions for vehicles that use hydrogen gas to generate electricity in fuel cells can also have very low GHG emissions. If the hydrogen is produced from methane by separating the carbon and hydrogen atoms but releasing the carbon in the form of CO_2, the fuel cell vehicle's emissions would be comparable to a plug-in hybrid. However, even if the hydrogen were produced from coal, the GHG impacts would be greatly reduced if the CO_2 emissions were captured and stored. At present, however, it is not known whether carbon capture and storage will be practical. Hydrogen produced from biomass or via electrolysis of water using either wind or solar power would also have nearly zero WTW GHG emissions.

Total GHG emissions by transportation depend on the distribution of passenger and freight activity among modes, the types of fuels used and their (fuel cycle) carbon intensity, and the efficiency with which each model transforms energy into useful work. Let Q_m be the quantity of transportation done by mode m, let e_{mf} be the energy efficiency of mode m when using fuel f, let S_{mf} be the share of mode m's activity that is powered by fuel f, and let C_f be the greenhouse gas intensity of fuel f, in CO_2 equivalents. Total GHG emissions are given by the following identity, known as the Kaya identity (Kaya & Yokobori, 1998).

$$C = \sum_m \sum_f Q_m S_{mf} e_{mf} C_f$$

The utility of this equation is that it neatly summarizes the options for reducing transportation's GHG emissions: (1) reduce the quantity of transportation, (2) shift activity to less carbon-intensive modes, (3) increase the efficiency of energy use and, (4) reduce the carbon intensity of the fuels used.

The GHG emissions of future vehicles and fuels will strongly depend on how and from what those fuels are produced. If the nations of the world decide to take serious steps to mitigate GHG emissions, future emission from all processes from electricity production to vehicle manufacturing will have much lower GHG emissions. As a consequence, battery electric and hydrogen fuel cell vehicles will have far lower emissions than they do today. In 2013, plug-in electric vehicles accounted for less than 1% of motor vehicle sales in the United States and far less than 1% of all vehicles on U.S. roads. Hydrogen fuel cell vehicles were not available to the general public. It will be decades before battery electric vehicles (BEVs) or fuel cell vehicles (FCVs) could become a large fraction of the total vehicle population, if they ever do. Therefore, it is the future carbon intensity of transportation fuels, not the present carbon intensity, that will determine the long-run impacts of these advanced technology vehicles.

ENERGY EFFICIENCY

Energy efficiency is not only a key determinant of energy use and GHG emissions; it has also proven to be the most effective target for public policies aimed at reducing transportation's petroleum dependence and GHG emissions. The main reason is the energy inefficiency of today's transportation system. Only about one-sixth of the chemical energy of the fuel in the tank of a modern automobile is transformed into power to move the vehicle. As shown in Plate 12.1, most of the energy is lost in the internal combustion engine. The way vehicles are designed and used compounds the energy inefficiency. The average weight of model year 2013 passenger cars and light trucks in the United States was 4,040 pounds (EPA, 2013b, Table 2.1). The average occupancy per mile was about 1.6 persons and the average weight of an American adult is about 175 pounds. Thus, the ratio of the passenger "payload" to the total weight of vehicle and passengers is $(1.6 \times 175)/(4040 + 1.6 \times 175) = 0.065$: 93.5% of the total weight moved in cars is in the vehicle itself. If we allocate the 16% of the energy in the fuel that powers the wheels in proportion to the weight of the payload, the overall system efficiency for moving passengers is $0.065 \times 0.16 = 0.01$, or 1%. Although freight transport is generally much more efficient on the basis of weight moved, there is obviously enormous potential for improving energy efficiency.

The Energy Paradox

There is not yet a consensus concerning the second reason public policy appears to have been very effective in raising energy efficiencies: a phenomenon called the "energy paradox." The energy paradox describes the fact that assessments of energy technology typically find a large potential to improve energy efficiency at a cost that is more than paid back by the energy saved over the life of the equipment (e.g., Gillingham & Palmer, 2013). The paradox arises from the expectation that markets will routinely adopt cost-effective technologies. If markets routinely fail to adopt cost-effective options to increase energy efficiency, government policies like fuel economy standards can produce substantial energy savings at little or no net cost. The nature of the energy paradox will be considered in greater detail later in this chapter because of its importance to public policies to promote energy efficiency.

BOX 12.2

Measuring Energy Efficiency

The energy efficiency of a process is the ratio of useful work output to the energy input. For an automobile, this could be the work done at the wheels divided by the chemical energy in the fuel. However, energy is also used producing, transporting, and refining oil and distributing petroleum fuels. A full fuel-cycle energy efficiency calculation includes all such "upstream" uses of energy. In the vehicle example, the energy services created were passenger miles, not vehicle miles. But not all passenger miles are equal. From an economic perspective the value of the travel produced is the correct measure of output. Different measures are useful for different purposes.

In the United States, vehicle efficiency is typically measured in miles per gallon (MPG). MPG is a measure of productivity: miles of output per gallon of input. Most countries measure the energy efficiencies of vehicles in liters of fuel per 100 kilometers, the inverse of fuel economy (in metric units). This measure is a better way to compare the fuel consumption and fuel costs of different vehicles. For example, a 20 MPG vehicle consumes 0.05 gallons per mile while a 40 MPG vehicle consumes half as much, saving 0.025 gallons per mile. A 60 MPG vehicle, however, will save only 0.00833 gallons per mile versus the 40 MPG vehicle. This would be easy to see if we measured energy efficiency in gallons per 100 miles. The three vehicles would be rated 5.0, 2.5, and 1.67 gallons per mile, respectively, and it would be clear that much less energy is saved by the increase from 40 MPG to 60 MPG than by the increase from 20 to 40. Apparently many consumers are misled by comparing vehicles in MPG rather than gallons per mile (Larrick & Sol, 2008).

Every new car sold in the U.S. has a fuel economy rating and a warning that "Your mileage may vary." The U.S. Environmental Protection Agency requires that manufacturers measure their vehicles' fuel economy over standardized test cycles under laboratory conditions. Prior to 2008, fuel economy was measured on two test cycles, the slower, stop-and-go city drive cycle and the faster, less variable highway cycle. Studies of real world driving have shown that these two cycles under-represent high-speed and rapid acceleration driving styles, as well as the impact of air conditioners and operation in cold temperatures. Since 2008, the fuel economy labels have been based on five test cycles that include all the missing elements. Still, no fixed test cycle will exactly match your traffic conditions or your individual driving style. The label is necessarily an estimate of the fuel economy of the average driver. Someday, you will be able to download your driving history from your vehicle's computer, upload it to a website like www.fueleconomy.gov or to an app on your mobile device and get a much more accurate, personal fuel economy estimate.

Energy Intensity

Energy intensity is the inverse of energy efficiency and measures how much energy is required to produce a certain amount of transportation. The average energy intensities of passenger modes in megajoules per passenger kilometer are shown in Figure 12.11. In the 1960s and early 1970s, air travel required more than twice as much energy per passenger kilometer as travel by automobile. The transition from turbo-jet to much more efficient turbo-fan engines, plus much higher load factors (passenger-kilometer/seat-kilometer) enabled by deregulation, reduced the energy intensity of commercial air travel to a level below that of automobiles. However, the order of

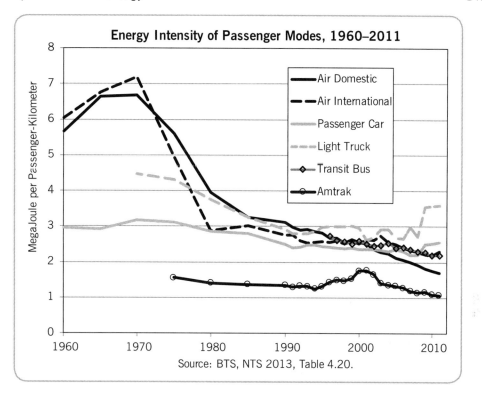

FIGURE 12.11. Energy intensities of passenger modes, 1960–2011. Source: Bureau of Transportation Statistics (2014, Table 4.20).

magnitude greater speeds of jet aircraft allow many more kilometers to be traveled in the same time budget. Intercity rail (Amtrak) is the least energy-intensive passenger mode.

On average, transit buses (and some rail transit operations) are no less energy-intensive than passenger cars. But averages don't tell the whole story because the energy intensity of mass transit modes is primarily determined by load factors. Empty seats increase energy intensity per passenger-kilometer. The energy intensities of urban heavy rail transit systems range from 1.2 MJ/P-kilometer in New York City where population densities are high and rail transit is extensive and well connected, to four times that energy per P-kilometer in San Juan and Baltimore. Similar differences can be seen across cities in light rail and commuter rail systems (Davis et al., 2013, Figures 2.2, 2.3, and 2.4).

Freight energy intensities are less well known because of a lack of data on ton kilometers transported by mode for trucks, ambiguities in allocating commercial air energy use between passenger and freight operations, and incomplete information for ships. Following the oil crisis of 1973–1974, rail energy intensity decreased sharply from 0.45 MJ/ton-kilometer to 0.25 in 1990 and to less than 0.2 in 2010, an average rate of decrease of 2% per year for four decades (Davis et al., 2013, Table 2.15). The change is partly due to increased car loading and longer length of hauls for raw materials like coal, and partly due to continuous improvements to locomotives, rolling stock, and operations. Heavy trucks have not achieved such dramatic improvements. Per vehicle kilometer, tractor trailers and heavy single-unit trucks reduced their energy use by only about 15% from 1970 to 2011 (from 16.4 to 14.2 MJ/kilometer). Trucks have become larger and heavier, however, so the decrease in energy use per ton-kilometer was probably larger.

U.S. Government Regulations and Fuel Economy

Over the past half century, fuel economy improvements to passenger cars and light trucks have achieved the greatest reductions in petroleum use and GHG emissions from the U.S. transportation sector. In 1975, less than 2 years after the oil crisis of late 1973, the average on-road fuel economy of a light-duty vehicle (passenger cars and light trucks) sold in the United States was 13.1 miles per gallon (MPG) (EPA, 2013b, Table 2.1). In December 1975 the Corporate Average Fuel Economy (CAFE) standards were enacted by the federal government and signed into law, to take effect in model year 1978. The standards required each manufacturer's new cars to average 27.5 MPG by 1985 (Figure 12.12). Light truck standards were established later by rule making by the Department of Transportation. For purposes of compliance with the CAFE standards, fuel economy was measured under laboratory conditions previously established by the Environmental Protection Agency (EPA) to measure air pollutant emissions. Despite large changes in gasoline prices (see Figure 12.5), new car fuel economy closely followed the fuel economy standards. After the collapse of

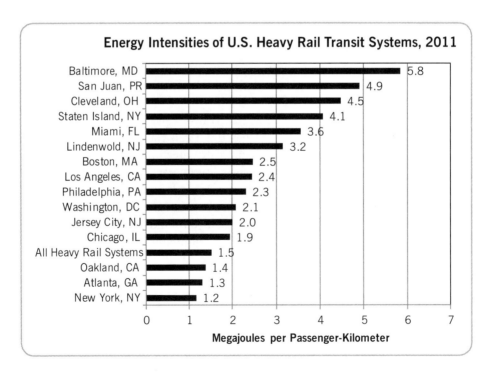

FIGURE 12.12. Energy intensities of U.S. heavy rail transit systems, 2011. Souce: Davis et al. (2013, Figure 2.3).

world oil prices in 1986, the CAFE standards were briefly lowered, then restored to the 1985 level and held there until 2005 when the Department of Transportation (DOT) raised the light truck standards via rule making. In 2007 the Congress enacted and President Bush signed the Energy Independence and Security Act, establishing a target of 35 MPG for passenger cars and light trucks by 2020. Also in 2007, the Supreme Court ruled that the EPA was obligated under the Clean Air Act to determine whether GHG were an air pollutant. In 2009 the agency ruled that GHG endangered the health and welfare of Americans and were therefore a pollutant covered by the Clean Air Act. Because CO_2 emissions from the combustion of petroleum fuels are the predominant GHG produced by motor vehicles, the authority to regulate motor vehicle and fuel economy, though not identical, are closely related. In 2010, President Obama directed the EPA and the National Highway Traffic Safety Administration (NHTSA) to jointly establish harmonized fuel economy and GHG emissions standards.

Prior to 1975, fuel use and vehicle miles traveled by light-duty vehicles increased in virtual lock step. Fuel economy improvements after 1975 broke the link creating a widening gap between vehicle use and fuel use (Figure 12.13). If the miles traveled in 2012 had been driven at the fuel economy of 1975, U.S. motorists would have had to purchase an additional 75 billion gallons of gasoline. Subtracting the motor fuel tax (approximately $0.42 per gallon) the fuel economy improvements are saving roughly one-quarter of a trillion dollars every year.

In 2012, the EPA and DOT jointly issued rules for regulating fuel economy and GHG emissions that raised the unadjusted fuel economy target for passenger cars

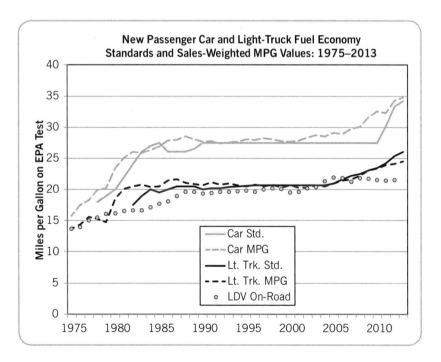

FIGURE 12.13. New passenger car and light-truck fuel economy standards and sales-weighted MPG values: 1975–2013. Source: Davis et al. (2013).

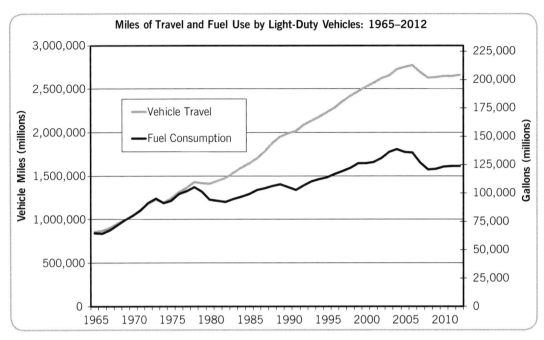

FIGURE 12.14. Miles of travel and fuel use by light-duty vehicles: 1965–2012. Source: U.S. Department of Transportation, Federal Highway Administration, various years, *Highway Statistics*, Table VM.1.

and light trucks to almost twice the 1985 standard. The federal agencies estimated that new passenger cars sold in 2025 would average 52.1 miles per gallon and new light trucks would average 37.6, for a combined average of 46.2 for all new light-duty vehicles (NHTSA, 2012, Tables 4a–c). In 2011, the EPA used its authority under the Clean Air Act to set the first GHG emissions standards for medium and heavy-duty vehicles. The EPA and DOT estimated that the truck and bus standards would cost-effectively reduce GHG emissions by 270 million metric tons and save 530 million barrels of oil over the life of vehicles built from 2014 to 2018 (EPA/DOT, 2011).

Before determining the fuel economy and GHG targets, the agencies compiled a list of technologies that could be used to increase vehicle fuel economy. They collected information from manufacturers and research organizations and conducted computer simulation modeling to estimate the energy consumption of vehicles equipped with multiple advanced technologies. The objective of these studies was to establish feasible levels of fuel economy rather than to predict how manufacturers will actually achieve the fuel economy standards through 2025. The agencies' assessment found that a near doubling of fuel economy could be achieved primarily by making conventional gasoline vehicles more energy-efficient. Engines can be downsized while maintaining power through better design and turbocharging. Transmissions can be made more efficient while increasing the number of gears to allow engines to run more of the time at their most efficient operating points. Tires can be developed with lower rolling resistance while maintaining traction. Vehicle bodies can be made more aerodynamic and lighter by substituting aluminum and high-strength steel for denser materials. The agencies' analyses indicated that all of these changes could be accomplished without requiring substantial numbers of plug-in electric or hydrogen fuel cell vehicles.

The federal agencies' assessment of fuel economy potential was yet another illustration of the energy paradox. The agencies estimated that the technologies needed to meet the 2025 fuel economy standards would increase the average price of a light-duty vehicle by approximately $1,900 but save the average driver four to five times that amount in fuel costs over the life of the vehicle for a net savings of $5,000–8,000 per model year 2025 vehicle.

The Rebound Effect

Increasing the energy efficiency of vehicles decreases the energy cost of driving a vehicle mile. Assuming for the moment that the increased energy efficiency did not increase the cost of the vehicle and did not affect its other attributes, the overall cost of driving would also be reduced. As vehicle travel is an economic good, decreasing its price should increase its consumption. More driving requires additional energy, thereby "taking back" some of the energy savings that would have occurred had vehicle use not increased. The increased demand for the services produced by energy caused by increased energy efficiency is known as the "rebound effect" (Greening, Greene, & Diffiglio, 2000). An increase in driving could be important because it diminishes the reduced energy use and negative externalities like GHG emissions brought about by improved energy efficiency. More driving would also tend to increase traffic congestion.

There is no question that the rebound effect occurs, but debate swirls around its magnitude. Although not all empirical studies agree, a reasonable consensus holds that the rebound effect for motor vehicle travel in the United States is relatively small, on the order of a 0.5 to 2% increase in vehicle travel for every 10% increase in fuel economy. Put another way, for every 10% increase in fuel economy, there is an 8 to 9.5% reduction in energy use (Greene, 2012; Hymel, Small, & Van Dender, 2010). The range is partly due to difficulties in measuring the rebound effect and partly due to the fact that the size of the rebound effect almost certainly increases as fuel prices increase. For the same reason, the rebound effect decreases as fuel prices decrease and as fuel economy increases. Today's average passenger car or light truck gets about 25 miles per gallon. If gasoline costs $3.50 per gallon, it takes $0.14 worth of fuel to drive a mile. To drive 10,000 miles in a year would mean a fuel bill of $1,400. A 10% increase in fuel economy would save $0.014 per mile and $140 per year, which might encourage a motorist to drive a bit more. But if cars could get 100 MPG, the cost of fuel to drive 1 mile would be only 3.5 cents. A 10% increase in fuel economy would save $0.0035, a fraction of a penny per mile or $35 per year, which would undoubtedly encourage a smaller increase in driving. In the limit, as fuel economy approaches infinity, fuel costs disappear and incremental reductions would have no effect on vehicle use.

Consumers and Loss Aversion

However, increasing vehicle fuel economy usually requires adding more costly technology, and increased costs are passed along to the customer. Because in the long run vehicle travel requires not only fuel but a vehicle, higher vehicle costs increase the long-run cost of vehicle travel, offsetting some of the fuel economy rebound effect.

While the rebound effect is reasonably well understood, consumer demand for energy efficiency is not. Given the importance of energy efficiency, this lack of knowledge is both surprising and disturbing. Advances in the field of behavioral economics may finally provide a satisfactory explanation. Basic economic theory says that economically rational consumers will be willing to pay for increased fuel economy up to the point where the increased cost of another mile per gallon equals the value of fuel saved by the additional fuel economy. Manufacturers, recognizing consumers' willingness to pay, will supply exactly the right amount. Since the fuel savings occur in the future, the economically rational consumer would discount the future fuel savings to their present value, but few households make any quantitative calculations to estimate the value of fuel economy when comparing vehicles (Turrentine & Kurani, 2007). However, applications for making such calculations are now available on the Internet (see, e.g., *www.fueleconomy.gov*). Why would consumers who spend $1,000–2,000 each year on gasoline not calculate and compare fuel costs before buying a new vehicle?

In fact, it is not possible to know the value of future fuel savings precisely. Future fuel savings will depend on the price of fuel, but what will the price of gasoline be 5, 10, or 15 years from now? The value of future fuel savings is uncertain, which makes paying more today to get future fuel savings a risky bet. The most firmly established theory of behavioral economics is that human beings are loss-averse; when faced with a risky bet, they count potential losses at twice the value of potential gains (Kahneman, 2011).

By measuring the uncertainties in fuel prices, fuel economy, and other factors, it has been shown that loss aversion could explain most or all of the energy paradox for fuel economy (Greene, 2011; Greene, Evans, & Hiestand, 2013). Although the theory of loss aversion seems to be a promising explanation for the energy paradox, it is not yet accepted, and the existence of the energy paradox remains controversial.

ALTERNATIVE ENERGY TECHNOLOGIES AND SOURCES FOR TRANSPORTATION

Even the most energy-efficient petroleum-powered cars, trucks, aircraft, ships, and locomotives will emit climate-changing GHG and continue transportation's dependence on petroleum. To achieve GHG reductions of 80% will almost certainly require substituting low-carbon energy for petroleum (NRC, 2013). The most comprehensive assessment of the global energy situation to date estimated that to avoid dangerous climate change petroleum use will have to be nearly eliminated by 2050 (GEA, 2012).

There are four main options for drastically reducing GHG emissions and petroleum use by transportation vehicles. The first and most important is increased energy efficiency. As discussed in the previous section, doubling or tripling the energy efficiency of transportation vehicles by 2050 appears to be feasible (NRC, 2013) and would not only cut petroleum use by half or more but also greatly reduce the amounts of alternative fuels that would be needed. The new standards, together with requirements to increase the use of biofuels to 36 billion gallons per year by 2022, would start the U.S. automotive sector on a path toward an 80% reduction in GHG emissions by 2050. However, to achieve that level of reduction by 2050 will require a transition to alternative, low-carbon forms of energy (NRC, 2013). The low-carbon alternatives include powering vehicles with electricity or hydrogen produced from renewable energy (wind, solar, hydropower, geothermal, biomass) or nuclear energy, or produced from fossil fuels with carbon capture and long-term storage, or fuels derived from biomass with minimal GHG emissions.

Natural Gas as an Alternative to Petroleum

Soaring domestic production and low prices have stimulated increased interest in natural gas as a transportation fuel. From 70 to 90% of natural gas is comprised of methane (CH_4) which has the highest hydrogen/carbon ratio of any fossil fuel and the lowest CO_2 emissions per joule (NRC, 2013). On the other hand, methane itself is a far more potent GHG than CO_2, with about 20 times the global warming effect per molecule. Methane leakage of only a few percentage points along the supply chain from well to tank could entirely offset its GHG advantage over petroleum fuels (Alvarez, Pacala, Winebrake, Chameides, & Hamburg, 2012). Natural gas is used around the world to power light- and heavy-duty vehicles, and is stored onboard in either compressed to 3,000–3,500 psi (pounds per square inch) or cooled to −260° F in liquefied form. As noted above, energy density and the cost of onboard storage tanks are the greatest disadvantages for natural gas as a transportation fuel. In many cases, these advantages are outweighed by its lower cost. Since the United States is expected to export natural gas for some years to come, using natural gas to power motor vehicles will reduce dependence on petroleum. Greater use of natural gas, especially in heavy-duty vehicles, would enhance energy security and probably contribute to reducing GHG emissions as well.

Biofuels as an Alternative Source

Biofuels could be an important part of the solution, but the supply of biofuels is limited. Today, almost every gallon of gasoline sold in the United States contains biofuel in the form of 10% ethanol by volume. The term "biofuel" refers to any liquid fuel made from biomass. Ethanol blended into gasoline is the most successful nonpetroleum source of energy for transportation since coal was phased out after 1950. Almost all of the ethanol used in the United States in 2012 was produced by fermentation of corn and distillation. Unfortunately, this is a relatively energy- and fertilizer-intensive method of making ethanol. As a consequence, the WTW GHG emissions from corn-derived ethanol use are only a little lower than gasoline from petroleum. In addition, making motor fuel from a food staple creates an undesirable conflict between food production and energy production (Hill, Nelson, Tilman, Polansky, & Tiffany, 2006). The same applies to biodiesel made from plant oils, although the GHG benefits tend to be somewhat greater.

Biomass can also be used to produce conventional hydrocarbon fuels like gasoline, diesel fuel, and jet fuel either by gasification and synthesis or by pyrolysis.

Gasification involves heating biomass in the presence of steam but not oxygen. This produces synthesis gas, a mixture of carbon monoxide (CO) and hydrogen (H_2) that can be recombined in a catalytic process to produce virtually any hydrocarbon fuel. Pyrolysis involves heating biomass in the absence of oxygen to produce a synthetic form of crude oil that can be refined to produce gasoline. The advantage of these processes is that they use the entire plant including cellulose and lignin, and produce a "drop-in" fuel compatible with existing internal combustion engines as well as with the existing distribution and refueling infrastructure (NRC, 2013). Gasification and synthesis of gasoline is proven technology but is also more expensive than conventional gasoline. Production by means of pyrolysis has yet to be proven at a commercially viable scale.

If biofuels are to be an important source of low-carbon energy for transportation in the future, they must be made from crops that do not compete significantly with food crops; they must be made primarily from cellulose and lignin rather than sugar, starch, or oils; and their production must not induce land use changes that significantly increase GHG emissions (NRC, 2013). Even so, the quantity of energy available from biofuels will be limited to much less than transportation's total energy needs. Even the most ambitious estimates of the potential to produce biofuels in the United States conclude that at most 50 billion gallons of gasoline equivalent biofuel could be produced (NRC, 2013), only about a quarter of the transportation sector's petroleum equivalent energy. Greatly increasing energy efficiency would reduce transportation's energy requirements so that a greater portion could be supplied by biofuel.

Electric Power as a Future Source of Fuel

Many of the first automobiles in the United States were powered by electric motors using energy stored on board in batteries (Wakefield, 1994). In 1997, the electric vehicle began a renaissance of sorts when Toyota Motor Company introduced the first mass-produced hybrid electric vehicle, the Prius, in Japan. Hybrid electric vehicles (HEVs) combine onboard storage of electricity with an electric motor that provides power to move the vehicle as a complement to an internal combustion engine. HEVs typically also use "regenerative braking" in addition to conventional braking to convert kinetic energy into electricity rather than waste heat. Having a powerful electric motor also allows the gasoline engine to be shut off when it would otherwise idle and to be restarted almost instantly when needed. HEVs are not alternative fuel vehicles because all of the energy to power the vehicle ultimately comes from gasoline. Nevertheless, the arrival of mass-produced hybrid vehicles has spurred improvements in batteries, electric motors, and control systems that benefit all types of electric drive vehicles.

Using their electric motors and internal combustion engines synergistically allows HEVs to downsize the internal combustion engine without loss of acceleration performance. These synergies, plus avoiding idling and recapturing energy via regenerative braking, give HEVs about 33% higher fuel economy than a comparable conventional vehicle (NRC, 2013). Despite their much improved energy efficiency, sales of hybrid vehicles have been disappointing. Fifteen years after they were first introduced to the market, hybrids comprise only 3% of total U.S. car and light truck sales (Figure 12.15). Because of their greater complexity (electric motors and gasoline engines, battery packs, regenerative braking, and engine-off at idle), hybrids cost more to manufacture and are generally priced higher than gasoline-powered vehicles by $2,000 or more. Still, in 2014 the government's fuel economy website listed 17 hybrids that would likely pay back their higher cost with fuel savings in less than 5 years. Except for the Toyota Prius, which has no conventional counterpart, all were selling poorly in comparison to their gasoline-only versions (perhaps owing to the energy paradox).

At present, only two types of vehicles qualify as zero (tailpipe) emission vehicles: electric-only vehicles with energy storage in batteries (BEV) and fuel cell electric vehicles (FCV) powered by hydrogen. Invented in the early 19th century and first used by NASA to provide onboard electricity for space missions, fuel cells produce electricity by reacting a fuel with oxygen. Both electricity and hydrogen can be produced from a variety of energy sources, including renewable energy. Electric-drive vehicles are highly efficient in terms of useful work produced per unit of energy stored in the vehicle's batteries. About 60% of the energy in a BEV's batteries becomes useful work moving the vehicle, making the electric powertrain about four times as efficient as

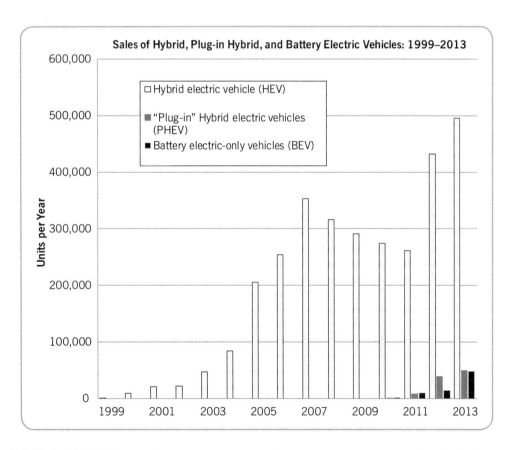

FIGURE 12.15. Sales of hybrid and plug-in electric vehicles in the U.S., 1999–2013. Source: Electric Drive Transportation Association (2012).

an internal combustion engine (see Plate 12.1; *fueleconomy.gov*, 2014).[5] Typically, about 10% of the electrical energy is lost charging the battery and another 10% during discharge. Electric motors are among the most efficient devices that exist for converting energy into work. Including losses for controls and transmission to the vehicle's wheel, 75–80% of the electrical energy to the motor becomes useful work. Hydrogen fuel cell vehicles are somewhat less energy-efficient. About 60% of the hydrogen in the vehicle's tank is transformed into electricity in the fuel cell. FCVs are also powered by efficient electric motors, giving them an overall energy efficiency of nearly 50%, about three times that of today's internal combustion engine vehicles (see Plate 12.1 in the color insert). The WTW efficiencies of BEVs and FCVs can be much lower, however, depending on how primary energy is converted to electricity or hydrogen.

The chief disadvantages of today's BEVs and FCVs in comparison to conventional vehicles are their costs and the difficulty of storing energy onboard the vehicle. The lack of recharging and refueling infrastructure is a much more serious problem for hydrogen FCVs than for BEVs. Electricity is available virtually everywhere in the United States, but recharging takes hours except when expensive fast chargers that can recharge a BEV in less than 30 minutes are available (NRC, 2013). The number of public recharging stations has increased rapidly in the past few years. In early 2014 there were almost 22,000 recharging stations in the United States. The geography of public charging was highly variable, however, with over 5,000 stations in California and only one in Montana (Alternative Fuels Data Center [AFDC], 2014a).

Battery costs are also a serious impediment to the market success of BEVs. In 2013, the lithium-ion batteries to power a modern BEV cost $400–500 per kilowatt-hour

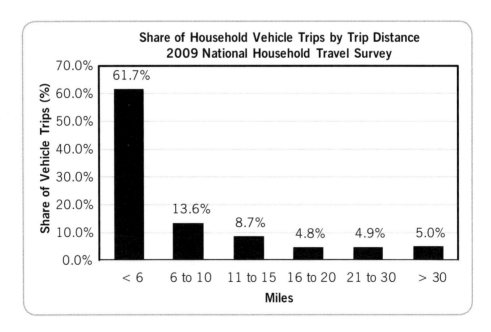

FIGURE 12.16. Share of household vehicle trips by trip distance. Source: Davis et al. (2013, pp. 8–15).

(kW).[6] At this rate, the 24 kWh batteries of a 2014 Nissan Leaf would cost between $9,600 and $12,000. Research goals call for reducing BEV battery costs to $150–200 per kWh. A large fraction of a BEV's battery cost is offset by much lower energy costs but how typical consumers will value the BEV's lower energy costs is not known.

Plug-in hybrid electric vehicles (PHEV) offer a possible solution to the limited range of electric vehicles by combining a larger electric motor and battery than a HEV with an internal combustion engine. Some PHEVs, like Chevrolet's Volt, use only their electric motor for power; the internal combustion engine serves as an onboard generator to extend the vehicle's range. Others use both the electric motor and gasoline engine to provide power. PHEVs are characterized by the distance they are able to travel on all-electric power, usually ranging from 10 to 40 miles. Because 90% of all household vehicle trips in the United States are less than 20 miles, PHEVs could potentially replace a large fraction of household gasoline use with electricity, especially if public recharging were widely available (Davis et al., 2013, pp. 8–15). The PHEVs chief disadvantage is cost. With larger batteries and electric motors than HEVs plus an internal combustion engine, PHEVs cost thousands of dollars more than comparable HEVs.

Hydrogen Power's Potential

Although not nearly as ubiquitous as electricity, the United States produces enormous quantities of hydrogen (about 20 billion kilograms) annually, enough to power 50 million hydrogen fuel cell automobiles (National Hydrogen Association, 2010). A tiny fraction of this hydrogen powers FCVs, however. About half the hydrogen produced is used in petroleum refining, with most of the rest going to the production of ammonia for fertilizer and other uses or to produce methanol. The U.S. Department of Energy counted just 10 publicly accessible hydrogen refueling stations in the United States in early 2014, nine of which were in California (AFDC, 2014b). Clearly, the lack of infrastructure is a much bigger problem for FCVs than for BEVs. The state of California has ambitious plans to promote hydrogen fuel cell vehicles in order to increase the number of zero-emission vehicles on the state's roads. By the end of 2015, California plans to have 68 public hydrogen refueling stations in operation (California Fuel Cell Partnership, 2012). The creation of hydrogen refueling infrastructure is timed to coincide with the announced plans of several major manufacturers to introduce FCVs in low-volume production between 2014 and 2017. Still, this is a very small number of hydrogen stations compared to the approximately 8,000 gasoline stations in the state and 150,000 gasoline stations nationwide. Coordinating the joint evolution of alternative fuel vehicle sales and fuel availability poses a major challenge for the transition to sustainable energy for transportation.

ENERGY FOR SUSTAINABLE TRANSPORTATION

The world's transportation systems are enormous and growing. Their appetite for energy is equally vast. Petroleum's dominance as the source of energy for transportation comes with economic, national security, and environmental costs. Most importantly, the continued dominance of internal combustion engine vehicles powered by fossil fuels is incompatible with avoiding dangerous climate change. Substituting other energy sources for petroleum has proven to be a difficult challenge. Yet it appears to be essential to create sustainable energy for transportation.

The essential concept of sustainability is that the current generation should not limit opportunities for future generations. Altering the world's climate in unpredictable, harmful, and potentially catastrophic ways is clearly inconsistent with the moral imperative of sustainability.

At present, it appears that sustainability will require an 80% global reduction in GHG emissions by 2050. Sustainability does not mean ensuring an infinite and prosperous future for the human race. Instead it is a dynamic concept that allows the current generation to use finite fossil energy resources as long as they drastically reduce GHG emissions and create new opportunities for future generations to use alternatives like renewable or safer nuclear energy. This process will require advancing technology, changing behaviors, and implementing new public policies.

Studies that have attempted to understand what would be necessary to reduce transportation's GHG emissions to sustainable levels have proposed the following steps (GES, 2012; Greene & Plotkin, 2011; McCollum & Yang, 2009).

1. Double or triple energy efficiency
2. Develop advanced, low-carbon renewable fuels
3. Make a transition from petroleum-fueled internal combustion engine motor vehicles to electric drive, battery, and/or fuel cell powered
4. Price energy to better reflect its full economic and social costs

At present, the technology to accomplish a cost-effective transition from petroleum to sustainable energy does not exist. There are no commercially competitive plants producing low-carbon biofuels from lignocellulosic biomass. Batteries for electric vehicles cost two to three times what would allow them to compete successfully against conventional gasoline vehicles, and hydrogen fuel cell systems cost at least twice as much as a comparable internal combustion engine.

Yet enormous progress has been achieved in all these areas over the past two decades, and there is every reason to be optimistic.

Two decades ago, an automotive fuel cell system used up all the available space in a full-sized van. It was impossible to start the systems quickly in winter weather and the best fuel cells wore out after only a few thousand miles of operation. Today, all these technical problems have been solved. Automotive fuel cells take up no more space than a conventional gasoline engine, start in seconds at temperatures down to −30° C, can travel 300 miles on a tank of compressed hydrogen, refuel in 3–5 minutes, and can operate for well over 100,000 miles. The first fuel cell vehicle to be produced on an assembly line, the Hyundai Tucson FCV was delivered to selected customers in 2013. Other manufacturers are preparing vehicles to sell in the 2015 and 2017 model years. But the first FCVs will cost between $50,000 and $100,000 each owing to low-volume production, high-cost carbon-fiber hydrogen storage tanks, and the still-too-large amount of platinum catalyst they require. On the other hand, today's costs are an order of magnitude lower than just a decade ago and promising research is ongoing in all areas. A similar story can be told for battery-electric vehicles. The development of lithium-ion batteries has reduced costs and greatly increased the capacity to store electricity. Further improvements in lithium-ion technology will come, and researchers around the world are investigating promising alternative battery chemistries.

Despite impressive progress and ongoing promising research and development, there is no guarantee of success for any of the key technologies necessary to achieve sustainable energy for transportation. Unfortunately, decades will be required to achieve a large-scale energy transition for transportation, and time is short (Intergovernmental Panel on Climate Change, 2014). To reduce transportation's GHG emissions by anything approaching 80%, energy efficiency must be steadily increased, and early efforts to deploy advanced vehicles and supporting infrastructure must begin despite the fact that some or all of the efforts might fail. Initiating a large-scale energy transition for transportation with uncertain success poses a new and difficult challenge for society.

NOTES

1. Moving information electronically from one place to another has not been considered transportation. Nevertheless, it should be and some day will be. It also requires using energy.

2. U.S. air and water energy use data generally exclude energy used by international travel and transport. Figure 12.1 includes fuel loaded onto international aircraft and ships based on Environmental Protection Agency data (2013b, Tables 3.53 and 3.54).

3. The fact that petroleum fuels can be synthesized from coal or natural gas is not included in these resource estimates.

4. The molecular weight of carbon is 12, while that of oxygen is 16. Thus, oxygen makes up 32/48, or two-thirds, of the mass of the CO_2 coming out of the tailpipe of a motor vehicle.

5. The energy used to power accessories, especially heating and air conditioning, comprises a much larger share of total use for an electric vehicle than for an internal combustion engine vehicle and is not included in the energy efficiency number cited.

6. KiloWatts (1 kW = 1,000 Watts) are a measure of power, like horsepower. One kW is the equivalent of 1.34 Hp. A kilowatt-hour is a measure of energy equal to an application of 1 kW for 1 hour. One kWh = 3,412 Btu or 3.43 MJ.

REFERENCES

Alternative Fuels Data Center. (2014a). *Energy efficiency and renewable energy.* Washington, DC: U.S. Department of Energy. Available at *www.afdc.energy.gov/data/10366*.

Alternative Fuels Data Center. (2014b). Energy efficiency and renewable energy. Washington, DC: U.S. Department of Energy. Available at *www.afdc.energy.gov/fuels/hydrogen_locations.html*.

Alvarez, R. A., Pacala, S. W., Winebrake, J. J., Chameides, W. L., & Hamburg, S. P. (2012). Greater focus needed on methane leakage from natural gas infrastructure. *Proceedings of the National Academy of Sciences, 109*(17), 6435–6440.

BP. (2013). BP statistical review of world energy, 2013: Historical data workbook. Available at *www.bp.com/en/global/corporate/about-bp/energy-economics/statistical-review-of-world-energy.html*.

Bureau of Transportation Statistics. (2014). *National transportation statistics 2014*. Washington, DC: U.S. Department of Transportation, Research and Innovative Technology Administration. Available at *www.rita.dot.gov/bts/sites/rita.dot.gov.bts/files/publications/national_transportation_statistics/index.html*.

California Fuel Cell Partnership. (2012). *A California road map: The commercialization of hydrogen fuel cell vehicles*. Sacramento, CA: Author.

Dargay, J. M., & Gately, D. (2010). World oil demand's shift towards faster growing and less price responsive products and regions. *Energy Policy, 38*(10), 6261–6277.

Davis, S. C., Diegel, S. W., & Boundy, R. G. (2013). *Transportation energy data book, edition 32* (ORNL-6989). Oak Ridge, TN: Oak Ridge National Laboratory. Available at *http://cta.ornl.gov/data/index.shtml*.

Deutch, J., Schlesinger, J. R., & Victor, D. (2006). *National security consequences of U.S. oil dependency* (Independent Task Force Report No. 58). New York: Council on Foreign Relations.

Electric Drive Transportation Association. (2012). Electric drive vehicle sales figures (U.S. market) EV sales. Available at *www.electricdrive.org/index.php?ht=d/sp/i/20952/pid/20952*.

Energy Information Administration. (2013a). Annual energy outlook 2014: Early release (DOE/EIA-0383ER2014). Available at *www.eia.gov/forecasts/aeo/er/index.cfm*.

Energy Information Administration. (2013b). *Electric Power Monthly*, Table 1.1. Net Generation by Energy Source: Total (All Sectors) 2003. Available at *www.eia.gov/electricity/monthly/epm_table_grapher.cfm?t=epmt_1_1*.

Energy Information Administration. (2016, September 27). Monthly energy review. Available at *www.eia.gov/totalenergy/data/monthly*.

Environmental Protection Agency. (2011, December). *Greenhouse gas emissions from a typical passenger vehicle* (EPA-420-F-11–041). Ann Arbor, MI: Office of Transportation and Air Quality.

Environmental Protection Agency. (2013a, April). *Inventory of U.S. greenhouse gas emissions and sinks: 1990–2011* (EPA 430-R-13–001). Washington, DC: Author.

Environmental Protection Agency. (2013b). *Light-duty automotive technology, carbon dioxide emissions and fuel economy trends, 1975–2013*. Ann Arbor, MI: Office of Transportation and Air Quality. Available at *www3.epa.gov/otaq/fetrends-complete.htm*.

Environmental Protection Agency. (2014). Overview of greenhouse gases: Emissions of fluorinated gases. Available at *http://epa.gov/climatechange/ghgemissions/gases/fgases.html*.

Environmental Protection Agency and Department of Transportation. (2011, September 15). Greenhouse gas emissions standards and fuel efficiency standards for medium and heavy duty engines and vehicles: Final rule. *Federal Register, 76*(179).

Environmental Protection Agency and National Highway Traffic Safety Administration. (2012, October 15). 2012 and later model year light-duty vehicle greenhouse gas emissions and corporate average fuel economy standards: Final rule. *Federal Register, 77*(199).

Gillingham, K., & Palmer, K. (2013). *Bridging the energy efficiency gap* (Discussion Paper RFF DP 13-02). Washington, DC: Resources for the Future.

Global Energy Assessment. (2012). *Global energy assessment—Toward a sustainable future*. Cambridge, UK: Cambridge University Press.

Greene, D. L. (2011). Uncertainty, loss aversion and markets for energy efficiency. *Energy Economics, 33*, 608–616.

Greene, D. L. (2012). Rebound 2007: Analysis of national light-duty vehicle travel statistics. *Energy Policy, 41*, 14–28.

Greene, D. L., Evans, D. H., & Hiestand, J. (2013). Survey evidence on the willingness of U.S. consumers to pay for automotive fuel economy. *Energy Policy, 61*, 1539–1550.

Greene, D. L., Lee, R., & Hopson, J. L. (2013). *OPEC and the costs to the U.S. economy of oil dependence: 1970–2010* (White Paper 1-13). Knoxville: Howard H. Baker Jr. Center for Public Policy, University of Tennessee. Available at *http://bakercenter.utk.edu/homepage/white-paper-1–13-opec-and-the-costs-to-the-us-economy-of-oil-dependence-1970–2010*.

Greene, D. L., & Plotkin, S. (2011). *Reducing greenhouse gas emissions from U.S. transportation.* Arlington, VA: Pew Center on Global Climate Change.

Greening, L. A., Greene, D. L., & Difiglio, C. (2000). Energy efficiency and consumption—The rebound effect—A survey. *Energy Policy, 28,* 389–401.

Hamilton, J. D. (2009, Spring). *Causes and consequences of the oil shock of 2007–08* (Brookings Papers on Economic Activity). Washington, DC: Brookings Institution. Available at *www.brookings.edu/about/projects/bpea/papers/2009/causes-consequences-oil-shock-hamilton.*

Hill, J., Nelson, E., Tilman, D., Polansky, S., & Tiffany, D. (2006). Environmental, economic and energetic costs and benefits of biodiesel and ethanol blends. *Proceedings of the National Academy of Sciences, 103*(30), 11206–11210.

Hymel, K. M., Small, K. A., & Van Dender, K. (2010). Induced demand and rebound effects in road transport. *Transportation Research Part B, 44,* 1220–1241.

Intergovernmental Panel on Climate Change. (2014). Climate change 2014: Mitigation of climate change. Available at *www.ipcc.ch/report/ar5/wg3.*

Kahneman, D. (2011). *Thinking fast and slow.* New York: Farrar, Straus & Giroux.

Kaya, Y., & Yokobori, K. (1998). *Environment, energy, and economy: Strategies for sustainability.* Tokyo: United Nations University Press.

Larrick, R. P., & Soll, J. B. (2008). The MPG illusion. *Science, 320,* 1593–1594.

Michalek, J. J., Chester, M., Jaramillo, P., Samaras, C., Shiau, C. N., & Lave, L. B. (2011). Valuation of plug-in vehicle life-cycle air emissions and oil displacement benefits. *Proceedings of the National Academy of Sciences, 108*(40). Available at *www.pnas.org/cgi/doi/10.1073/pnas.1104473108.*

McCollum, D., & Yang, C. (2009). Achieving deep reductions in U.S. transport greenhouse gas emissions: Scenario analysis and policy implications. *Energy Policy, 37*(12), 5580–5596.

National Highway Traffic Safety Administration. (2012). *Corporate average fuel economy for MY2011-MY2025 passenger cars and light trucks* (Final Regulatory Impact Analysis). Washington, DC: U.S. Department of Transportation.

National Hydrogen Association. (2010). *Hydrogen and fuel cells: The U.S. market report.* Washington, DC: Author.

National Research Council. (2012). *Climate change: Evidence, impacts and choices.* Washington, DC: National Academies Press.

National Research Council. (2013). *Transitions to alternative vehicles and fuels.* Washington, DC: National Academies Press.

Smil, V. (2010). *Energy transitions.* Santa Barbara, CA: Praeger.

Turrentine, T. S., & Kurani, K. S. (2007). Car buyers and fuel economy? *Energy Policy, 35,* 1213–1223.

U.S. Department of Energy and Environmental Protection Agency. (2014, April). All-electric vehicles. Available at *www.fueleconomy.gov/feg/evtech.shtml.*

U.S. Department of Transportation, Federal Highway Administration. (various years). *Highway statistics,* Table VM-1. Glastonbury, CT: Author.

Wakefield, E. H. (1994). *History of the electric automobile.* Warrendale, PA: Society of Automotive Engineers, SAE International.

Yergin, D. (1992). *The prize.* New York: Simon & Schuster.

CHAPTER 13

Social Equity and Urban Transportation

EVELYN BLUMENBERG

Throughout U.S. history, low-income families and minorities have faced numerous inequities related to the provision of transportation infrastructure and services. In the years following the Civil War, free African Americans were prohibited from sharing rail cars with white passengers (Bullard, 2003). Well into the 20th century, the legacy of segregation continued on buses, streetcars, and trains. Prior to the civil rights movement, black passengers in the South were required to sit at the back of the bus and give up their seats to white passengers upon demand; bus and train stations had race-segregated waiting rooms, bathrooms, and concession stands (Bullard, 2003). Inequality was associated not only with the provision of transit services but also with the development of the U.S. highway system. During the period of urban renewal in the 1950s and 1960s, planners routed interstate highways through low-income and minority neighborhoods, demolishing large swaths of affordable housing and severing the social and economic fabric of already vulnerable communities (Mohl, 2004).

Certainly times have changed. Civil rights struggles helped to dismantle racial segregation on public transit, ensuring that African Americans no longer had to use separate transit vehicles or sit at the back of the bus. Technological innovation greatly lowered the costs of automobiles, making them available to families who could not previously afford them. Consequently, automobiles became the dominant mode of travel among all population groups, even the poor. As for highway construction, the Federal-Aid Highway Act of 1956 provided the federal dollars and incentives to spur the freeway construction boom of the 1950s and 1960s (Mohl, 2004).

Despite these changes, inequality remains deeply embedded in the planning and operations of urban transportation systems. Compared to higher-income households, the poor bear a disproportionate share of the burdens while receiving fewer of the benefits of transportation. Disparities in mobility influence individuals' access to opportunities—to employment, healthy food, and health care. The costs of travel—whether they are associated with automobiles or public transit—heavily burden low-income families, squeezing already constrained budgets and further limiting their mobility. These inequities assume a spatial dimension since a high percentage of low-income and minority families remain concentrated in inner-city neighborhoods. Compared to higher-income communities, these urban neighborhoods are more likely to experience transit service cuts and the negative externalities of transportation-related pollutants and less likely to capture the benefits of new transit investments.

In this chapter, therefore, I begin by discussing these inequities organized around the following four themes: spatial access to opportunities, the transportation expenditure burden, public transit service and finance, and the externalities of transportation infrastructure. The content of this discussion draws broadly from current research on transportation inequality; I supplement these studies with U.S. data from the 2010 American Community Survey as well as from the most recent National Household Travel Survey (NHTS; 2009).[1] The chapter concludes with a discussion of poverty as it relates to transportation policy.

ISOLATION OF HOUSEHOLDS WITHOUT ACCESS TO AUTOMOBILES

Given the evolving spatial structure of metropolitan areas, there is compelling evidence that, under most circumstances, adults without access to automobiles are increasingly disadvantaged. In the postwar years, employment steadily decentralized such that only 21% of employees in the largest 98 metropolitan areas now work within 3 miles of the central business district; in contrast, 45% of jobs are located more than 10 miles away from the city center (Kneebone, 2009). The metropolitan population has also suburbanized over time. As Figure 13.1 shows, currently 38% of the population lives in the central city, defined by the U.S. Census Bureau as the principal (largest) cities within metropolitan areas. A higher percentage of the poor (49%) remain in central-city neighborhoods largely to take advantage of the availability of affordable housing and—for those without automobiles—access to high levels of public transit service (Glaeser, Kahn, & Rappaport, 2008). Minority poverty is particularly concentrated in the central city where 63% of poor African Americans and 61% of poor Asians live (U.S. Census Bureau, 2016).

Advocates of the spatial mismatch hypothesis, an idea originally developed by Kain (1968), contend that owing to racial discrimination in housing, low-income residents have been left behind in urban areas, distant from suburban employment opportunities. Kain actually advanced not one but three hypotheses. The first said simply that residential segregation helped shape the geographic distribution of the African American population. Race-based housing discrimination limited African Americans' access to suburban housing, contributing to their concentration in central-city neighborhoods. The second said that residential segregation increased black unemployment, and the final argument was that both of these conditions were exacerbated by the decentralization of employment. It is this final idea—that surplus labor in minority inner-city areas leads to either joblessness or the need to travel

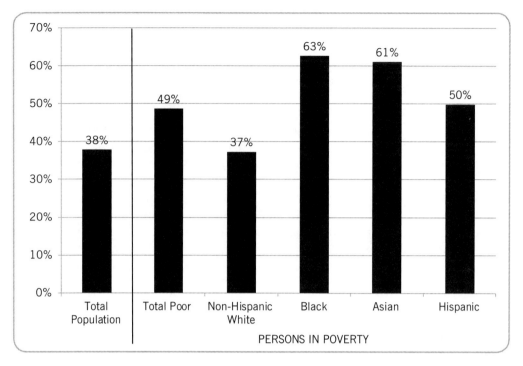

FIGURE 13.1. Metropolitan population in central city by poverty status, race, and ethnicity. Source: U.S. Census Bureau, Current Population Survey, 2016 Annual Social and Economic Supplement.

longer distances (which results in lower net earnings because of travel costs)—that has grown into the modern idea of the spatial mismatch. In Kain's initial investigation, the focus was on African American men in the central city, but since then scholars have applied the concept to a wide range of the urban poor including teens, women, immigrants, other racial and ethnic groups, welfare recipients, and African Americans living in the suburbs. There is a large literature on the spatial mismatch hypothesis; while the specific results vary across studies, a numerical majority is generally supportive of the concept (e.g. Gobillon, Selod, & Zenou, 2007).

A number of scholars suggest that rather than facing the classic "spatial mismatch," low-income, inner-city residents suffer from a modal mismatch, a drastic divergence in the relative advantage between those who have access to automobiles and those who do not (e.g., Taylor & Ong, 1995). In 1992, Kain (p. 392) noted, "None of the spatial mismatch studies, including my original 1968 study, does a good job of dealing with mode choice." Since that time, scholars have worked to correct this omission.

As Figure 13.2 shows, the vast majority of adults in the United States—rich or poor—have access to household vehicles. As of 2014, 95% of all adults lived in households with at least one automobile. Nearly 80% of low-income adults now live in households with vehicles, an increase from just over 50% in 1960. For the remaining 21%, reliable access to vehicles remains out of reach since the costs of automobile ownership can be prohibitive. As of 2010, over 6.5 million poor adults lived

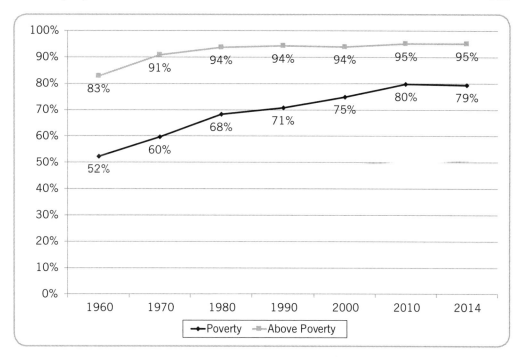

FIGURE 13.2. U.S. adults (16+) in households with vehicles by poverty status (1960–2010). Source: Steven Ruggles, Katie Genadek, Ronald Goeken, Josiah Grover, and Matthew Sobek (2015). *Integrated Public Use Microdata Series: Version 6.0* [Machine-readable database]. Minneapolis: Minnesota Population Center, University of Minnesota.

in households without automobiles. Furthermore, access to vehicles can be limited even in low-income households *with* automobiles since on average there is less than a one-to-one ratio between cars and drivers. Data from the 2009 NHTS show that approximately 25% of the poor live in households with fewer cars than drivers compared to 12% of nonpoor households. Low-income adults, therefore, often compete with other drivers for use of a household vehicle, and therefore may not be able to use a car whenever they might need one. Other population groups are less likely to live in households with automobiles. For example, African American adults are more than three times as likely to live in zero-vehicle households (16%) than non-Hispanic whites (5%). Seniors, a rapidly growing population group, are also less likely to have access to automobiles. Six percent of working-age adults (16–64) live in households without cars compared to 8% of seniors 66 to 84 and more than 26% of seniors 85 and older.

In almost all metropolitan areas in the United States, individuals without reliable access to automobiles can reach far fewer opportunities within a reasonable travel time compared to those who travel by automobile. Even in cities with good transit service, transit travel times are, on average, much longer than automobile travel times—a result of walking to and from transit stops, waits at stops and for transfers, and frequent vehicle stops along the way. Data from the NHTS show that on average transit commutes are 53 minutes, more than twice as long as commutes by private vehicle (23 minutes) (Santos, McGuckin, Nakamoto, Gray, & Liss, 2011).

Consequently, studies of employment access show that automobile commuters have a substantial access advantage over transit commuters in metropolitan areas as diverse as Boston, Detroit, Los Angeles, the San Francisco Bay Area, and Tokyo (e.g., Kawabata & Shen, 2006). For example, in the Los Angeles Pico–Union neighborhood, a high-poverty neighborhood located adjacent to job-rich downtown, residents with cars have access to five times as many low-wage jobs within a 30-minute commute than residents who travel by public transit (Blumenberg & Ong, 2001). A longitudinal analysis of employment access in the Twin Cities shows that the accessibility gap between cars and public transit grew from 1990 to 2000; access by automobile increased at the same time as access by public transit declined (El-Geneidya & Levinson, 2007).

Automobiles offer other important employment advantages. They enable job seekers to search more widely for employment and to travel more easily to multiple, unfamiliar destinations in a single day. For women, particularly single women, who have primary responsibility for the functioning of their households, the ability to sustain employment rests on access to a variety of household-supporting destinations, only one of which is work (Blumenberg, 2004). Cars offer women numerous advantages in balancing both home and work responsibilities. They enable flexibility in trip making, which allows women to more easily and safely manage their multiple responsibilities as heads of households. Low-income women are more likely than men to work nonstandard schedules requiring travel on nights and weekends; cars enable women to travel safely during off-peak hours when transit service is limited and after dark when women's concerns for their personal safety are highest. Compared to public transit, cars also enable women to more easily "trip chain," that is, make multiple stops in a tour. Given the advantages of cars, working mothers—particularly those with young and/or many children—are more likely to drive to work at all income levels than are comparable men or other women.

Given these access advantages, it should come as no surprise that private automobiles are pivotal in improving economic outcomes for low-income and minority adults. Cars can make it easier to search and regularly commute to jobs, and employment can provide households with the resources to purchase automobiles. Regardless, the importance of automobiles to employment persists even when controlling for the simultaneity of the car ownership and employment decision. Automobile ownership is associated with higher employment rates, weekly hours worked, and hourly earnings (Raphael & Rice, 2002). Using a variety of data sources, a number of studies have examined the effect of automobile ownership on outcomes for welfare participants—largely poor, female-headed households. These studies show a positive relationship between the presence of household automobiles and employment rates, the likelihood of exiting welfare, and earned income (e.g., Cervero, Sandoval, & Landis, 2002). Automobile ownership also reduces racial disparities in employment rates (Raphael & Stoll, 2001) and unemployment duration (Dawkins, Shen, & Sanchez, 2005). Finally, studies show automobile ownership to be a far more powerful determinant in job seeking and job retention than public transit (Cervero et al., 2002).

While there is an extensive literature on automobiles and their relationship to employment, there is also a small—but growing—body of scholarship on the role of automobiles in facilitating other healthy behaviors and outcomes. This research runs counter to the recent emphasis in public health and urban planning on the negative health effects of driving associated with living in low-density, auto-oriented environments. Overall, the evidence suggests that cars can make it easier for low-income families to purchase healthier food and travel to health care providers.

Scholars have characterized some low-income and minority neighborhoods in the United States as "food deserts," low-income communities where the availability of healthy and affordable food is limited. Although specific measurements of food deserts differ across studies, research in this area tends to focus on spatial disparities in the location of food stores by the racial and income composition of neighborhood residents. (See, e.g., Walker, Keane, & Burke, 2010, for a review of this extensive body of literature.) A number of studies support the existence of food deserts, revealing racial and income neighborhood disparities in access to food in the United States, particularly access to large full-service supermarkets. Consistent with this finding, residents of some low-income neighborhoods must travel farther to access the same number of supermarkets as residents of nonpoor neighborhoods (e.g., Zenk et al., 2005), and therefore spend more time traveling to grocery stores (19.5 minutes) than the national average (15 minutes) (U.S. Department of Agriculture, 2009).

However, other studies find that many low-income neighborhoods have greater—not less—access to supermarkets than higher-income neighborhoods (e.g., Leete, Bania, & Sparks-Ibanga, 2011). For example, in their analysis of food deserts in Portland, Oregon, Leete and colleagues (2011) found that only a small percentage of the poor (12%), elderly (4%), and those without access to automobiles (9%) live in food deserts—low-income neighborhoods with limited access to grocery stores. However, access to food can still be a challenge for disadvantaged households who are increasingly spatially dispersed in outlying areas. Leete and colleagues found that between five and six times as many poor persons live in what they term "food hinterlands," neighborhoods where food access levels are below the area average.

Whether in food deserts or not, currently six million low-income households reside more than a half mile from a grocery store and do not have access to household vehicles (U.S. Department of Agriculture, 2009). Some studies show that distance is inversely associated with the purchase of healthy foods such as fresh fruit (e.g., Rose & Richards, 2004), although the link to particular health outcomes (obesity, etc.) is difficult to support. Automobiles make it easier to travel longer distances, allowing low-income shoppers to frequent grocery stores outside of their neighborhoods and enabling them to buy and carry large quantities of groceries at one time (e.g., Fuller, Cummins, & Matthews, 2013). A few recent studies suggest that automobiles also may contribute to healthier food consumption. For example, in a study of low-income African American adults, D'Angelo, Suratkar, Song, Stauffer, and Gittelsohn (2011) found that driving to a "food source" was associated with obtaining healthier foods than the use of other modes of travel. Finally, cars allow low-income families to more easily reach health care providers. In general, many families cite transportation as a major barrier to the receipt of health care and, in particular, note the difficulty of traveling to appointments without use of a vehicle (e.g., Guidry, Aday, Zhang, & Winn, 1997). Although the research is sparse in this area, studies show a positive relationship between the lack of an automobile and the greater likelihood of missed appointments (Yang, Zarr, Kass-Hout, Kourosh, & Kelly, 2006); in rural areas those without driver's licenses tend to have fewer visits for chronic care and regular check-ups than those with licenses (Arcury, Presser, Gesler, & Powers, 2005).

THE CONSEQUENCES OF SPENDING TOO MUCH ON TRANSPORTATION VERSUS SPENDING TOO LITTLE

A number of advocacy organizations claim that transportation expenditures—particularly auto-related expenses—are overly burdensome and therefore restrict the economic opportunities of low-income households. These analyses largely find that low- and moderate-income families spend a higher percentage of their incomes on transportation compared to higher-income households and that their transportation expenditure burden has increased over time in conjunction with continued sprawl, growing automobile usage, and rising gas prices.

The evidence, however, does not support this conclusion. Data from the U.S. Consumer Expenditure Survey (U.S. Bureau of Labor Statistics, 2016) show that low-income households devote a slightly smaller percentage of their budgets to transportation than do higher-income households. Figure 13.3 presents expenditure data for 2015 and reveals that households in the bottom income quintile spent 15% of

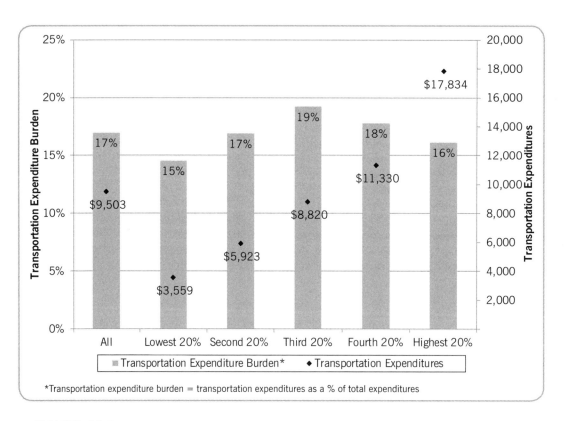

*Transportation expenditure burden = transportation expenditures as a % of total expenditures

FIGURE 13.3. Transportation expenditures by income quintile. Source: U.S. Bureau of Labor Statistics. (2016). *Current expenditure tables, 2015: Quintiles of income before taxes.* Consumer Expenditure Survey. Washington, DC: U.S. Department of Labor.

their total expenditures on transportation compared to an average of 17% for all households. As discussed previously, lower-income households are less likely than higher-income households to own automobiles. Lower-income households *with* automobiles own fewer and less expensive vehicles and travel fewer miles. Poor adults drive vehicles that are, on average, 11 years old compared to 7 years old among non-poor adults (2009 NHTS, author's calculations). The median vehicle miles traveled (VMT) of poor households is 11,600 miles, 65% that of nonpoor households (2009 NHTS, author's calculations)

Low-income households *do* face an expenditure burden, just not as characterized by many scholars and advocates. Low-income families—particularly those with limited access to automobiles—are more likely to live in dense urban areas where they can more easily access public transit (Glaeser et al., 2008). Consequently, for many households a lower transportation expenditure burden is more than offset by the higher cost of housing. The combined housing and transportation burden (H + T) of households in the bottom income quintile is 55%, compared to 50% for all households (Bureau of Labor Statistics, 2016).

Furthermore, low-income households pay more than higher-income households for comparable levels of transportation, a finding that holds true for both automobiles and public transit expenditures. Most low-income households purchase used vehicles, a market fraught with abuse including the sale of vehicles in poor condition, unfair financing arrangements, deceptive sales practices, junk products with fees that add to the vehicle's cost, and frequently fraud (Van Alst, 2009). A high percentage of low-income households are unbanked, and therefore do not have established relationships with mainstream financial institutions (Caskey, 2005). Furthermore, African Americans, Hispanics, young adults, and individuals residing in low-income or minority neighborhoods, on average, have lower credit scores than other population groups (Board of Governors of the Federal Reserve System, 2007). Combined, these factors result in the overuse of alternative financial products and with them, the high costs of payment services (Sutton, 2007). Consequently, low-income and minority buyers tend to pay high prices for lower-quality vehicles, including excessive interest rates and fees (Sutton, 2007).

The poor also face a mix of regressive auto-related operating expenses that contribute to the high costs of auto ownership. Flat fees for driver's licenses, smog checks, and automobile registration, among other things, by their very nature comprise a higher percentage of the budgets of low- than higher-income households. Other fees—while perhaps less regressive than flat fees—also disproportionately burden low-income households. For example, most states levy a tax on motor vehicles based on total purchase price or fair market value of the vehicle. In a study of California's vehicle license fees, Dill, Goldman, and Wachs (1999) showed that these fees comprise a higher percentage of the income of lower-income households—despite having fewer and less expensive vehicles. Automobile insurance rates also place a disproportionate burden on low-income households due to the widespread use of flat rates as well as redlining in low-income and high-minority neighborhoods (Ong & Stoll, 2007). Finally, low-income families have higher vehicle operating costs since they tend to drive older cars that are less fuel-efficient, a problem that can be particularly onerous with increased gas prices. Data from the 2009 NHTS show that within class (e.g., car, van, SUV, truck), the poor own vehicles that are less fuel-efficient than nonpoor households.

In recent years, the financing of transportation systems has shifted from user fees—charges related to the actual use of the system—to local instruments such as local option sales taxes (Goldstein, Corbett, & Wachs, 2001; Wachs, 2003). A sales tax can be politically acceptable as it has the potential to generate enormous revenues by charging a small fee over a large population base; it also allows jurisdictions to collect revenues from nonresidents of tax districts (Goldstein et al., 2001). Local sales taxes—used to fund both road and transit projects—are regressive (Goldstein et al., 2001; Schweitzer & Taylor, 2008). While taxable expenditures are positively related to income, the tax burden—the tax as a percentage of income—is higher for low- and higher-income households than middle-income households (Schweitzer & Taylor, 2008). Furthermore, retail purchases rise more slowly than income, and therefore, once again, disproportionately burden low-income households (Goldstein et al., 2001). If low-income travelers benefit from the locally funded transportation projects, then accessibility benefits may compensate them for their higher costs. Otherwise, if the benefits accrue largely to higher-income households, low-income families will be doubly burdened, first by the tax and then again by the investment itself.

Similar cost disparities occur with respect to public transit. On average, low-income transit users pay more per mile than higher-income transit users. Most transit agencies (except for longer distance, commuter rail systems) have adopted a flat fare structure in which riders pay the same fee regardless of distance or time of day. Compared to higher-income transit riders, the poor travel shorter distances and are more likely to use transit during off-peak hours (Giuliano, 2005). With flat fares—much like flat-rate insurance premiums—low-income transit riders pay significantly more per mile than higher-income transit riders who tend to travel longer distances. Moreover, transit costs are higher during the peak when loads are heavy (Cervero, 1990) yet riders who travel during off-peak—many of them low-income—pay the same fare, thereby cross-subsidizing higher-income, peak-period commuters. Additionally, low-income riders often have difficulty taking advantage of discount pass programs since many of these programs require a lump-sum payment at the beginning of the month, when families have numerous competing expenses (Blumenberg & Agrawal, 2014)

Overall, transportation expenditures can be financially onerous for income-constrained families who struggle to carefully manage their household budgets. In response, they adopt a variety of strategies to cover the costs of their transportation. They limit their transportation expenditures by taking fewer trips and traveling shorter distances than higher income households (e.g., Blumenberg & Agrawal, 2014; Giuliano, 2005). They minimize their auto expenses by finding inexpensive parking, gas, and insurance, buying used cars, driving vehicles until they are no longer operational, and avoiding auto-related fees such as for licensing and insurance (see, e.g., Blumenberg & Agrawal, 2014). Individuals without the resources to own cars often rely on their social networks for rides or to borrow vehicles (e.g. Blumenberg & Agrawal, 2014).

Finally, while transportation expenditures certainly can heavily burden low-income families, the hidden cost is likely not how *much* low-income households spend on transportation but rather just how *little*. Figure 13.3 shows that on average households in the bottom income quintile spend $3,559 on transportation, only 37% of that spent by all households (U.S. Bureau of Labor Statistics, 2016). Transportation is a cost but also provides numerous benefits including the ability to travel to needed destinations. It is likely that low-income households spend so little on their transportation—almost $6,000 less per year than the average household—that their lack of resources results in unmet, or latent, travel demand. In other words, high transportation costs may lead to forgone travel as families attempt to make ends meet by minimizing trip making and vehicle miles traveled. Reduced travel—particularly for nondiscretionary purposes—likely influences the well-being of households, perhaps reducing employment and negatively affecting health and other outcomes.

CURRENT FUNDING PRIORITIES FAVOR HIGHER-INCOME HOUSEHOLDS

Between 1990 and 2000, transit ridership increased by 14%, from approximately 9 billion riders to just over 10 billion riders (American Public Transit Association, 2012). However, the growth in transit ridership has not kept up with the increase in the population; consequently, transit's mode share has continued to decline over time. In urban areas, approximately 2.5% of all trips—and 4.5% of commute trips—are now made on public transit (NHTS, 2009; author's calculations). Public transit works best in dense urban areas where origins and destinations are proximate and where road congestion and high parking fees make driving less attractive. As urban areas have dispersed, transit agencies have had trouble maintaining and attracting discretionary riders, many of whom find it faster and more convenient to travel by automobile. The one remaining market for higher-income transit users is commute travel into downtown areas where concentrated employment, expensive parking, and congestion make transit a reasonable substitute to driving.

However, for low-income households—particularly those without reliable access to automobiles—public transportation is essential. Data from the 2009 NHTS show that in metropolitan areas public transit comprises 10% of all trips among low-income adults—about three times that of all adults. As Figure 13.4 shows, transit use is highest among low-income African Americans who are almost six times as likely to use transit as all adults. In addition to serving a heavily poor and minority ridership base, bus transit also plays an especially crucial role in the transportation lives of poor and working-class women and immigrants. Women have less access to cars than men in households with fewer vehicles than drivers, are frequently burdened with childcare responsibilities, and can be especially vulnerable to safety issues brought on by infrequent bus service and overcrowding. Immigrants are also highly dependent on public transit particularly during their first few years in the United States when they are less likely to have driver's licenses and the resources necessary to purchase vehicles (Blumenberg & Evans, 2010).

For adults without cars, higher levels of transit-based employment accessibility can increase the probability of employment and of working 30 hours or more per week (Kawabata, 2003; Yi, 2006). Similarly, studies of welfare recipients show that

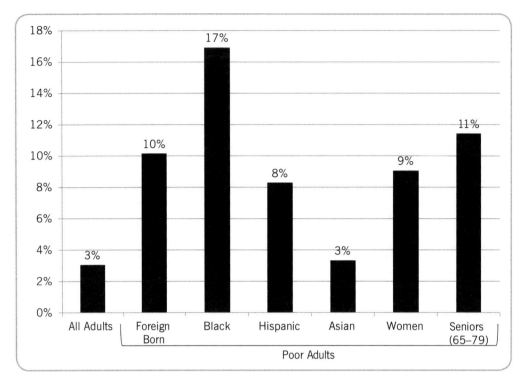

FIGURE 13.4. Percentage of trips taken using mass transit—all adults and poor adults by race/ethnicity (NHTS, 2009). Source: National Household Travel Survey (2009).

transit service can result in improved economic outcomes (Ong & Houston, 2002). Yet, despite the importance of public transit to the lives of the poor, transit users face a set of equity concerns with respect to the divergent properties of bus and rail transit systems, which differ in the extent to which they serve disadvantaged populations and the extent and manner in which they are subsidized.

While nonwhite and poor travelers make disproportionately heavy use of public transportation, they are much more likely to rely on bus than on rail transit. As Figure 13.5 shows, 73% of all bus transit trips are made by nonwhite riders compared to 55% of rail transit trips (NHTS, 2009; author's calculation). Beyond these disparities in the racial and ethnic composition of their respective ridership, bus and rail transit also differ in terms of the economic status of their riders. Forty-three percent of bus trips are made by poor riders compared to only 12% of rail trips, and the mean family income of bus riders is approximately a third that of rail riders (NHTS, 2009; author's calculation).

Between 2002 and 2011, fixed-route bus transit accounted for 31% more trips than did rail transit, 52.3 billion versus 40.0 billion (National Transit Database, 2002–2011). Over this same time period, however, rail transit garnered a slightly higher overall subsidy, an investment largely justified by efforts to attract discretionary riders and in so doing to reduce automobile travel and its negative externalities (Giuliano, 2005). In constant 2011 dollars, bus transit subsidies over this period totaled $163.1 billion, compared to $167.6 billion for rail transit subsidies. Thus, rail transit was subsidized at $4.19 per trip, a figure 34% higher than the comparable bus transit subsidy of $3.12 per trip (National Transit Database; author's calculations). These disparities in aggregate nationwide funding levels translate to higher per-passenger subsidies of rail users within individual metropolitan areas, such as Los Angeles and Atlanta (Bullard, Johnson, & Torres, 2004; Grengs, 2002). Given the average socioeconomic disparities between bus and rail passengers, these funding priorities lead to greater per-trip transit subsidies accruing to wealthier, nonminority transit users (Bullard et al., 2004; Grengs, 2002).

Beyond the disproportionate level of subsidies going to rail travel, federal transit funding structures can also leave local bus systems imperiled during economic downturns. As discussed previously, bus transit systems rely to a great extent on local and state funds to maintain operations; when these funding sources become constrained, local bus systems can be forced to reduce service and pass a greater cost burden onto their already economically distressed users. During the 2009 calendar year, over 80% of large public transportation agencies experienced a reduction in local or regional funding, and over 70% saw a reduction in state funding (American Public Transit Association, 2010). In 2009 and 2010 the decline in funding forced many agencies to cut service and/or to increase fares. Some bus advocates argue that there has been an implicit bias in the implementation of these service cuts. They contend

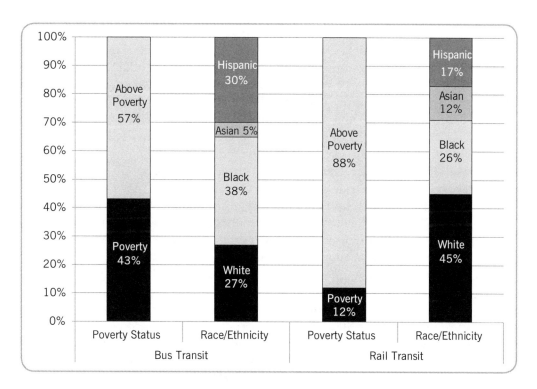

FIGURE 13.5. Poverty and racial/ethnic composition of transit trips. Source: National Household Travel Survey (2009).

that transit agencies have cut cost-effective transit lines that provide needed service to and from low-income communities while continuing to invvest in expensive rail transit service (Romann, 2011).

Although the use of public transit is far less expensive to the user than automobile ownership, low-income transit riders tend to be very poor and therefore more concerned about the costs of public transit than higher-income, discretionary riders (Blumenberg & Agrawal, 2014; Giuliano, 2005). With limited access to other modes of travel, captive riders have little choice but to endure fare increases, albeit unhappily. For many transit agencies, increases in transit fares compound already inequitable fare structures, such as the flat fares discussed previously.

Title VI of the Civil Rights Act of 1964 prohibits discrimination on the basis of race, color, and national origin in programs and activities receiving federal financial assistance. The federal government, through the Federal Transit Administration, provides financial assistance to transit agencies and metropolitan planning organizations to develop new transit systems and to improve, maintain, and operate existing systems. As such, these agencies must comply with Title VI. In 2001, however, the U.S. Supreme Court in *Alexander v. Sandoval* held that there was no private right of action to enforce Title VI's disparate impact regulations. This ruling made it much more difficult for individuals, who must now prove discriminatory intent rather than simply disparate treatment.

Income and racial disparities in transit subsidies have motivated civil rights lawsuits and/or complaints against a number of public transit agencies in Atlanta; Boston; Chicago; Houston, King, Pierce, and Snohomish Counties in Washington; Livonia; Los Angeles; Memphis; New York; St. Louis; and the San Francisco Bay Area (e.g., Grengs, 2002; Mercantonio, 2011). In most of these cases, the Court ruled that there was no violation and, therefore, took no action. In two of these cases, however, the court ruled in favor of the plaintiffs. In *Labor/Community Strategy Center v. Los Angeles County Metropolitan Transportation Authority* (LACMTA) (1994), the Bus Riders Union filed an administrative complaint against the LACMTA in which they argued that the agency's decision to eliminate over 387,500 hours of bus service while preserving existing levels of rail transit service had a discriminatory impact on minorities and low-income riders since they were substantially more likely to use the bus system (Grengs, 2002). In 1996 the district court approved a consent decree in which LACMTA was instructed to improve its bus system by reducing fares and overcrowding, and providing improved transit service to major centers of employment, education, and health care. When LACMTA did not meet the service improvement goals set forth in the decree, in 1999 the district court entered an order requiring the LACMTA to acquire 248 additional buses to reduce passenger overcrowding. The appeals court concurred, stating that consent decree imposed an "obligation" on the LACMTA "to meet the scheduled load factor targets."

Similarly, in *Metropolitan Atlanta Transportation Equity Coalition* (MATEC) *v. MARTA,* MATEC alleged that MARTA, the Metropolitan Atlanta Rapid Transit Authority, provided disparate service to white and minority communities—less transit service, fewer clean compressed natural gas buses and bus shelters, and inadequate security at rail stations in minority neighborhoods (Bullard, 2003). Additionally, they were concerned about fare increases and transit access for disabled riders. MARTA worked with MATEC to achieve a mediated settlement that included a negotiated agreement regarding fares and vehicle assignment (Ward, 2005).

Despite the resolutions in these two cases, the tension between bus riders and transit agencies—and between investments in bus and rail systems—continues. For example, in Los Angeles, the consent decree was lifted in 2006; since that time, the LACMTA has reduced bus service to, as they state, cut costs and make the system more efficient. In November 2010, the Bus Riders Union prepared a Title VI administrative complaint against the LACMTA alleging that the agency's decision to eliminate over 387,500 hours of bus service while expanding rail service, once again, has a discriminatory impact on minority and low-income riders. While the FTA dismissed the complaint, it conducted a Title VII audit of the agency and instructed the agency to address a number of civil rights violations.

LOW-INCOME NEIGHBORHOODS AND NEGATIVE EXTERNALITIES OF TRANSPORTATION INVESTMENTS

Not only do low-income individuals and their households face transportation inequities, but also so too do the low-income neighborhoods in which many of them live. Poverty has suburbanized over time (Garr & Kneebone, 2010). Yet low-income and minority households still disproportionately reside in high-poverty and racially segregated neighborhoods, a trend that—at least among the poor—worsened during the economic crises of the 2000s. Figure 13.6 shows the distribution of the population by race/ethnicity and poverty across neighborhoods of varying levels of poverty. Poor, African American, and Hispanic residents are significantly more likely to live in neighborhoods with poverty rates of 20% or higher. A recent study shows that 28% of low-income households live in neighborhoods that are majority low-income, up from 25% in 2000 (Frey & Taylor, 2012). Additionally, the population living in extreme-poverty neighborhoods rose during this period such that 10.5% of the poor now live in neighborhoods where at least 40% of residents live below the poverty line (Kneebone, Nadeau, & Berube, 2011). The poor are highly segregated, in large part due to the spatial concentration of minorities (Massey & Fisher, 2000). Despite the steady decline in black–white segregation since 1990, in 2010 segregation—measured as the percentage of a minority group that would need to move to be distributed like the white population—remains highest for African Americans (55%) and is still sizeable among Hispanics (43%) and Asians (40%) (Frey, 2011).

Among the many issues facing low-income neighborhoods, residents have suffered from the disproportionately high and adverse effects of transportation facilities located in their communities. Historically, as part of broader efforts to remake urban neighborhoods, transportation agencies routed freeways through low-income, African American neighborhoods; in so doing, thousands of low-income families were displaced and many inner-city communities devastated (Mohl, 2004). Highway policies have shifted as some cities—such as Boston, Oakland, and San Francisco—have torn down elevated, inner-city expressways to "reclaim portions of cities decimated by highway construction, and to restore more human-scale streetscapes, neighborhoods, and commercial districts" (Mohl, 2012, p. 90). While these examples are

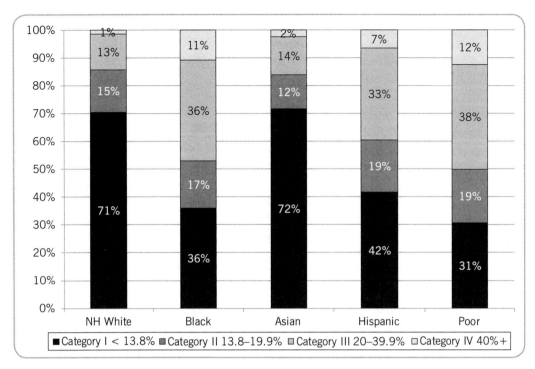

FIGURE 13.6. Distribution of racial/ethnic population and poor by poverty rate of census tracts. Source: Bishaw, Alemayehu (2011). *Areas with Concentrated Poverty: 2006–2010.* American Community Survey Briefs. Washington, DC: U.S. Census Bureau.

promising, they represent only a small fraction of the expressways that wind their way through metropolitan areas; and, at least for now, these teardowns are likely to remain the exception rather than the rule (Mohl, 2012). Consequently, low-income communities continue to bear the harmful environmental and health effects of previous transportation facility siting. Furthermore, the evidence suggests that these communities are negatively affected by new infrastructure investments, particularly rail development. It is to these two topics that I now turn.

Adverse Environmental Effects

Although transportation investments increase accessibility, they also expose adjacent residents to a host of negative externalities including air, water, and solid waste contaminants, as well as noise and accidents. Low-income and minority communities are exposed to greater environmental hazards from transportation infrastructure than higher-income, white neighborhoods (Schweitzer & Valenzuela, 2004). Low-income and nonwhite families are more likely to live in high-traffic neighborhoods, adjacent to busy freeways where they are exposed to much higher levels of particulate matter than other residents (see, e.g., Schweitzer & Zhao, 2010). Exposure to particulate matter is magnified in low-income and minority neighborhoods surrounding seaports from heavy-duty diesel trucks (HDDT), ship traffic, as well as the concentration of industries that rely on these waterways. Figure 13.7 shows that individuals in poverty, renters, and minorities are much more likely to live within 300 feet of a 4+ lane highway, railroad, or airport than nonpoor individuals (U.S. Department of Housing and Urban Development & U.S. Census Bureau, 2011).

Air pollution has multiple determinants. Therefore, it is difficult to specifically link exposure to particulate matter from mobile sources such as cars and diesel trucks to health outcomes. However, the evidence suggests that the relationship exists and that residents of low-income and minority communities suffer disproportionately. Mobile sources account for more than half of all air pollution in the United States. Numerous single- and multicity studies find that air pollution is positively associated with reduced lung function and premature mortality with effects that increase with long-term exposure (Pope & Dockery, 2006). A subset of these studies link air pollution from mobile sources to disparities in health outcomes by race and income (see, e.g., Chakraborty, 2009).

In addition to air pollution, a number of other negative externalities concentrate in low-income and minority neighborhoods. For example, if low-income families are more likely to live near heavily-trafficked roads and highways, they also are more likely to experience noise pollution, that is, unwanted or detrimental sound. All else being equal, noise levels are higher in neighborhoods with high levels of truck

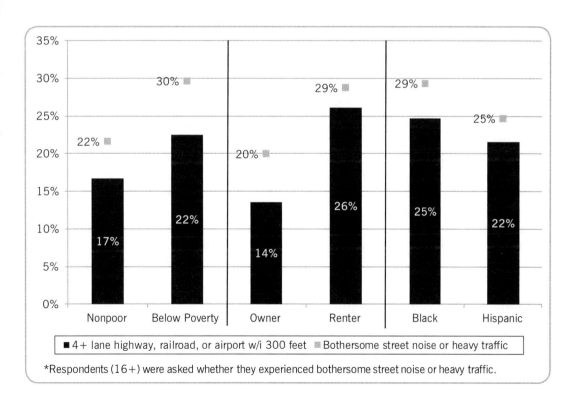

*Respondents (16+) were asked whether they experienced bothersome street noise or heavy traffic.

FIGURE 13.7. Proximity to highway, railroad, or airport, and to bothersome noise. Source: Office of Policy Development and Research and U.S. Census Bureau (2011). *American Housing Survey for the United States: 2009.* Washington, DC: U.S. Department of Housing and Urban Development and U.S. Department of Commerce.

traffic as one truck produces as much noise at 55 mph as 28 automobiles (Forken-brock, 1999). Noise levels are also problematic in neighborhoods surrounding major airports; in these neighborhoods the negative externalities outweigh the benefits of increased employment accessibility (Nelson, 2004). As Figure 13.7 shows, the same population groups that live close to highways, railroads, and airports also report higher levels of bothersome street noise or heavy traffic (U.S. Department of Housing and Urban Development & U.S. Census Bureau, 2011).

Finally, disadvantaged communities tend to be overrepresented with respect to transportation-related accidents from transport spill events as well as pedestrian and auto crashes (e.g,. Loukaitou-Sideris, Liggett, & Sung, 2007; Schweitzer, 2006). With respect to pedestrian-auto crashes, rates are largely determined by high levels of use—both pedestrian and auto traffic. However, some aspects of the built environment can play a role in increasing risks including the presence of multiple driveways, visual impairments, low lighting levels, and fewer safe spaces for off-street recreation (Loukaitou-Sideris, Liggett, & Sung, 2007).

The associations among race, income, and the presence of hazards may be due to the increased likelihood of minorities moving into neighborhoods adjacent to transportation infrastructure rather than discrimination in siting. The negative externalities of transportation investments are typically capitalized into lower housing prices, attractive to lower-income households. Yet, hazardous facilities are also more likely to be sited in low-income and minority neighborhoods than in other locations (Morello-Frosch, Pastor, Porras, & Sadd, 2002). There is a need for additional research to further examine this issue.

Regardless of the reason, residents of low-income communities suffer from the negative environmental effects of transportation projects that reduce their quality of life and contribute to health inequities. Disadvantaged neighborhoods are often subjected to multiple hazards, although far less is known about the cumulative risks facing low-income and minority residents. Finally, low-income travelers receive fewer benefits from the transportation infrastructure located in their communities, further exacerbating income and racial inequities. Transportation investments provide increased accessibility. Yet low-income residents are less likely than higher-income residents to use some of these facilities. For example, freeways cut through many low-income neighborhoods facilitating the commutes of higher-income commuters into the central city; they may not be the best choice for making the short-distance trips often taken by low-income travelers.

Transit-Oriented Development

From 1990 to 2010, rail systems expanded at a rapid rate; total vehicle miles of rail (commuter, heavy, and light rail) grew by 43% compared to a 9% decline for buses (American Public Transit Association, 2012). There is mounting concern that improving transit services and concentrating growth around transit stops and stations may have unintended social equity impacts. While transit-oriented development (TOD) has many benefits, it may contribute to gentrification and, in so doing, potentially price low-income families out of these neighborhoods, forcing some families to move to more affordable but less desirable locations. Low-income families who remain in TOD neighborhoods may enjoy improved neighborhood amenities yet also face higher rents and housing prices, further straining their already limited budgets.

Research on the effect of TOD on low-income households is limited. In general, transit investments can increase land value by improving regional accessibility to employment and other destinations and also by its association with TOD design and mixed-use amenities. Bartholomew and Ewing (2011) note the mixed findings of this body of literature; however, the research suggests a positive relationship between transit investments and land value, particularly as a consequence of increased accessibility. They find that the positive effect varies by type of transit investment (light rail, heavy rail, bus rapid transit), housing type (multifamily, single family), levels of transit service, and by the presence of pedestrian-oriented design. Some transit investments, however, are associated with lower land values, particularly in properties immediately adjacent to transit station areas; this finding is likely due to the negative externalities discussed previously (Bowes & Ihlanfeldt, 2001).

A small but growing body of research focuses specifically on the relationship between transit investments and outcomes for low-income households. Pollack, Bluestone, and Billingham (2010) examined patterns of neighborhoods change in 42 neighborhoods in 12 metropolitan areas first served by rail transit between 1990 and 2000. In many of these neighborhoods, the changes mirrored those of the larger metropolitan area in which they were located. However, in neighborhoods where changes in neighborhood characteristics were significantly different from the regional average, the addition of transit was positively related to more expensive housing stock, wealthier neighborhood residents, and higher levels of vehicle ownership (Pollack et al., 2010). They did not find a systematic relationship between new transit stations and changes in neighborhood racial composition. Transit investments can differ across some important dimensions, likely with diverse outcomes for low-income residents. For example, Kahn (2007) finds that gentrification occurs more rapidly in communities with new "walk and ride" stations compared to those that received "park and ride" stations. Transit investments and TOD are intended to improve accessibility by modes other than the automobile; yet by attracting households with automobiles, gentrification in TOD neighborhoods may actually reduce neighborhood transit ridership. Rail systems are generally structured to provide quick and convenient access to downtown job centers. Therefore, in TOD neighborhoods, if rail transit is a substitute—rather than a complement—to bus transit, low-income transit users, most of whom will be traveling to destinations other than downtown, will experience reduced accessibility.

POVERTY AND TRANSPORTATION POLICY

The issue of transportation and poverty has only sporadically received national policy attention. In the aftermath of the Watts riots in Los Angeles, the McCone Commission reported that inadequate and costly transportation limited the employment opportunities of African Americans and, among their policy recommendations, proposed an expansion of public transit services. Following closely on its heels, in 1968, the Kerner Commission reported on racism in America and offered three potential solutions for better linking the African American community to jobs. One of these solutions was "creating better transportation between ghetto neighborhoods and new job locations."

In the 1990s the issue of transportation and poverty again resurfaced in two separate contexts—environmental justice (EJ) and welfare reform. Despite these efforts, progress toward addressing the equity issues facing low-income families and communities remains limited. In the 1980s and 1990s, a number of studies highlighted the adverse effects of government and corporate environmental decisions on low-income and minority communities. These initial studies focused on issues largely unrelated to transportation such as the location of unpopular land uses, toxic waste sites and landfills. In response, in 1994 President Clinton issued Executive Order 12898 in which he directed every federal agency to make EJ part of its mission.

> Each Federal agency shall make achieving environmental justice part of its mission by identifying and addressing, as appropriate, disproportionately high and adverse human health or environmental effects of its programs, policies, and activities on minority populations and low-income populations. (President of the United States, 1994)

Subsequent to this order, the U.S. Department of Transportation (2011) adopted the following three guiding principles:

- "To avoid, minimize, or mitigate disproportionately high and adverse human health and environmental effects, including social and economic effects, on minority populations and low-income populations.
- To ensure the full and fair participation by all potentially affected communities in the transportation decision making process.
- To prevent the denial of, reduction in, or significant delay in the receipt of benefits by minority and low-income populations."

These principles include a *distributive* objective, to ensure that low-income and minority communities do not accrue fewer benefits and bear greater costs than higher-income, white communities; a *procedural* objective related to full and fair participation in transportation decision making; and a *proactive* mandate, one that requires agencies to prevent potential harm to vulnerable communities. Over time, the scope of EJ in transportation has widened as state departments of transportation (DOTs), metropolitan planning organizations (MPOs), and other agencies have responded to EJ mandates (Chakraborty, 2006). However, while many agencies have developed methods to assess the burdens of transportation investments and to facilitate public involvement in transportation decision making, at least so far, very few agencies "assess the equity of benefits, fewer access outcomes of EJ actions, and fewer still link EJ analysis outcomes with future funding and policy decisions" (Amekudzi, Smith, Brodie, Fischer, & Ross, 2012, p. 1).

President Clinton also signed into law the 1996 Personal Responsibility and Work Opportunity Reconciliation Act, or welfare reform. As part of this act, Congress created the Temporary Assistance to Needy Families Program (TANF), a welfare block-grant program that, among other provisions, includes strict work requirements. In response to welfare reform, policymakers again turned to transportation as a strategy for rapidly transitioning welfare recipients and other low-income adults into the labor market. To aid in this process, Congress passed the Job Access and Reverse Commute (JARC) program, one component of the Transportation Equity Act of the 21st Century (U.S. Congress, 1998). JARC specifically notes the uneven spatial pattern in job growth and allocates federal funds for projects that connect inner-city residents to suburban employment centers (49 USC § 5309).[2] Additionally, other federal agencies—U.S. Departments of Health and Human Services, Housing and Urban Development, and Labor—made resources available to provide transportation for welfare recipients and other low-wage workers (U.S. General Accounting Office, 1998).

Efforts to aid low-income individuals largely have centered on the role of public transit in meeting the work-related transportation needs of the poor and, in particular, the importance of strengthening public transit connections from center cities to suburbs. However, public transportation has had difficulty accomplishing these objectives. Following the McCone Commission, the State of California funded the Transportation-Employment Project, which ran a new bus line through underserved parts of Los Angeles (Governor's Commission on the Los Angeles Riots, 1965, p. 9). The Transportation-Employment Project was intended to gauge the effect of better transit on low-income workers, and its results were sadly ambivalent. Although ridership on the new bus line was consistently strong, and the new route made some existing commutes far less circuitous, there was little evidence that the bus contributed meaningfully to any decline in unemployment (California Transportation-Employment Project, 1970; Sanchez, 2008). Similar experiments in St. Louis, Boston, and New York also yielded disappointing results (e.g., Falococcio & Cantelli, 1974). Recent studies of the effect of JARC-sponsored transit programs show that participants benefit from these programs; however, once again, their overall effect on welfare usage and employment rates appears modest (Sanchez & Schweitzer, 2008; Thakuriah, Sriraj, Sööt, & Persky, 2008).

Transit functions most effectively in the inner city, where both employment and residents are clustered and where employment is frequently located adjacent to transit stops. These conditions are increasingly rare for entry-level work, and they occur with particular infrequency for reverse commutes as suburbanized employment is highly dispersed. Furthermore, work travel comprises a minority of trips, just 19% (Santos et al., 2011). Therefore, transit programs that center on alleviating work-related transportation barriers fail to address the hardships associated with travel for other purposes such as for food shopping, health care, school, and social and recreational purposes. A focus on fixed-route public transit may be particularly problematic for low-income women who have complex and demanding travel schedules and for seniors with physical limitations that make walking to and from transit stops and stations difficult.

Despite the mounting evidence of the importance of automobiles to low-income adults, relatively few policy efforts have focused on increasing automobile access and use among the poor. Many, if not most, policymakers loathe policies and programs that might be perceived as promoting automobile use, thus contributing to traffic congestion, air pollution, and sprawl. Policymakers ought to address the negative externalities associated with automobiles. However, until public transit and other alternatives can compete effectively with the automobile, it is difficult to justify efforts to limit automobile access among low-income households, a population group that faces significant accessibility barriers, travels the least, and is the most economically vulnerable.

A few policy efforts have attempted to expand automobile ownership and use among the poor. The Food Stamp program, now called the Supplemental Nutrition Assistance Program (SNAP), is a federal program that provides food to low-income families. The Food Stamp Act of 1977 mandated that cars be included as resources if they had a fair market value (FMV) in excess of $4,500. More recent federal regulations allow states flexibility in their food stamp vehicle asset policies; some states have exempted at least one vehicle from consideration in determining eligibility for benefits. Some of these same states have raised or lifted their vehicle asset limitation for welfare recipients set at $1,500 prior to welfare reform. Although the evidence is somewhat mixed, in most studies raising the vehicle asset limitation appears to increase automobile ownership among participating low-income families (Bansak, Mattson, & Rice, 2010). Finally, many social programs provide user-side subsidies for both public transit as well as automobile travel. For welfare recipients, these subsidies include reimbursement "in whole or part to TANF-eligible individuals for work-related transportation expenses (e.g., mileage, gas, public transit fare, auto repairs/ insurance, or a basic cash allowance for transportation needs)" (U.S. Departments of Transportation, Health and Human Services, & Labor, n.d., p. 4).

According to data from the 2010 American Community Survey, only a fraction of the poor (30%) receive any form of public assistance; the remaining, therefore, will not benefit from these policy changes (Ruggles et al., 2010; author's calculation). However, two other government programs may make it easier for some low-income families to purchase vehicles. Individual development accounts (IDAs) are matched savings accounts that are intended to help low-income families build assets, typically homes, businesses, and postsecondary education or training. A subset of IDA programs allows participants to use their savings for the purchase of automobiles. Among these programs, vehicle purchases are often the most frequent use of these funds (Hein, 2006). Furthermore, not typically on the list of automobile ownership programs, tax rebate programs also contribute to automobile purchases by providing the lump-sum payments needed to either purchase the vehicle outright or for the down payment. One of the largest subsidies for low-income families is the federal Earned Income Tax Credit (EITC), a refundable tax credit for low- and moderate-income families. Families whose credit exceeds the amount of taxes that they owe receive a tax refund in February or March. The payment enables families to purchase large-ticket items that do not fit into their monthly budgets. Goodman-Bacon and McGranahan (2008) find that EITC families are more than six times as likely as non-EITC families to purchase cars in February when they are most likely to receive refund checks.

Because policies and programs to help low-income families purchase automobiles are few and fall short of meeting the high demand for vehicles, nonprofit organizations have stepped in to fill some of the gap. As of 2006, there were 151 low-income automobile programs around the country (National Economic Development Law Center, 2006). In addition to selling or donating vehicles to low-income families, these programs often provide ongoing car repair support, consumer protection from price gouging on car purchases, repair, insurance, or interest rates, referrals to job training or family support services, and financial literacy training. Evaluations of these programs show that they are effective in increasing automobile ownership rates among participants (e.g. Brabo, Kilde, Pesek-Herriges, Quinn, & Sanderud-Norquist, 2003).

TRANSPORTATION EQUITY

For low-income households, transportation is a critical link to opportunities. Yet, as this chapter shows, neither the costs nor the benefits of transportation are equitably distributed. Low-income and minority families and neighborhoods on average bear higher costs and accrue fewer benefits than do middle- and higher-income households. Unfortunately, public policies rarely focus explicitly on these inequities. Policymakers often are torn by competing views of transportation equity, which as Chapter 11 shows can center on individuals and households, various groups, or geographies. How, then, can we fairly allocate the costs for and benefits from transportation investments? This seemingly simple question regarding distributive justice is surprisingly difficult to define, as equity means very different things to different people at different times. Some policymakers emphasize market equity to bring those who pay for transportation in line with those who benefit from it. Others focus on whether programs and policies treat people, groups, or places equally. And still others are concerned with outcomes, such as who has mobility or access and who does not.

One common way to evaluate competing claims is in terms of horizontal and vertical equity, concepts often used in discussions of equity, particularly transportation policy and finance (Taylor & Norton, 2009). Both concepts should inform transportation policies to aid the poor. *Horizontal equity* means that individuals in the same group—comparable in ability and need—receive the same level of transportation services (the same benefits) and also bear the same costs. As discussed in this chapter, low-income families bear different costs and benefits depending on the neighborhoods in which they live. For example, families who live in the central city typically have greater access to public transit than low-income families who live in the suburbs. One policy response may be to provide comparable levels of transit service across regions, a very expensive proposition given the costs associated with providing transit service in low-density environments. However, a focus on outcomes—access to opportunities, for example—rather than service distribution could mean adopting programs and policies that vary across neighborhoods. In dense central-city neighborhoods, low-income travelers may benefit from enhanced public transit service. Conversely, in more dispersed environments, policies ought to focus on increasing automobile access and ownership.

In contrast, *vertical equity* centers on the equitable distribution of costs and benefits *across* groups, in this case by income, race, or ethnicity. A transportation policy would be "proportional" if the net benefits are proportional to the size of the group, "regressive" if the net benefits are larger for the rich than the poor, or "progressive" if the net benefits are larger for the poor than the rich. Vertical equity lies at the center of targeted transportation and environmental justice initiatives to protect and advance opportunities and the quality of life for low-income and minority families. It is also associated with a set of policies to engender proportional costs/benefits with respect to transportation expenditures. For example, gas taxes, emission fees, and road tolls are mechanisms to ensure that individuals pay for the costs of driving commensurate with the extent to which they travel. In this scenario, people who drive less (who tend to be lower income) would pay less than people who drive more (who tend to be higher income). Other examples include pay-as-you-drive automobile insurance and distance-based transit fares (because low-income transit users tend to take shorter trips).

Transportation policies and programs cannot be held accountable for poverty and racial discrimination. Numerous factors contribute to income and racial inequality, of which transportation is only one. However, targeted transportation policies can help increase access to opportunities, improve environmental quality in low-income and minority neighborhoods, and perhaps, by drawing attention to these issues, ensure that future transportation investments and policies will not have adverse and disproportionate effects on vulnerable communities and their residents.

NOTES

1. The U.S. government relies on absolute thresholds by which to determine poverty. Thresholds based on pretax income are intended to reflect the amount of money necessary to meet basic needs such as food, clothing, and shelter. They vary by family size and composition (but not geography) and are updated for inflation using the Consumer Price Index. For analysis of the NHTS data, I use adults living in urban areas. To determine whether these adults live in poverty households, I match both household income and size to the income thresholds developed by the U.S. Census Bureau, *www.census.gov/hhes/www/poverty/data/threshld/thresh08.html*.

2. As of 2012, applicants may apply for up to 3 years of funding per project with a maximum of $400,000 per year. Eligible projects may include activities such as late-night weekend service, guaranteed ride home service, shuttle service, demand-responsive service, supporting the administration and expenses related to voucher programs, subsidizing the costs associated with adding reverse commute bus, train, carpool van routes, or service from urbanized areas and nonurbanized areas to suburban work places, intelligent transportation systems (ITS), transit vehicles, and mobility management activities. See *www.dot.ca.gov/hq/MassTrans/Docs-Pdfs/Jarc-NF/Cycle%207/2012.jarc.5316.factsheet.pdf*.

REFERENCES

Amekudzi, A. A., Smith, M. K., Brodie, S. R., Fischer, J. M., & Ross, C. L. (2012). Impact of environmental justice on transportation applying environmental justice maturation model to benchmark progress. *Journal of the Transportation Research Board, 2320*, 1–9.

American Public Transit Association. (2010). Impacts of the recession on public transportation agencies: Survey results. Available: *www.apta.com/resources/reportsandpublications/Documents/Impacts_of_Recession_March_2010.pdf*.

American Public Transit Association. (2012). Public transportation fact book, Appendix A. Available at *www.apta.com/resources/statistics/Pages/transitstats.aspx*.

Arcury, T. A., Presser, J. S., Gesler, W. M., & Powers, J. M. (2005). Access to transportation and health care utilization in a rural region. *Journal of Rural Health, 21*(1), 31–38.

Bansak, C., Mattson, H., & Rice, L. (2010). Cars, employment, and single mothers: The effect of welfare asset restrictions. *Industrial Relations, 49*(3), 321–345.

Bartholomew, K., & Ewing, R. (2011). Hedonic price effects of pedestrian- and transit-oriented development. *Journal of Planning Literature, 26*(1), 18–34.

Bishaw, A. (2011). *Areas with concentrated poverty: 2006–2010* (American Community Survey Briefs). Washington, DC: U.S. Census Bureau.

Blumenberg, E. (2004). En-gendering effective planning: Spatial mismatch, low-income women, and transportation policy. *Journal of the American Planning Association, 70*(3), 269–281.

Blumenberg, E., & Agrawal, A. (2014). Getting around when you're just getting by: Transportation survival strategies of the poor. *Journal of Poverty, 18*(4), 355–378.

Blumenberg, E., & Evans, A. E. (2010). Planning for demographic diversity: The case of immigrants and public transit. *Journal of Public Transportation, 13*(2), 23–45.

Blumenberg, E., & Ong, P. (2001). Cars, buses, and jobs: Welfare participants and employment access in Los Angeles. *Transportation Research Record, 1756*, 22–31.

Board of Governors of the Federal Reserve System. (2007). *Report to the Congress on credit scoring and its effects on the availability and affordability of credit.* Washington, DC: Author.

Bowes, D. R., & Ihlanfeldt, K. R. (2001). Identifying the impacts of rail transit stations on residential property values. *Journal of Urban Economics, 50*(1), 1–25.

Brabo, L. M., Kilde, P. H., Pesek-Herriges, P., Quinn, T., & Sanderud-Norquist, I. (2003). Driving out of poverty in private automobiles. *Journal of Poverty, 7*(1–2), 183–196.

Bullard, R. D. (2003). Addressing urban transportation equity in the United States. *Fordham Urban Law Journal, 3*, 1183–1210.

Bullard, R. D., Johnson, G. S., & Torres, A. O. (2004). Dismantling transit racism in metro Atlanta. In R. D. Bullard, G. S. Johnson, & A. O. Torres (Eds.), *Highway robbery: Transportation racism and new routes to equity* (pp. 49–74). Cambridge, MA: South End Press.

California, Transportation Employment Project. (1970). *A research project of the State of California to determine and test the relationship between a public transportation system and other opportunities of low income groups.* Sacramento, CA: Business and Transportation Agency.

Caskey, J. P. (2005). Reaching out to the unbanked. In M. Sherraden (Ed.), *Inclusion in the American dream: Assets, poverty and public policy* (pp. 149–166). Oxford, UK: Oxford University Press.

Cervero, R. (1990). Transit pricing research—A review and synthesis. *Transportation, 17*(2), 117–139.

Cervero, R., Sandoval, O., & Landis, J. (2002). Transportation as a stimulus of welfare-to-work private versus public mobility. *Journal of Planning Education and Research, 22*(1), 50–63.

Chakraborty, J. (2006). Evaluating the environmental justice impacts of transportation improvement projects in the US. *Transportation Research Part D: Transport and Environment, 11*(5), 315–323.

Chakraborty, J. (2009). Automobiles, aid toxics, and adverse health risks: Environmental inequities in Tampa Bay, Florida. *Annals of the Association of American Geographers, 99*(4), 674–697.

D'Angelo, H., Suratkar, S., Song, H., Stauffer, E., & Gittelsohn, J. (2011). Access to food source and food source use are associated with healthy and unhealthy food-purchasing behaviours among low-income African-American adults in Baltimore City. *Public Health Nutrition, 14*(9), 1632–1639.

Dawkins, C. J., Shen, Q., & Sanchez, T. W. (2005). Race, space, and unemployment duration. *Journal of Urban Economics, 58*(1), 91–113.

Dill, J., Goldman, T., & Wachs, M. (1999). California vehicle license fees: Incidence and equity. *Journal of Transportation and Statistics, 2*(2), 133–148.

El-Geneidya, A., & Levinson, D. (2007). Mapping accessibility over time. *Journal of Maps, 3*(1), 76–87.

Falococchio, J., & Cantelli, E. J. (1974). *Transportation and the disadvantaged.* Lexington, MA: D. C. Heath.

Forkenbrock, D. J. (1999). External costs of intercity truck freight transportation. *Transportation Research Part A: Policy and Practice, 33*(7–8), 505–526.

Frey, R., & Taylor, P. (2012). *The rise of residential segregation by income.* Washington, DC: Pew Research Center.

Frey, W. H. (2011, March). *Analysis of 1990, 2000, and 2010 censuses.* Washington, DC: Brookings Institution and the University of Michigan Social Science Data Analysis Network. Available at *www.psc.isr.umich.edu/dis/census/segregation2010.html.*

Fuller, D., Cummins, S., & Matthews, S. A. (2013). Does transportation mode modify associations between distance to food store, fruit and vegetable consumption, and BMI in low-income neighborhoods? *American Journal of Clinical Nutrition, 97*(1), 167–172.

Garr, E., & Kneebone, E. (2010). *The suburbanization of poverty: Trends in metropolitan America, 2000 to 2008* (Metropolitan Policy Program). Washington, DC: Brookings Institution.

Giuliano, G. (2005). Low income, public transit, and mobility. *Journal of the Transportation Research Board,* No. 1927, 63–70.

Glaeser, E. L., Kahn, M. E., & Rappaport, J. (2008). Why do the poor live in cities?: The role of public transportation. *Journal of Urban Economics, 63*(1), 1–24.

Gobillon, L., Selod, H., & Zenou, Y. (2007). The mechanisms of spatial mismatch. *Urban Studies, 44*(12), 2401– 2427.

Goldstein, T., Corbett, S., & Wachs, M. (2001). *Local option transportation taxes in the United States: Issues and trends* (Research Report UCB-ITS-RR-2001-3). Berkeley, CA: Institute of Transportation Studies.

Goodman-Bacon, A., & McGranahan, L. (2008). How do EITC recipients spend their refunds? *Economic Perspectives, 32*(2), 17–32.

Governor's Commission on the Los Angeles Riots. (1965). *Violence in the city: An end or a beginning? Final report.* Los Angeles: Author.

Grengs, J. (2002). Community-based planning as a source of political change: The transit equity movement of Los Angeles' Bus Riders Union. *Journal of the American Planning Association, 68*(2), 165–178.

Guidry, J. J., Aday, L. A., Zhang, D., & Winn, R. J. (1997). Transportation as a barrier to cancer treatment. *Cancer Practice, 5*(6), 361–366.

Hein, M. L. (2006). *The Office of Refugee Resettlement's Individual Development Account (IDA) program: An evaluation report.* Washington, DC: ISED Solutions.

Kahn, M. E. (2007). Gentrification trends in new transit-oriented communities: Evidence from 14 cities that expanded and built rail transit systems. *Real Estate Economics, 35*(2), 155–182.

Kain, J. (1968). Housing segregation, Negro employment, and metropolitan decentralization. *Quarterly Journal of Economics, 82*(2), 175–197.

Kain, J. (1992). The spatial mismatch hypothesis: Three decades later. *Housing Policy Debate, 3*(2), 371–460.

Kawabata, M. (2003). Job access and employment among low-skilled autoless workers in US metropolitan areas. *Environment and Planning A, 35*(9), 1651–1668.

Kawabata, M., & Shen, Q. (2006). Job accessibility as an indicator of auto-oriented urban structure: A comparison of Boston and Los Angeles with Tokyo. *Environment and Planning B: Planning and Design, 33,* 115–130.

Kneebone, E. (2009). *Job sprawl revisited: The changing geography of metropolitan employment.* Washington, DC: Brookings Institution.

Kneebone, E., Nadeau, C., & Berube, A. (2011). *The re-emergence of concentrated poverty: Metropolitan trends in the 2000s.* Washington, DC: Brookings Institution.

Leete, L., Bania, N., & Sparks-Ibanga, A. (2011). Congruence and coverage: Alternative approaches to identifying urban food deserts and food hinterlands. *Journal of Planning Education and Research, 32*(2), 204–218.

Loukaitou-Sideris, A., Liggett, R., & Sung, H. (2007). Death on the crosswalk: A study of pedestrian–automobile collisions in Los Angeles. *Journal of Planning Education and Research, 26,* 338–351.

Massey, D. S., & Fischer, M. J. (2000). How segregation concentrates poverty. *Ethnic and Racial Studies, 23*(4), 670–691.

Mercantonio, R. A. (2011). Just transportation planning: Lessons from California. *Progressive Planning, 186,* 4–7.

Mohl, R. A. (2004). Stop the road: Freeway revolts in American cities. *Journal of Urban History, 30,* 674–706.

Mohl, R. A. (2012). The expressway teardown movement in American cities: Rethinking postwar highway policy in the post-interstate era. *Journal of Planning History, 11*(1), 89–103.

Morello-Frosch, R., Pastor Jr., M., Porras, C., & Sadd, J. (2002). Environmental justice and regional inequality in southern California: Implications for future research. *Environmental Health Perspectives, 110*(Suppl. 2), 149–154.

National Economic Development Law Center. (2006). Low-income car ownership programs: 2006 survey. Available at *www1.insightcced.org/uploads/publications/assets/LICO2006survey-report.pdf.*

National Transit Database. (2002–2011). Washington, DC: Federal Transit Administration.

Nelson, J. P. (2004). Meta-analysis of airport noise and hedonic property values. *Journal of Transport Economics and Policy, 38,* 1–27.

Ong, P. M., & Houston, D. (2002). Transit, employment, and women on welfare. *Urban Geography, 23*(4), 344–364.

Ong, P. M., & Stoll, M. A. (2007). Redlining or risk?: A spatial analysis of auto insurance rates in Los Angeles. *Journal of Policy Analysis and Management, 26*(4), 811–830.

Pollack, S., Bluestone, B., & Billingham, C. (2010). *Maintaining diversity in America's transit-rich neighborhoods: Tools for equitable neighborhood change.* Boston: Northeastern University, Dukakis Center for Urban and Regional Policy.

Pope III, C. A., & Dockery, D. W. (2006). Health effects of fine particulate air pollution: Lines that connect. *Journal of the Air and Waste Management Association, 56*(6), 709–742.

President of the United States. (1994, February 11). Executive Order 12898. *Federal Register, 59*(32).

Raphael, S., & Rice, L. (2002). Car ownership, employment, and earnings. *Journal of Urban Economics, 52,* 109–130.

Raphael, S., & Stoll, M. A. (2001). Can boosting minority car-ownership rates narrow inter-racial employment gaps? In W. G. Gale & J. R. Pack (Eds.), *Brookings–Wharton Papers on Urban Affairs* (pp. 99–145). Washington, DC: Brookings Institution Press.

Romann, E. (2011). *Transit civil rights and economic survival in Los Angeles: A case for federal intervention in LA Metro.* Los Angeles: Bus Riders Union.

Rose, D., & Richards, R. (2004). Food store access and household fruit and vegetable use among participants in the US Food Stamp Program. *Public Health Nutrition, 7*(8), 1081–1088.

Ruggles, S., Genadek, K., Goeken, R., Grover, J., & Sobek, M. (2015). *Integrated public use microdata series: Version 6.0* [Machine-readable database]. Minneapolis: University of Minnesota.

Sanchez, T. W. (2008). Poverty, policy, and public transportation. *Transportation Research Part A: Policy and Practice, 42*(5), 833–841.

Sanchez, T. W., & Schweitzer, L. (2008). *Assessing federal employment accessibility policy: An analysis of the JARC Program.* Washington, DC: Brookings Institution.

Santos, A., McGuckin, N., Nakamoto, H. Y., Gray, D., & Liss, S. (2011). *Summary of travel trends: 2009 National Household Travel Survey.* Washington, DC: U.S. Department of Transportation, Federal Highway Administration.

Schweitzer, L. (2006). Environmental justice and hazmat transport: A spatial analysis in southern California. *Transportation Research Part D: Transport and Environment, 11*(6), 408–421.

Schweitzer, L., & Taylor, B. (2008). Just pricing: The distributional effects of congestion pricing and sales taxes. *Transportation, 35,* 797–812.

Schweitzer, L., & Valenzuela Jr., A. (2004). Environmental injustice and transportation: The claims and the evidence. *Journal of Planning Literature, 18,* 383–398.

Schweitzer, L., & Zhou, J. (2010). Neighborhood air quality, respiratory health, and vulnerable populations in compact and sprawled regions. *Journal of the American Planning Association, 76*(3), 363–371.

Sutton, R. (2007). *Car financing for low and moderate income consumers.* Baltimore: Annie E. Casey Foundation.

Taylor, B. D., & Norton, A. T. (2009). Paying for transportation: What's a fair price? *Journal of Planning Literature, 24*(1), 22–36.

Taylor, B. D., & Ong, P. (1995). Spatial mismatch or automobile mismatch?: An examination of race, residence, and commuting in U.S. metropolitan areas. *Urban Studies, 32*(9), 1453–1473.

Thakuriah, P., Sriraj, P. S., Sööt, S., & Persky, J. (2008, June 20). *Economic benefits of employment transportation services* (Final Report). Chicago: University of Illinois at Chicago, Urban Transportation Center.

U.S. Bureau of Labor Statistics. (2016). *Current expenditure tables, 2015: Quintiles of income before taxes.* Consumer Expenditure Survey. Washington, DC: U.S. Department of Labor.

U.S. Census Bureau. (2016). *2016 annual social and economic supplement.* Washington, DC: Current Population Survey.

U.S. Congress. (1998). Transportation Equity Act for the 21st Century. Public Law 105-178, 105th Congress.

U.S. Department of Agriculture. (2009). *Access to affordable and nutritious food: Measuring and understanding food deserts and their consequences.* Report to Congress. Washington, DC: Economic Research Service.

U.S. Department of Housing and Urban Development and U.S. Department of Commerce. (2011). *American Housing Survey for the United States: 2009.* Washington, DC: Office of Policy Development and Research and U.S. Census Bureau.

U.S. Department of Transportation. (2011). *Memorandum of understanding on environmental justice and Executive Order 12898.* Washington, DC: Departmental Office of Civil Rights. Available at *www.transportation.gov/sites/dot.gov/files/docs/Enviroment_Executive_ Order12898_0.pdf.*

U.S. Department of Transportation, Federal Highway Administration. (2009). *2009 National Household Travel Survey.* Washington, DC: U.S. Department of Transportation.

U.S. Departments of Transportation, Health and Human Services, and Labor. (n.d.). *Use of TANF, WtW, and job access funds for transportation.* Washington, DC: Author.

U.S. General Accounting Office. (1998). *Welfare reform: Transportation's role in moving from welfare to work* (Report to the Chairman, Committee on the Budget, House of Representatives, GAO/RCED-98-161). Washington, DC: Author.

Van Alst, J. (2009). *Fueling fair practices: A road map to improved public policy for used car sales and financing.* Boston: National Consumer Law Center.

Wachs, M. (2003). *Improving efficiency and equity in transportation finance.* Washington, DC: Brookings Institution.

Walker, R. E., Keane, C. R., & Burke, J. G. (2010). Disparities and access to healthy food in the United States: A review of food deserts literature. *Health and Place, 16*(5), 876–884.

Ward, B. G. (2005). *Case studies in environmental justice and public transit Title VI reporting.* Tampa: University of South Florida, National Center for Transit Research.

Yang, S., Zarr, R. L., Kass-Hout, T. A., Kourosh, A., & Kelly, N. R. (2006). Transportation barriers to accessing health care for urban children. *Journal of Health Care for the Poor and Underserved, 17*(4), 928–943.

Yi, C. (2006). Impact of public transit on employment status: Disaggregate analysis of Houston, Texas. *Journal of the Transportation Research Board,* No. 1986, 137–144.

Zenk, S. N., Schulz, A. J., Israel, B. A., James, S. A., Bao, S., & Wilson, M. L. (2005). Neighborhood racial composition, neighborhood poverty, and supermarket accessibility in metropolitan Detroit. *American Journal of Public Health, 95*(4), 660–667.

CHAPTER 14 | # Looking to the Future

GENEVIEVE GIULIANO
with SUSAN HANSON

MANAGING THE AUTO

The previous 13 chapters have described the many and complex dimensions of the contemporary urban transportation system in the United States. Until very recently, we have seen consistent historical trends, especially rising private vehicle ownership and use, accompanied by declining market share for public transit and nonmotorized modes. These trends have gone hand-in-hand with rising per capita income, the decentralization of metropolitan areas, and changes in economic structure that make households and firms ever more mobile. Furthermore, these fundamental trends are observed throughout the Western world.

The decline in U.S. vehicle miles traveled (VMT) from 2008 to 2011 attracted a lot of attention and speculation that travel behavior is at last changing as more households choose more central residential locations, drivers licensing loses its importance as a rite of passage among the millennial generation, and the baby boom generation retires. However, the most recent data shows VMT rising again (though at a slower rate than in previous decades), while public transit's share for the journey to work remains stagnant. Although fundamental changes in travel behavior may be on the horizon, we do not yet see significant change in the long-term trend data.

Nevertheless, the United States is an extreme case when we look at travel patterns around the world. The United States has the highest rate of private vehicle ownership, the highest level of daily miles traveled, and the lowest rates of trip making by modes other than the auto. The transit mode share in Canada is far larger than that of the United States, despite comparable per capita income and geography. Countries of northern Europe also have comparable per capita income, yet private vehicle ownership is much lower, and VMT per capita is about two-thirds that of the Unitrd

States.[1] These differences are explained by a combination of history, culture, and policy. We have seen how the contemporary U.S. situation developed from decisions and actions taken by many actors, including households, government, and business.

The private vehicle-oriented system—some would say the private vehicle-*dependent* system—that has evolved in the United States provides unprecedented mobility to those people who have access to an auto, and that mobility has created many benefits for both individuals and society. But the U.S. system also has generated many problems. Those who do not have access to a private vehicle have suffered declining mobility as other modes have become less-effective substitutes. The transport sector consumes enormous amounts of energy, accounts for more than 30,000 accidental deaths each year[2], affects human health and the well-being of the environment, and is bound up with urban sprawl. Understandably, the current urban transportation system in the United States is and has been the target of a great deal of criticism (see, e.g., Grescoe, 2012; Mitchell, 2010; Speck, 2013).

What can be done to change things? It is important to recognize that the urban transportation system we now experience is the result of choices made by individuals and institutions; it is not the outcome of some relentless natural or necessary process. As several authors in this book have noted, the transportation planning process itself has contributed significantly to the current situation of urban transportation in the United States. The traditional four-step modeling process, for example, emphasized highway building over other approaches to providing accessibility. Through the ongoing decisions and actions of a variety of actors (including planners, firms, citizens, and government), the transportation system is constantly changing. How would *you* like to see it changed? What new policies might bring about the kinds of changes *you* would like to see? How can citizens and activist groups help to bring about positive change in the urban transportation system?

Because the automobile is usually identified as *the* source of contemporary transportation-related problems, it is the focus of our thoughts in this chapter on potentials for change. We consider various approaches to managing the auto. Our focus is on policy: what governments at all levels can do, or how citizens can influence what governments do. We acknowledge at the outset that policies aimed at getting people out of their cars are unlikely to be broadly successful, simply because the mobility cars provide is of great value to most people. Nevertheless, change is both possible and necessary.

The remainder of this chapter discusses the context of transportation policy decision making, strategies for making positive change, and examples of how change is accomplished. We present our "Top Eight" strategies for improving the transportation system. They are based on their potential effectiveness as well as their implementation feasibility. We close with some success stories of positive change.

THE POLICY CONTEXT

Developing effective strategies to solve the problems generated by our contemporary urban transportation system requires an understanding of context, of what works and what doesn't. We make four observations related to context.

There Is No Magic Bullet

Transportation policy history is littered with failures. Many of these failures are the result of thinking that there is a "magic bullet"—a single, painless solution (meaning one that is low in cost and requires no big change in individual behavior) to the urban transportation problem. In the 1950s, urban leaders saw the problem as congestion and the solution as the circumferential beltway. The Interstate Highway Program offered 90/10 funding (whereby the federal government provided 90% of the funds, the states 10%), making highway building a bargain from the perspective of states and local leaders. From the 1970s, rail transit has been seen as the answer to an expanding list of problems. Again, the provision of 80/20 (federal/local) capital funding made rail transit attractive to local leaders. Decision makers believed that better and faster transit would surely draw people out of their cars. In the 1980s, many claimed that transportation system management (TSM) was the answer. TSM proponents argued that major highway building was over (highways are too costly, disruptive, and environmentally damaging); we would simply use existing capacity more efficiently. Needless to say, increasing efficiency was no match for increasing VMT. Today's magic bullets include automated vehicles, bicycle facilities, high speed trains, information technology substitutes for travel, and more rail transit investment.

No single policy can solve the urban transportation problem. Transit, bicycles, information technology, automation, and a host of other strategies all have a role to play. As with many areas of public policy, many strategies, each contributing incrementally, may collectively result in significant change. Effective approaches are based on a suite of complementary and mutually reinforcing measures and policies. We describe a range of possible approaches that need to be considered as a whole so that the synergies among them can be realized.

Effective Solutions Are Tailored to Specific Places and Circumstances

Many authors in this book have emphasized the distinctiveness of place. Each urban area combines resources, governance structures, and geography in a different way. Communities and neighborhoods within urban areas are distinctive as well. Furthermore, transportation policy decision making is increasingly local (see Handy & Sciara, Chapter 6, this volume, on the transportation planning process, and Taylor, Chapter 11, this volume, on transportation finance). It is therefore not surprising that what works in one place does not necessarily work well in another, or that a given policy will require different implementation strategies in different places to be effective. In addition, the synergies among a suite of policies will come together and work differently in different places.

Many examples illustrate that what works in one place may not work in another. An obvious example is rail transit. New York City is utterly dependent upon its extensive subway system, yet the rail systems in Miami, Sacramento, and Atlanta have no significant impact on accessibility or travel patterns. Deep discounts on monthly transit passes seem to be an obvious strategy for increasing transit ridership. People with extremely low income, however, cannot accumulate the funds to buy the monthly pass, so the discount is effectively unavailable.

The environmental impacts of roads, transit lines, or bicycle paths provide a good example of the need for solutions that match local conditions. Impacts depend on design, location, characteristics of the adjacent residents and firms, and a host of other factors.

This is not to say that we cannot generalize about strategies for solving transportation problems or that lessons learned in one community are not transferable to another. From the perspective of the transit customer, for example, lower fares and higher service quality are always better. Parking charges provide a second example. Extensive evidence shows that hefty parking charges promote carpooling and transit use wherever they are imposed (Shoup, 2011); increases in the use of these modes depend, however, on local circumstances. Our point about uniqueness and generalization is simply that analysts and policymakers need to be sensitive to local differences when thinking about creating or adopting various policy options.

Policymaking Is Complex, Contentious, and Increasingly Local

This volume has identified many problems with our urban transportation system. Readers may legitimately ask why we seem to make so little progress in solving them. Urban transportation is subject to policy decisions at all levels of government. Infrastructure projects are often funded from many different sources, thus requiring approvals from multiple levels of government. However, complexity alone does not explain the slowness and contentiousness of transportation policy decisions. The more fundamental explanation is that there is no broad consensus on how transportation problems should be solved, or on who should pay for them.

At every level of geography, from the nation to the neighborhood, policy action is dependent upon the democratic process. Both houses of Congress and the president must agree in order to fund the federal surface transportation system, establish fuel economy standards, or allocate subsidies to urban transit systems. At the local level, mayors and city council members must agree in order to approve changes in parking policy or the location of a new transit stop. In all cases, elected officials are subject to the interests of their constituents.

Different and often conflicting perspectives exist at every geographic level. Public transit advocates may promote replacing on-street parking with exclusive lanes for buses, while merchants resist any loss of on-street parking, claiming it will adversely affect business. In cities where council members are elected by district, each member represents the interests of that community. Taylor's geographic equity (Chapter 11, this volume) operates at every scale. In order for policy action to take place, compromises and trade-offs must be made across these diverse and often conflicting policy objectives.

There are good reasons why urban transport choices are difficult and contentious. First, transportation infrastructure has a large footprint and is very durable, and impacts are typically both positive and negative. Chicago's elevated rail lines increased accessibility, drawing new businesses to station areas and increasing land values. They also changed the urban landscape, allowing screeching trains to travel within a few feet of apartment windows, and darkening the streets and sidewalks below. Building the urban routes of the Interstate Highway System literally destroyed neighborhoods—mostly poor and minority neighborhoods—while greatly improving travel speeds for those who used them. Decades later these facilities continue to operate. Given the potential for both positive and negative impacts over a long period of time, it is understandable that local residents can be very concerned about the proposed location of the next rail station.

Second, there are more stakeholders and interest groups involved in policymaking, and they have more ways to influence decisions (see Handy & Sciara, Chapter 6, this volume, for the case of regional planning). As these groups increase in number and narrow in perspective, it becomes more difficult to achieve consensus on any given issue. The downsides of this rich complexity of stakeholders and participants are inflated costs, wasteful investments, or abandoned projects. Costs are inflated as project proponents face pressure to pay off those who can stop the project (Giuliano, 2007). Geographic equity considerations may require making costly investments where they are not justified in terms of travel demand or efficiency criteria, as all communities demand their "fair share." We have many examples of transportation projects that are mired in lawsuits for decades, and of policy proposals that are abandoned for lack of consensus. While some of these projects are deservedly stalled, others are worthy initiatives that would benefit society as a whole.

Finally, transportation policymaking is becoming more local as a result of two trends. The first is the increasing number of special authorities, government units established by state legislation to perform some function over a given area. Although special authorities have been around for a long time (as urban transport systems shifted to public ownership, most were organized as special authorities; the Port Authority of New York and New Jersey (PANYNJ), a bi-state special authority, was established in 1921), special authorities are increasingly used as vehicles for specific projects or activities.

The second trend is the rise of local option taxes as a revenue source for transportation. For many decades the federal Highway Trust Fund, supported by federal fuel and excise taxes, was the major source of highway (and later transit) funding. However, demands for funding have exceeded funding supply. The many explanations for this problem are beyond the scope of this chapter. State and local governments have responded by generating new revenue sources. These include some limited increases in user fees (fuel taxes, toll roads), but mostly increases in nonuser fees, including bonding and local option sales taxes (Sciara & Wachs, 2007). As more of the funds are derived from local sources, so are the policy decisions. The result is increasing geographic policy diversity. As we shall see, innovative policies are emerging in this increasingly "bottom-up" process of policy and decision making.

A final point to be made on the policy process is the mismatch between the slowness of public policy decision making and the rapidity of economic and technological change. One example is the emergence and rapid growth of carsharing services such as Uber and Lyft. Are these services a new type of carpooling, or a new type of taxi? We have a regulatory framework for taxis. They are "public conveyances" because a financial transaction takes place between the customer and the taxi company. They are therefore subject to state or local regulation, because of the government's responsibility to protect the safety of the public. Carpools are not public conveyances because the choice to carpool is made among private individuals, and there is no (formal) financial transaction. In the carsharing model, individual travelers seek a ride, and individual drivers (under contract to Uber or Lyft) offer to fulfill the request for a given fee. Uber claims that it is a technology platform to facilitate matches between drivers and riders, not a transportation company. Others claim that it is a taxi service, and hence should be subject to the same regulations as other taxi firms. Governments throughout the United States and Europe are struggling to develop a regulatory model for these services. Meanwhile, as of 2015, Uber is operating in over 150 U.S. cities.[3]

Beware of Unintended Consequences

It is, of course, impossible to know in advance how people will react to changing circumstances, so there is always some uncertainty when new policies are implemented. In some cases, however, policies have unintended consequences that create new problems. One example is prioritization of bike lanes. Cities around the United States are embracing the redesign of streets to facilitate pedestrian, bicycle, and transit travel, rather than focusing on throughput of auto traffic.[4] Under Mayor Bloomberg, New York City launched an expansive bike lane program as part of his larger effort to promote alternative modes as a means to address the city's legendary mobility problems. Some of the bike lanes were placed on streets that are designated truck routes. These streets have heavy truck traffic and large volumes of deliveries. In some cases, bike lanes replaced loading zones; in other cases traffic lanes were eliminated. In the absence of legal places to park, trucks more frequently park illegally, in this case in the bicycle lanes. Research on the impacts of these changes show that truck–bicycle crashes are concentrated along truck routes with heavy truck traffic, and a much higher-than-expected share of all bicycle-involved crashes occur on these routes (Conway, Cheng, Peters, & Lownes, 2013). In this case, the bike lanes had the unintended consequence of increasing accident risk to bicyclists on these routes.

Local truck deliveries are essential to the functioning of a city (see Dablanc & Rodrigue, Chapter 2, this volume). Food must be delivered to grocers and restaurants, retail stores must be stocked, and parcels must be delivered to businesses and residents. Parking prohibitions are ineffective, because carriers are willing to pay the fine rather than search for more distant legal parking. Efforts to redesign local streets without taking into account the demand for local deliveries leads to more congestion and safety risk, and ultimately discourages the behavior promoted by the bike lane policy.

Unintended consequences are sometimes the result of conflicting policies. U.S. policies on public transit and the private vehicle are clearly conflicting. Free parking, low vehicle and fuel taxes, and easy access to driver's licenses contribute to the attractiveness of the private vehicle, making it more difficult to attract more travelers to public transit, despite the deep subsidies that are provided. Thus our policies regarding the private vehicle are reducing the effectiveness of our public transit policies.

A more interesting case is when the results of conflicting policies are less predictable. The emergence of the SUV is a good example. The CAFE standards provided a loophole for supplying the U.S. market with large vehicles via the lower fuel efficiency standards for light trucks. Light trucks were not subject to regulation until 1982, and the standard was set at 20.7 mpg, compared to 27.5 mpg for passenger cars. The 1980s and 1990s were a period of declining real fuel price. Travelers have little incentive to economize on fuel when its real price (i.e., price adjusted for inflation) is dropping. When a consumer considers the extra cost of fuel compared to the comfort, power, passenger capacity, and perceived safety of a larger vehicle, the larger vehicle becomes more attractive as the price of fuel declines. SUV sales peaked in 2004. After a precipitous decline during the Great Recession, SUV sales are once again increasing.[5] This trend is likely to continue as long as fuel prices remain relatively low. The price of fuel is itself the result of a policy decision to keep fuel taxes as low as possible. Most other countries see the fuel tax as an important lever to discourage both fuel consumption and private vehicle use.

The Role of Government in Transportation Supply

It is difficult to overstate the role of government in the urban transportation system, even in an era of devolution and privatization. In addition to the role of policy in influencing the decisions of individuals and firms through prices and rules, the government is the supplier of transportation infrastructure. Federal, state, and most local streets and roads are collectively funded and maintained. Even when roads are built by private entities (as in planned communities), they are approved by public agencies and must comply with government standards. Public transit is almost entirely a public enterprise, funded by taxes and built and operated by public agencies. Bicycle and pedestrian facilities are also typically publicly funded and maintained.

The transportation systems operating in U.S. metropolitan areas are the result of decades of public decisions made by federal, state, and local agencies. Differences in transportation infrastructure investment explain some of the major differences among metro areas. Houston has one of the nation's highest rates of carpooling and one of the lowest rates of transit use among large metro areas, in part because of its extensive investment in high-occupancy vehicle (HOV) lanes and lack of investment in a high-quality public transit service. The extreme auto orientation of metro areas like San Diego and Atlanta is explained in part by freeway investments made before these areas entered their phase of rapid growth. The lack of sidewalks in the suburbs of Tallahassee, Phoenix, and Dallas is the result of local government decisions not to require them. The crumbling sidewalks of Los Angeles are the result of a lack of consensus on who should pay to fix them. Portland, Oregon, is conducting a major experiment in using transportation infrastructure to try to change growth patterns by making large investments in public transit and no investment in new highways. Thus transportation infrastructure decisions are a major element in the public policy toolbox.

POLICY APPROACHES

Many books have been written on how we might improve the U.S. urban transportation system or on how we might move toward a more sustainable transportation system. Before we discuss specific strategies for achieving these goals, we discuss two broad categories of policy: rules based and market based.

Rules-based policies are those that mandate specific outcomes. They may be applied at any level of government and to either firms or individuals. Examples include municipal parking requirements for new development (a local policy applied to firms), EPA/NHTSA fuel economy and GHG reduction standards (a national policy applied to auto manufacturers), and seat belt use laws (a national policy applied to individuals). *Market-based policies* use price incentives or disincentives to promote desired outcomes. They also may be applied at any level of government and to either firms or individuals. Examples include emissions-and-trade programs (national or state policy applied to firms), city parking taxes (a local policy applied to individuals), or off-peak discounts on transit fares. Less obvious policies are various forms of subsidies, such as transit capital subsidies or the support of the local road system by property taxes rather than by direct user charges.

The two policy approaches have contrasting advantages and disadvantages. First, because a rules-based policy has a specific standard or requirement that must be met, some believe that the outcome is more predictable than that of a market-based policy. For example, emissions standards on new vehicles require manufacturers to produce vehicles that meet the standard, and the standard can be enforced via random testing. If we were to impose instead an "emissions tax" on vehicles—for example, a charge per mile—individuals would respond by some combination of driving less and purchasing cleaner vehicles. In theory, we can set the tax to generate the same outcome as the regulation, but in practice people's response to the tax is much harder to predict.

Second, many claim that rules-based policies are more politically acceptable. Regulation appears to impose the same burden on everyone (e.g., all auto manufacturers face the same emission standards), whereas pricing generates equity concerns (e.g., an emissions tax would fall disproportionally on the poor because the poor drive the oldest and dirtiest cars). Of course, those who are regulated have different capabilities and resources for complying with the regulation, and regulation has its own inequities (e.g., why should people without a car pay the additional costs the developer passed on to provide the required parking spaces?). Moreover, pricing inequities can be addressed via targeted subsidies or reductions in other taxes. Rules-based policies are more politically palatable also because costs are hidden. New car buyers do not know how much they are paying for state-of-the-art emissions control systems, but an emissions tax would be an obvious "extra cost" of using a car. As the vast majority of voters are also car owners, it is not surprising that direct taxes on car use are not popular among elected officials.

In addition, industry may prefer rules-based policies, as regulatory regimes are more easily "captured." Some observers, for example, argue that the U.S. auto industry much prefers the emissions regulatory regime to an emissions tax because it protects them from competition. With an emissions tax, auto producers would have to compete to produce the cleanest car at the lowest cost, whereas today they need only meet the federal standard and pass the associated costs on to the consumer. The auto industry's incentive is therefore to minimize the burden of the regulation but not to eliminate it.

The major advantage of pricing policy is that it is more efficient: the policy objective—less congestion, less pollution, or less energy consumption—can be achieved at lower cost because those who are creating the problem have choices on how to respond. To continue with the emissions discussion, a carbon tax would impose a fee per mile based on the vehicle's carbon emissions rate. Hence SUV drivers would pay more per mile than hybrid-vehicle drivers, and drivers of new cars would pay less than would drivers of old cars. As noted above, drivers would have the choice of driving fewer miles, driving a cleaner car, or some combination of both. Such a policy would generate the following: (1) old cars would be retired more quickly from the fleet; (2) demand for cleaner cars would increase; (3) industry would compete to produce cleaner cars; (4) demand for car use would decline. Hence we could reduce vehicle carbon emissions much faster and more efficiently with a carbon tax than by continuing the current regulatory approach.

EIGHT STRATEGIES FOR BETTER URBAN TRANSPORTATION AND MORE LIVABLE COMMUNITIES

We noted earlier that there is an extensive literature on improving urban transportation. Here we focus on the strategies that in our judgment are most likely to be both effective and feasible to implement; we provide our "Top Eight" choices for policies that can make a real difference. These policies pay attention to specific market segments and places, address the source of the problem, acknowledge financial costs and equity concerns, and at least recognize the political challenge of fostering change.

1. Selectively Implement Pricing Strategies to Reduce Problems of Auto Use

The fundamental problem of the U.S. urban transportation system is that private vehicle users do not pay the full costs of vehicle use. We automobile travelers do not fully pay for the health and environmental damages resulting from auto emissions, for the additional congestion we cause, or for the many other social and environmental problems discussed in this volume in Chapters 11 (Le Vine & Lee-Gosselin), 12 (Greene), and 13 (Blumenberg). In addition, we pay only indirectly for parking and other services. This causes two problems. First, we use private vehicles more than we should from the perspective of society as a whole. Second, all other competitors to private vehicle travel are placed at a disadvantage. In economic terms, the private vehicle is underpriced.

As long as the private vehicle remains underpriced, it will be very difficult to develop viable alternatives. This is clear from recent history. We have made large investments in public transit and HOV lanes, yet transit market share remains flat and carpooling continues to decline. Why? The low price and high convenience of private vehicle travel makes it the preferred mode of travel for most people most of the time. We engage in wasteful public policy when we invest in alternative modes while doing nothing to correct for the underpricing of the private vehicle.[6] Therefore, despite the difficulties involved, our top-priority strategy is to "level the playing field" by broader implementation of pricing strategies.

Any effort to broadly implement a significant increase in the price of private vehicle travel in the United States is bound to fail. For example, the last increase in the federal per-gallon fuel tax took place in 1993, in the amount of 4.3 cents per gallon, raising the federal total to 18.4 cents. Since the fuel tax is charged per gallon, increased fuel efficiency has resulted in proportionately less tax revenue per VMT. Since the tax is not indexed to inflation, each dollar of tax revenue has less buying power as other prices increase. Efforts over the past two decades to raise the federal tax have met with failure. Insolvency of the Highway Trust Fund since 2008 has been addressed by transfers from the general fund rather than raising the fuel tax or imposing any other new charges on highway users.

The combination of rising congestion, limited public funding, and public resistance to new roads is, however, creating new opportunities. We now have many examples of pricing projects around the United States. Over 30 states have some type of toll facility in operation, and 10 states have high occupancy toll lanes, which allow vehicles with less than the required occupancy to use the lanes and pay a toll.[7] Tolls range from fixed fees, such as the Orange County toll roads described in Chapter 9, this volume (Giuliano & Agarwal), to tolls that vary by time of day. Examples of variable pricing include the SR 91 toll road in Orange County, California, the I-10 Katy Freeway managed lanes in Houston, Texas, the I-25 Expressway in Denver, Colorado, and the I-394 HOT lanes in Minneapolis. Variable pricing allows prices to fluctuate in response to demand so that the flow of vehicles is maximized and congested conditions are avoided. Toll facilities typically are either new highways or added lanes, or are facilities for which no-toll alternatives are readily available. In the first case, a new road (or additional lanes) could be constructed only with debt financing, and the tolls are justified as a means for paying off the debt. In the second case, tolls are justified as a way to make better use of existing capacity, while still offering travelers a "free" alternative. You can see that these types of facilities are far more politically feasible than, say, a conversion of all existing highways to toll roads.

Political feasibility explains why we have no examples of cordon pricing in the United States, despite London's success with its downtown pricing scheme (see Box 14.1). *Cordon pricing* imposes a fee on any vehicle entering the targeted zone, for example, a downtown area. Cordon pricing has been proposed for areas including New York City and Los Angeles. In New York City, a pricing plan proposed by then Mayor Bloomberg in 2006, approved by the city, and supported by the USDOT with over $350 million in funding was ultimately defeated in the New York State legislature in 2008. No other cordon pricing proposals have yet to reach the planning stage in the United States.

Parking policy can be used as another form of pricing. Donald Shoup (2011), the world's leading expert on parking, has pointed out that the subsidy represented by free parking at the workplace is typically greater than the price of fuel for the commute. That is, free parking is a more powerful incentive to drive alone to work than paying for an employee's gasoline would be! Shoup and others have shown that charging employees for parking, *even when combined with a "transportation allowance" of the same amount,* leads to significant reductions in drive-alone commuting. When offered a choice of free parking or the equivalent amount of cash, some employees will choose the cash and change commute mode, because the additional cash is worth more to them than the free parking. Other parking strategies include limiting the number of parking spaces permitted for new development (i.e., setting a maximum rather than a minimum number, as is now the case), parking taxes, or special reduced parking requirements in areas with high transit accessibility.

BOX 14.1
London's Congestion Pricing Scheme

In February 2003 London became the first Western city to use pricing to reduce downtown traffic congestion. London has long been notorious for its congestion. By 2000 congestion had reached such a state (average travel speed was down to less than 10 miles per hour within central London; Impacts Modeling Group, 2003) that residents and businesses were demanding relief. Mayor Ken Livingstone proposed a cordon pricing scheme: charge private vehicles £5 (about US$8) for entry into an 8-square-mile area of central London between 7:00 A.M. and 6:30 P.M. on weekdays. In order to make the plan politically acceptable, many exemptions were included, notably a 90% fee reduction for those living within the cordoned area and a total exemption for taxis.

Early results of the program were positive. Traffic within the cordon area declined by about 20%, resulting in increased average speeds of about 15%. Traffic also declined in areas just outside the cordon zone, suggesting that the new fee did not simply shift traffic from one area to another. Perhaps the biggest beneficiary of the scheme was public transit. Reduced traffic on the roads greatly increased reliability on the bus system, and public transit carried about 20,000 more daily trips. In addition, net revenues from the fee (fee revenue less costs associated with operation the pricing system) are reserved for public transit improvements.

In 2007 an extension of the charge was implemented in some of London's western neighborhoods. Similar benefits were experienced: congestion fell by 20% while transit and nonmotorized vehicle usage increased. However, in 2008 the newly elected mayor, Boris Johnson, conducted a public consultation and found growing opposition to the Western Extension. In 2011 the extension was removed. At the same time the fee for the congestion charge was increased to £10, and again in 2014 to £11.50.

What made implementation of the London congestion fee possible? We suggest the following. First, there was a widespread perception that congestion was a serious problem. Second, a rather small share of people traveling into central London traveled by car; the public transit mode share for commuters is 75%. Very high parking charges were already a deterrent for most commuters. Third, key stakeholder groups that could have defeated the proposal were neutralized by exemptions to the fee. Fourth, the fee provided a new source of funding for badly needed public transport improvements. Finally, Mayor Livingstone was an adept entrepreneurial leader. Once he left office, there was less support for the program, and the plans for expansion were not realized.

Sources: *https://tfl.gov.uk/corporate/publications-and-reports/congestion-charge*; Impacts Modeling Group (2003).

A discussion of pricing requires a discussion of equity. What about the low-income traveler, who is already spending a disproportionate share of his or her income on transportation? There are four responses. First, we should consider pricing effects relative to the current situation. In many metropolitan areas a major source of transportation funding is the sales tax, which is highly regressive. Pricing policies that reduce reliance on sales taxes could reduce inequities. Second, pricing policies generate revenue, and this revenue can be used to provide subsidies to low-income households (imagine a "lifeline transportation allowance"), to reduce other taxes that fall disproportionately on the poor, or to subsidize alternative modes. Third, pricing will increase demand for alternative modes, thereby increasing public transit revenues and enabling expanded transit services. Finally, when pricing is used to reduce congestion, public transit users receive additional benefits because transit travel times are reduced, as illustrated in the case of London (see Box 14.1).

2. Selectively Increase and Improve Public Transit Service

Many advocates of sustainable transportation are convinced that investment in public transit—particularly rail transit—is the key to solving transportation problems and building more livable communities. Such advocates argue that if we stop building roads, limit suburban development, build extensive rail transit networks, and promote higher density, mixed-use development around the rail network, we can transform the urban landscape and reduce auto dependence. Unfortunately, this is a tall order that is unlikely to be fulfilled; transit cannot and should not be viewed as the "magic bullet" that will solve urban problems.

The last four decades of transit investment in the United States illustrate how difficult it is to achieve significant change in mode choice by massive investment in public transit. From 1980 to 2013, total unlinked passenger trips by transit increased from 8.6 billion to 10.6 billion, or about 24%. Over the same period, total vehicle miles of service increased by more than 135% (from 2.3 to 5.4 billion) (American Public Transportation Association, 2015). Attracting additional passengers required a much greater than proportional addition in transport supply, resulting in reduced productivity and growing operating subsidy requirements. As noted earlier, we cannot expect public transit to compete effectively with underpriced private vehicles.

Today's metropolitan areas are a mix of higher-density clusters of activities and broad expanses of moderate- to low-density development. Jobs and the activities they represent are dispersed largely in parallel with the population. Traditional central cities, the prime market for fixed-route transit, make up an ever-smaller proportion of metropolitan areas (see Muller, Chapter 3). It is simply not financially feasible to build extensive rail systems in places that not only do not have the density and clustering of activity to support them but are unlikely ever to have it, even under the most optimistic scenarios.

Moreover, carsharing and ridesharing services are changing the passenger market in ways we cannot yet foresee. Carsharing makes car use cheaper by eliminating the costs of purchasing, owning, and garaging. For some people, carsharing will be more convenient for some trips than transit, but at the same time, not owning a car should promote more transit use. Rideshare services provide a cheaper alternative to the taxi, and in all but the most dense parts of U.S. metro areas will be a better option than transit.

Given these conditions, we should focus on making public transit as attractive as possible in places where transit is competitive. There are many possibilities, as Schweitzer (Chapter 8) describes. Extensive research tells us that people want high-quality service and low fares. High-quality service means frequent, reliable, and safe service to and from a wide variety of origins and destinations. Frequent service reduces wait and transfer times, and therefore travel time. Buses with wide doors or multiple doors reduce the time required to board and off-load passengers, reducing dwell time. Fare cards or smart phone apps reduce fare payment transaction time. The various "bus rapid transit" services are good examples of how higher-quality service can increase transit use. A notable example is Los Angeles's Metro Orange Line, a transitway connecting the suburbs to the northwest with the current terminus of the Metro Red Line subway. Opened in 2005, the 18-mile line carries 8.6 million annual passengers.[8]

Reliable and safe public transit is fundamental. If buses or trains do not keep to their schedule or if buses are so crowded that they don't stop, the transit user has more uncertainty about how long the trip will take. The traveler must then build in some extra time to account for the possibility of the bus being late, adding an even greater time penalty to using transit. People with a choice will avoid the uncertainty and inconvenience. New information technology provides transit managers with the tools to ensure service reliability by real-time tracking of vehicles and wireless communications with drivers and dispatchers. Real-time information via smart phone gives travelers more options for routing and adjusting to service schedules. Similarly, few of us are willing to put ourselves in personal danger. If passenger safety cannot be guaranteed, the transit system will be avoided.

High-quality transit also means a transparent route network that is easy to understand and negotiate. Transit use has a high cost for the new user; one must learn the route structure and timetables, how to pay the fare, how to transfer, and so on. In our judgment, part of the attractiveness of the San Francisco BART or the Washington METRO is due to the small number of routes, coded by color, and the availability of clear information about the system at every station. Transit systems should be organized so that the effort required to use the system is as small as possible. Strategies for large systems would include naming routes based on main streets or final destinations, using color coding for major routes, and providing route information at every stop and interactive maps at major stops. Physical location guides, along with virtual (smart phone) guides, would greatly improve transparency, reduce uncertainty, and hence make transit more attractive.

The United States has been slow to adopt practices that have proven successful in other countries. While most large U.S. transit systems have a variety of fare options (trip pass, day pass, multiday pass), there are few examples of bundling transit passes with hotel reservations and air tickets to attract tourists. Group fares can be negotiated, so that greatly reduced fares can be offered. An increasingly common example is the university pass program, where the university negotiates a fixed price per student per year with the transit provider, and students get unlimited-use transit passes. Typically, the program is funded via student fees. Student pass programs have resulted in significantly greater transit use (Brown, Hess, & Shoup, 2003).

One of the greatest challenges facing U.S. public transit is how to provide service at a reasonable cost outside traditional central cities. Fixed-route service in low- or moderate-density areas is both ineffective (not enough origins and destinations can be served) and expensive (not enough demand). It is clear, however, that some form of mobility service must be available; many transportation-disadvantaged households live in low- or moderate-density areas. Although we can expect emerging carshare and rideshare services to address some of the gaps, we cannot expect private-sector solutions to eliminate the need for public subsidies and services.

Some models worth considering include shared-ride services and user-side subsidy programs. Shared-ride services serve two or more different trips at the same time. When provided by the public sector, taxi companies are typically contracted to offer the service, and rides are by advance reservation. On-demand private shared ride services are appearing in the cores of major cities: Via in Manhattan and UberPool in San Francisco. These are smart phone-based, cashless services that do real-time trip matching—a much more convenient service than one that requires advance reservations and cash fares. However, at $5–10 per ride, they are not necessarily a solution for low-income travelers, and these services typically do not serve disabled passengers.

User-side subsidies provide funding directly to the traveler, to be used on any mode for any travel purpose. Unlike a transit pass program, user-side subsidies can be used on public transit, taxi, private bus, or car, giving the traveler the ability to choose the best alternative for any given trip. One example is a "transportation voucher" that would be good for anything, including gas for a neighbor's or a relative's car. The major barriers to such programs are cost and perceptions. Because the transportation-disadvantaged have so much unmet demand, user-side subsidies would be widely used and hence quite costly to provide. Also, user-side subsidies are perceived as a form of welfare.

3. Make Walking and Biking Safer

Despite the well-documented health, environmental, and congestion reduction benefits of walking and cycling instead of driving, walking and biking still account for a small proportion of urban travel in the United States. In 2009, 10.5% of all trips were made on foot and just 1% were made by bicycle. These figures compare with a nonmotorized mode share of 33% in Germany and 45% in the Netherlands. Reliance on motorized travel need not be a hallmark of an industrialized economy. Davis, California, and Boulder, Colorado—where, respectively, 15.5 % and 9.6% of work trips are made by bicycle, compared with 0.6% nationwide—show that even in the United States people can forgo the car for a substantial portion of their travel.

One way of improving the sustainability of urban transportation is to substantially increase the proportion of travel conducted by nonmotorized modes. Meeting such a goal in the United States would mean making walking and cycling attractive to large segments of the population as a means of everyday transport to a variety of destinations. Currently a major reason people give for not traveling by foot or by bike is concern about the dangers posed by motorized traffic (Nuworsoo & Cooper, 2013; Winters, Davidson, Kao, & Teschke, 2011). So bicycle and pedestrian safety—real and perceived—is essential for expanding the use of these nonmotorized modes and making urban transportation more sustainable; moreover, the goal of making walking and bicycling safer is equivalent to the goal of significantly increasing the portion of travel done on foot or by bike.

The vulnerability of pedestrians and cyclists to motor vehicle traffic in the United States is evident in the data: each year motor vehicles kill about 1,000 cyclists and 6,100 pedestrians and injure an additional 120,000 cyclists and 160,000 pedestrians seriously enough to require medical treatment (National Safety Council, 2015a, 2015b). Pucher and Buehler (2012) contrast cycling fatality and injury rates per kilometer traveled in the United States with rates in several European countries: U.S. cyclists suffer 5.5 fatalities per kilometer, compared with fatality rates of 1.1 for the Netherlands, 1.6 for Denmark and Germany, and 3.3 for the United Kingdom; the injury rate per kilometer for cyclists in the United States is 33.5, 20 times higher than the injury rate in Denmark and the Netherlands and seven times higher than in Germany.

Countries where cycling injuries and fatalities are the lowest are precisely those countries that have the highest percentage of all trips made by bicycle: 26% in the Netherlands, 18% in Denmark, and 10% in Germany (compared to 1% in the United States) (Pucher & Buehler, 2012). How have places succeeded in simultaneously improving safety and increasing the share of travel by bicycle?

Furth (2012) stresses the importance of two key infrastructure strategies:

1. Reduce traffic speeds on low-volume streets where bicycles and motor vehicles share the road; in addition to setting speed limits (e.g., below 30 mph (~50km/h)), install infrastructure changes such as speed bumps to enforce traffic calming;

2. On higher-volume roads physically separate motorized from nonmotorized traffic by providing networks of integrated walkways and bikeways; particularly important is the treatment of intersections, whether grade-separated by mode and/or instrumented with separate signals for each mode.

Pucher and Buehler (2012) include these two strategies in a comprehensive summary of lessons learned from the many urban areas around the world that have successfully managed to increase cycling in everyday travel. Many of these measures also work to improve pedestrian safety. These authors stress the importance of adopting a multipronged, integrated set of policies and programs to promote cycling. In addition to the infrastructure changes already described, their recommendations include, but are not limited to, providing bicycle parking, developing bikesharing programs, improving education and traffic enforcement for cyclists and motorists, restricting the use of autos, and pursuing a number of community-engagement strategies to support the successful implementation of all of the recommendations. To mitigate the adverse effects of motor traffic on pedestrians and bicycles, Loukaitou-Sideris (2003) outlines a similar integrated package of approaches that spans education and training, reductions in motorized traffic volume, regulation of motor traffic, and changes to physical infrastructure.

The high proportions of trips made by bicycle in the Netherlands, Denmark, and Germany—and the accompanying low fatality and injury rates in these countries—have been achieved via comprehensively integrated policies and programs such as these. The slowing of traffic on shared roads and the separation of bikeways and walkways on roads where traffic travels at speeds in excess of 30 mph (~50 km/h) have together been at the core of programs to make cycling safer and more attractive in these countries. Such approaches have been adopted in several American cities, such as Davis, California, and Boulder, Colorado, and are gradually being introduced in other cities around the United States. The evidence suggests that improving bicycle and pedestrian safety and increasing nonmotorized mode share go hand in hand; they are essential strategies for attaining sustainable urban transportation and livable communities. That nearly half of all trips in the United States are less than 3 mi (5 km)—an easily bikable distance—makes moving toward this goal feasible.

4. Take Advantage of New Technology

New technology has the potential to solve some of our most critical transportation problems: air pollution and GHG emissions, safety, and mobility. Vehicle emissions reductions and fuel efficiency savings are primarily the result of technological advances. Through "technology-forcing" regulation—imposing performance standards that cannot be met with current technologies—vehicle manufacturers have been "forced" to develop lighter-weight and more aerodynamic vehicle designs, more efficient engines, and so on. Indeed, vehicle emissions have declined dramatically despite the almost doubling of VMT over the past 30 years. Emissions standards for heavyduty trucks have been gradually tightened through 2007. And in California, a set of "optional" standards call for a tenfold reduction in NOX and PM from the 2007 standards by 2015.

Technology-forcing regulation has been the policy of choice because it focuses on the vehicle manufacturer rather than the vehicle user, and therefore does not require a change in people's behavior. Efforts to change behavior (e.g., encourage more transit and ridesharing, or purchase of more fuel-efficient vehicles) have been quite unsuccessful, as noted earlier, because we as a society have been unwilling to change the relative price of private vehicle travel sufficiently to induce significant changes.

We can expect technology to continue to provide solutions. Although efforts in the 1990s to force the development of a zero emissions vehicle (ZEV) market were unsuccessful, the hybrid electric vehicle (HEV) emerged instead. By 2013 HEV annual sales were approaching a half million,[9] and Washington (state) had the highest market share (1.6%) for EVs.[10] As of the 2013 sales year, over 20 different makes of electric or plug-in hybrid electric vehicles (PHEVs) were offered in the retail market.[11] In the longer term, hydrogen fuel cells, solar energy, or some technology yet to be invented will give us a clean and renewable fuel source.

Technology has great potential for improving public transit through improvements to existing services and development of new services. Automated guidance systems will smooth bus rides, provide for shorter headways, and allow vehicles to travel both on and off of exclusive guideways. Ubiquitous information will enhance the passenger experience. Technology should help to make public transit service "seamless," coordinating all modes and providers to perform as a single system.

The automation revolution in transport has begun. Although a fully automated road system is still decades away, each year brings more automation to conventional passenger cars and trucks. Today the average car is equipped with antilock brakes, cruise control, GPS-guidance, and other technologies. High-end cars have lateral guidance (lane keeping), adaptive cruise control (adjusts to the speed of the vehicle in front), automated braking (sounds alarm and begins braking should the driver not respond), and the capacity to park themselves. There are also breathalyzers, sensors that notice when a driver is falling asleep, and night vision enhancements. Flashing cross-walks and smart signals that sense pedestrians in the roadway further increase pedestrian safety. Information and telecommunications technologies will improve surveillance and emergency response for both private and public transport. These technologies increase safety for all travelers—drivers, passengers, bicyclists, and pedestrians.

Along with automation comes "vehicle-to-vehicle" and "vehicle-to-infrastructure" communications: vehicles and infrastructure communicating in real time and adapting as roadway conditions change. With communication, information can be pushed to vehicles, informing them of a red light or traffic accident ahead, and slowing traffic in advance of the problem. These technologies will increase the efficiency and safety of the transport system, hence accommodating more motorized travel.

The service revolution in transport is also well under way. The traditional 20th-century model of rigid modes, private vehicles, and (mostly) fixed-route public transit, is giving way to a spectrum of modes that promise to greatly enhance mobility while reducing the inherent inefficiencies of the private vehicle. One of the early entrants to this new marketplace was carsharing, or shared ownership of a fleet of vehicles available for use on a prereservation basis. The fleet is centrally managed; participants pay a monthly fee for access to the service, and typically an hourly fee for use of the vehicle. A more recent model is peer-to-peer carsharing: individuals make their private cars available on a short-term rental basis. These vehicle-sharing arrangements depend on information technology to match drivers and vehicles (see Box 14.2).

BOX 14.2
Using Cars More Efficiently

Carsharing is a good example of what technology makes possible. The carsharing concept emerged in Europe in the late 1980s as a highly localized, grassroots effort. Carsharing was introduced in the United States about a decade later. The typical U.S. carshare program requires a one-time membership fee plus a monthly subscription. Car use is charged by the hour and by the mile. Members reserve cars in advance via the Internet and are notified where to pick up the car. A smart key-card is used to gain access to the assigned car. The car is returned to the site at the end of the trip and locked with the key-card. All key-card transactions are logged, so the central manager knows at all times the availability of each vehicle in the fleet.

Carsharing is concentrated in large cities. High residential densities with ample transit and walkable neighborhoods complement and facilitate forgoing car ownership and encourage car share membership. Carsharing is more difficult in neighborhoods with lower population densities, because of the need for more vehicles (trips are more frequent and longer) and therefore higher costs. The peer-to-peer (P2P) carsharing model uses existing car owners to temporarily rent their cars to other individuals. This model maximizes the use of private vehicles that sit idle for the vast majority of their operating life. Several new service companies are dedicated to P2P carsharing including Getaround, Go-op, RelayRides, Spride Share, and WhipCar.

Each new shared car added to existing carsharing fleets in the market has been shown to remove 4.6 to 20 private vehicles from the road (Hampshire & Gaites, 2011). This reduction is due to members deferring car purchases or even selling their current vehicles once they join the car share system. Carsharing eliminates the fixed costs of car ownership and allows individuals to pay only for what they consume, thereby reducing travel as travelers no longer feel pressured to maximize their investment on their automobile. Car sharers have been shown to reduce their adjusted VMT by 67%. (Hampshire & Gaites, 2011).

The new transportation marketplace has many other options. Rideshare services match demand for a specific trip with a driver available to serve the trip. This model, pioneered by Uber and Lyft, provides what is essentially a lower-cost taxi service. The "Uber model" has been extended to specialty services, including shuttling children to school and delivering small packages. Private suppliers are exploring new transit market niches. Chariot in San Francisco offers customized fixed-route service that is based on crowd sourcing: travelers place requests for commute trips, and when sufficient demand is expressed, Chariot offers a point-to-point service with a fare of about $5 per trip. Bridj in Boston and Washington, DC, offers on-demand shared-ride van service between specific areas of the city, with stops determined by the origins and destinations of passengers. These services offer new options for travelers and hold promise for both increasing mobility and reducing VMT.

5. Remove Barriers to Flexible Use of the Transportation System

If people could use our existing transportation system more flexibly, that is, if they could easily exercise more choices in terms of when and how they traveled, the efficiency of the system would be increased. Many constraints, including those of scheduling, system design, and the nature of the workplace in the United States, prevent more flexible use.

Scheduling constraints include the weekday concentration of work and school start times in the morning and end times in the afternoon, the major source of peak-period congestion. If workers could choose their work hours, for example, they might start earlier so as to have more time with family or friends at the end of the day, reducing peak congestion. They might take the bus instead of drive if they could schedule work around the transit schedule and have some day-to-day flexibility in arrival time at work. Parents with young children have particularly restrictive schedules, complicated by childcare that may not be available early enough or late enough to allow a change in work schedule. Moreover, few people enjoy the privilege of being able to change their work arrival and departure times from day to day, even by small increments of time, especially on short notice; this constraint makes it difficult for people to deal with the inevitable vagaries of everyday life, such as bus riders who encounter congestion because of an accident or parents whose children become ill.

Scheduling constraints appear to be at the root of many truck-related congestion problems. Deliveries to wholesale and retail establishments are determined by the receiver; many restrict delivery to daytime hours. Consequently, truck traffic adds to peak-hour congestion. Some restrictions are due to labor constraints (afternoon- or night-shift workers typically are paid a shift premium); others are due to local ordinances that limit delivery hours. Again, lifting these restrictions in places where local residents would not be adversely affected would smooth out the flow of truck traffic and make better use of existing capacity.

A second constraint on flexible use of our transportation system has to do with the design of the transportation system itself, in particular the difficulty, in many, if not most, places, of making trips that involve combining different travel modes. Making sure that travelers can readily combine modes (e.g., airports and public transit; bikeways and train or bus lines, together with bike racks on transit vehicles; sidewalks and transit; adequate parking at transit stations for cars and bikes) would enhance the flexible—and therefore the efficient—use of the transportation system.

Bikesharing systems also increase the flexible use of the transportation system by enabling travelers to combine biking with primarily walking and transit. Subscribers pay to access a bike from any bikeshare station and return it to any station in the bikeshare system. In part because of the enhanced flexibility it offers, bikesharing's popularity has increased rapidly in the past decade, such that by mid-2015, 880 cities worldwide had some kind of bikeshare system; the largest portion (80%) of the world's 1 million bikeshare bikes are in China (Shaheen, 2015).

Perhaps the ultimate form of flexibility is eliminating the need for travel entirely, via the use of information technology (IT), but the work process in most U.S. workplaces together with employees' enjoyment of the sociability of working with others preclude the easy and widespread substitution of IT for commuting. Telecommuting and home-based work allow people to work at home at least part of the time. Telephones, video conferences, and web-based collaborative worksites make it possible for people to collaborate across the globe. Yet, as described in Chapter 4, this volume (Circella & Mokhtarian), telecommuting is feasible for only a small share of all jobs, and almost all people who telecommute do so for only a minority of their workdays. Employees recognize the importance of "presentism" (i.e., being present in the workplace) to an upward career trajectory, and are therefore unwilling to become invisible by telecommuting extensively. Few are interested in working in isolation at home full time.

IT makes networked production possible, but management of these processes requires face-to-face contact. The dramatic increase in long-distance air transport in recent decades is consistent with these ideas. Studies of telecommuting indicate that whereas telecommuting can reduce the number of work trips, total travel may or may not be reduced as work trips may be replaced by other trips (Helling & Mokhtarian, 2001). And, of course, Internet shopping allows shopping from the comfort of home, but a truck delivers the goods either to the home, a neighborhood dropbox, or, in Japan, to the nearest convenience store. Like Circella and Mokhtarian (Chapter 4, this volume), we conclude that the likelihood for extensive trip reduction via IT is limited, despite its promise for increased flexibility.

6. Selectively Increase and Improve Highway Capacity

It may be surprising to the reader that we suggest increasing highway capacity as an appropriate strategy for solving transportation problems. Most advocates of sustainable transportation would argue just the opposite: highways are blamed for urban sprawl, loss of open space, deterioration of urban cores, air pollution and other environmental damages, global warming, and the obesity epidemic. While some of these claims may have merit, we must face facts. U.S. metropolitan areas are utterly dependent upon the highway system; 83% of all person trips and about two-thirds of all goods movements are made in private vehicles on roadways. Although passenger vehicles will surely be used more efficiently as the technology revolution in transport proceeds, we do not see any reduction in the value of personal mobility. The mobility provided by the private vehicle is superior to other alternatives under most circumstances. Outside the cores of the largest metro areas, car ownership is the key to job access, as well as to medical care, recreational opportunities, and reasonably priced groceries. Even if the mode share for transit and nonmotorized modes were to double or triple, about 75% of all person trips would still be in private vehicles. Thus, while we must reduce the damage cars generate, we must also recognize that cars—whether privately owned or shared, and whether self-driving or driven by people—are here to stay.

One might argue that the efficiencies resulting from technology changes described earlier should allow us to accommodate increased travel demand; this is likely the case in existing, slow-growing metro areas, where the costs (financial and environmental) of acquiring additional rights of way are prohibitive. However, in metropolitan areas where population and employment are growing more rapidly, some additional road capacity will be required. Capacity can be added in ways that avoid the pitfalls of the past. New roads can and should be toll roads, as discussed in Strategy 1. Willingness to pay is the best indicator for determining whether additional capacity is justified. Environmental damage can be reduced by context-sensitive design. There is no reason why every new highway must meet federal interstate standards. Some might be modern versions of the parkway, allowing for tighter curves and steeper grades and requiring less rights of way. Such facilities would be more consistent with the fabric of local development and topography and perhaps less likely to be opposed by local residents.[12] Limited-access facilities have higher capacity per lane and therefore require less land to serve a given volume of traffic than an equivalent arterial street. Thus adding capacity in the form of limited-access highways is the most effective option for fast-growing areas.

Somewhat less controversial are strategies to increase the capacity of existing roadways. Traffic signalization technology is one of the big success stories of traffic engineering. It is estimated that optimization of signal timing can increase arterial capacity by 5–10% and greatly reduce delay. Reduced delay means reduced emissions, as running emissions are greatest at very low and very high speeds. Reduced delay at intersections means less exposure of pedestrians to carbon monoxide "hot spots," particulates, and other toxics. It also means greater reliability and shorter running times for public transit. The technology advances discussed in Strategy 4 extend the benefits of traffic management via real-time, interactive management strategies.

7. Promote More Flexible Land Development and Redevelopment

It is clear that transportation and land use patterns are interdependent. It's also clear that travel characteristics (e.g., trip frequency, travel distances, and modes used) are related to both personal/household circumstances (e.g., income, employment status, gender, age, number of children) and land use features (e.g., density of population and potential destinations, kinds of potential destinations, roadway configurations, availability of sidewalks and bike paths). But because personal/household attributes have more influence over travel behavior than do land use patterns, we can't expect land use strategies—however needed and important—to transform cities overnight, or even over several years. Land use change has to be a long-term strategy; most of the built environment of 2030 or even 2050 is already here.

Creating urban environments that are conducive to nonmotorized and transit travel, for example, is desirable for a number of reasons including increased mode choice, physical activity, and sustainability. Constructing such environments requires land use strategies that include increasing the density of population and establishments; building networks of walkways and bikeways that link residential areas with transit and potential destinations, including workplaces; and ensuring that the networks for cyclists and pedestrians are integrated with the rest of the transportation system.

Not only is land use change a long-term strategy; it can also be difficult to carry out. Substantial evidence shows that local land use controls often stymie or prohibit the kinds of land use changes needed to increase residential and travel choices, to make walking and biking attractive to large numbers of people, or to raise the efficiency of the existing transportation system.

High-density mixed-use redevelopment near transit stations in urban areas or moderate-density mixed-used development near suburban employment centers are often blocked by local zoning ordinances that set building height restrictions, for example, or prohibit mixing residential and commercial land uses. Such policies warrant changing for many reasons, among them the need to take maximum advantage of the investments we have already made in rail transit systems. Higher-density, mixed-use development around rail transit stations and elsewhere can be encouraged by reducing parking requirements, allowing development density above current zoning limits, creating public–private partnerships to share infrastructure costs, and selectively offering tax reductions. A first step is to remove the planning and zoning barriers that can prevent such development. A second step is to convince existing residents that such developments will benefit the entire community by providing more housing, activity, and mobility choices.

New development is equally important. Land use policies that restrict apartment houses, condominiums, apartments over stores, or mother-in-law apartments in suburban houses limit choice and often result in longer commutes or poorer employment opportunities. Policies that do not require sidewalks or bike paths as part of new development give a further advantage to car travel. At a minimum, local land use policy should be mode-neutral. Ideally, new development should be structured to allow for growth and densification over time, so that today's suburbs can gracefully mature into tomorrow's satellite cities.

If we provide safe choices, those who prefer to walk to the local coffee shop or bike to school will do so. As noted above, we have examples of communities where the nonmotorized share is quite large. However, nonmotorized modes are substitutes for short, local trips and therefore by themselves cannot be expected to reduce congestion or improve job access for the disadvantaged. Building communities with abundant safe walking and biking opportunities can increase bike and walk mode shares on the work trip, depending on the locations of jobs and residences, but in other places the availability of such facilities will be more about improving livability than solving transportation problems.

Adaptive reuse is another land use strategy. Proponents argue that instead of developing land at the fringe of urban areas, thereby contributing to sprawl, we should be making better use of the existing buildings, streets, and infrastructure at the core of U.S. cities. Deteriorated commercial zones often include warehouses or office buildings with sufficient architectural attractiveness to merit refurbishment and reuse. Factories and warehouses can be rebuilt to provide loft apartments; office buildings can be restructured for mixed use with street-level retail spaces and apartments above. These areas typically have high levels of public transit access and high accessibility to employment and services, making them ideal candidates for in-town transit-oriented developments.

Older downtown districts in cities throughout the United States have been pursuing adaptive reuse projects. Such redevelopment projects in former industrial inner-city areas, however, must first clean the site of toxic wastes deposited there by earlier industrial activity. A serious barrier to the redevelopment of such areas (known as "brownfields") is the complexity surrounding who is responsible for the cleanup costs as in some cases the liability may be shared among previous owners. Uncertainties around legal liabilities can dampen the attractiveness of brownfield sites to potential buyers, who may be responsible for cleanup if the land proves to be contaminated.

Finally, as mentioned in Strategy 3, cities can also employ adaptive reuse of existing transportation infrastructure to enhance livability and travel by nonmotorized modes. These strategies include traffic-calming measures such as speed bumps or narrowing roadways to slow traffic and creating auto-free streets or zones on certain days of the week or permanently.

8. Promote Reinforcing Suites of Strategies That Are Appropriate for Local Conditions

Our last point has to do with successful outcomes. If we are to solve urban transportation problems, we must develop policies that work together synergistically rather than conflict with one another, and we must create policies that are tailored to the specific circumstances of a specific location. We have noted that many public policies work at cross-purposes, as in the provision of large subsidies to public transit in an effort to attract travelers out of their cars, while also providing free parking almost everywhere.

What would a suite of strategies be? Let's take a truly challenging and difficult problem: increasing the mobility of low-income people residing outside the urban core, where public transit is a poor substitute for private vehicle travel. Providing more fixed-route transit would be both costly and ineffective; very large increases in transit service would not materially improve accessibility. Strategies might include (1) a transportation "lifeline allowance" to qualifying households; (2) neighborhood-based carsharing; and (3) redevelopment of local commercial areas to expand availability of basic goods and services. This combination of strategies would provide more choice for residents, offer new mobility options, and increase access.

How might we implement this suite of programs? The first challenge is cost: lifeline allowances require subsidies. Possibilities include some form of tax credit (e.g., refunds of sales taxes to low-income households), or state or federal demonstration grants. A carshare provider might require subsidies during the start-up period; these could be covered via cross-subsidies from already profitable services in other parts of the city. A business improvement district (which allows local tax revenues to be used for local improvements, including public safety), an overlay zone providing development incentives, increased access to small business loans, and permitting street vendors are possible strategies for fostering economic redevelopment.

Our focus on place-specific solutions leads us to suggest strategies that limit the private vehicle in some places, while accommodating it in others. It makes sense to foster alternatives to the car, and the best incentive is to correct the current underpricing of the private vehicle. If we did so, all the other strategies we have discussed would be more effective. If Atlanta, Miami, and Los Angeles continue to grow at their historical rates, this growth cannot be accommodated through infill and densification; new development at the edges of these metropolises will continue. Growing population and employment will require more roads, as well as more public transit. Therefore, it also makes sense to develop better ways to build those roads—ways that consider the local context and place the cost of these facilities on the people and firms who use them.

HOW DO WE FOSTER CHANGE?

Given the scale of urban transportation problems and the array of forces that reinforce the status quo, it may seem quixotic to expect that we can achieve significant change. But change does happen. It happens as people's perceptions change, and perceptions often change as a result of the persuasion of key individuals or groups. We close this chapter with some examples of how a few major changes have been accomplished and how you might make a difference.

California's Greenhouse Gas Reduction Policy

The State of California has a long history of being at the leading edge of air pollution control policy. Air pollution control districts were authorized in 1947, and in 1959 California established the nation's first air quality standards for criteria pollutants. The first vehicle emissions control technology was mandated in 1961, and in 1966 the first new vehicle auto tailpipe emissions standards were set for HC and CO. California continued to lead the effort to combat air pollution, and when the federal government established air quality standards in 1970, California was (and still is) permitted to continue to enforce its more stringent standards. The campaign to reduce air pollution has generated efforts to shift to cleaner fuels, and California currently has the most ambitious clean fuels program in the nation.

With its history of air pollution regulation, it is perhaps not surprising that California was among the first states to recognize greenhouse gases (GHG) as a pollutant to be regulated. In 2002 the state legislature passed AB 1493 mandating reduced GHG emissions from passenger cars (the law was not implemented until 2009 due to industry opposition). The California Global Warming Solutions Act, AB32, was signed in 2006. It requires that GHG emissions be reduced to 1990 levels by 2020, a reduction of approximately 25% (Schmidt, 2007). The target reductions are to be achieved by a combination of market-based and rules-based policies. The main market-based policy is a cap-and-trade program that, when fully implemented, will apply to about 85% of all GHG emissions sources. The cap-and-trade program is operated by the state's Air Resources Board (ARB). Cap and trade involves establishing a quantity of "pollution credits" that together reflect the target amount of GHG to be emitted by the covered industry sectors in a given year. The credits are either allocated or sold at auction, and firms use the credits to offset their emissions. The total amount of credits is reduced each year in order to achieve the emissions reduction target. Credits may be exchanged between firms or banked for use or sale in later years. The California program initially applied only to large emitters (power generators and heavy industry). Additional industries are phased in over time; transportation fuels were added in 2015. Early results are positive; emissions in the covered industries decreased, and the first nine auctions generated over $900 million in auction revenue.[13]

AB32 is supported by many other policies. SB 375, the Sustainable Communities Act, sets regional GHG reduction targets. MPO's are required to prepare "sustainable communities strategies" along with their regional transportation plans, and demonstrate how these targets will be met. Local governments and developers are incentivized to conform by receiving certain exemptions from the state environmental review requirements (CEQA) for compliant projects. Other policies include: (1) the low carbon fuel standard, which requires a 10% reduction in the carbon intensity of transportation fuels by 2020; (2) tax credits for purchase of clean fuel vehicles; and (3) mandates for a 50% increase in fuel efficiency and a 50% share for renewable energy sources by 2030. California's GHG reduction policy is a good example of the "suite of policies" approach.

Although California has been a leader in air quality, GHG reduction is a more difficult problem, given that the benefits of GHG reductions are global rather than local and will take place decades into the future. What was the motivation for the passages of AB32? Hanemann (2007) identifies the following factors: the long history and expertise of ARB as a regulatory agency; extensive research within California on global climate change and its projected impacts on California; the electricity crisis of 2001 resulting from a clumsy deregulation of the electric utilities that motivated calls for energy conservation; and the governor's need for a political win after some major setbacks on other policy issues. AB 32 cannot be ascribed to one individual or group. There were many players: state political leaders, environmental advocacy groups, and university researchers.

The San Francisco Freeway Revolt

San Francisco was one of the cities where grassroots opposition to proposed urban freeways resulted in dramatic action. Plans developed in the 1940s called for a grid of freeways in San Francisco. Communities throughout the city strongly opposed the freeway plans as they became publicized in the 1950s. Citizen groups organized, held meetings, and signed petitions. In 1959 the San Francisco Board of Supervisors responded by voting to cancel seven of the 10 planned routes. Particularly hated was the Embarcadero Freeway, which was to link the Golden Gate and Bay Bridges via a route along the city's waterfront at Fisherman's Wharf. About 1 mile of the Embarcadero Freeway was built by 1959 before the full force of the revolt was felt. In the face of widespread public opposition, the project was stalled, and nearly a decade later the board of supervisors voted to stop construction.

For many years the unfinished freeway stood (or, more accurately, hung in mid-air) as a monument to the highway revolt that spread around the country and marked the beginning of the end for many of the urban segments of the interstate system. Decades later, the Embarcadero freeway suffered structural damage in the 1989 Loma Prieto earthquake. In 1985 the San Francisco County Board of Supervisors voted to tear down the freeway, but city leaders were opposed, citing the potential for more traffic congestion. The earthquake damage forced the closure of the freeway, and after a short time no major traffic problems were apparent. City leaders became convinced, and in 1991 approved a plan for demolishing the freeway and replacing it with a boulevard. Demolition of the freeway made possible extensive redevelopment, and today the corridor includes retail and office development, new residential neighborhoods, and a new baseball stadium.

The Embarcadero Freeway revolt demonstrates the potential power of ordinary citizens. If, rather than opposing the freeway, the residents of San Francisco had accepted it as inevitable (given its strong support by the federal and state governments), the Embarcadero would probably have been completed and be standing to this day. Instead, the demolition of the freeway became a model for reclaiming urban space in other cities.

Individuals Who Made a Difference

Sometimes individuals provide the leadership for policy change. Here we briefly describe two who were instrumental in developing and promoting new mobility options.

Our first example is Jaime Lerner, who served three terms as mayor of Curitaba, Brazil, during the 1970s and 1980s. Lerner's background is in architecture and planning. Curitaba, like many cities in developing countries, was experiencing rapid growth but lacked the infrastructure to accommodate an ever growing population. Growth was haphazard, congestion was pervasive, and environmental pollution was extensive. Lerner focused on "simple solutions" that could be implemented quickly. His solution for transportation was to use buses as a rapid transit system: provide exclusive right of way, use large capacity vehicles, and collect fares off the vehicle. The concept became known as bus rapid transit (BRT) and has been replicated in cities around the world. BRT has many advantages. First, at-grade construction is orders of magnitude less costly than subway construction, allowing for many more miles of route to be constructed with the available funds. Second, construction can be accomplished in a relatively short period of time—most of the Curitaba system was built within a few years. Third, buses on exclusive rights of way can be operated at short headways (2 minutes), providing very high frequency service. Fourth, large buses reduce labor costs, as more passengers can be carried per driver. Finally, low capital costs make the total costs of operating and maintaining the system lower, permitting lower fares, or less subsidy. The BRT system greatly increased accessibility and mitigated demand for private vehicle use. Discount fare programs for low-income travelers makes the system more affordable.

Lerner also used BRT as a framework for land development to improve the overall livability of Curitiba by promoting new growth around the BRT stations, so that all residents could be within 400 meters of a bus stop (the BRT is supported by a feeder system). Downtown areas are car-free zones, making streets public space for pedestrians. As of 2014 the Curitiba system was carrying 1.75 million daily passengers. Although motivated by the transportation problems of Curitiba and other similar cities, BRT has proven effective in high-income countries as well. Nearly 200 cities around the world now have BRT systems. Lerner's influence has become truly global.

Our second example is Robin Chase, founder of ZipCar. Chase, having had a successful professional career in a variety of jobs, decided that she would like to start a company. A trip to Europe exposed her to the private car-share activity in Germany. Chase believed that there was demand in the United States for a mobility service that was shorter term than traditional car rental and allowed more freedom and flexibility than taxis. The concept was to provide the mobility benefits of hassle-free car ownership on demand to noncommuting individuals who need an occasional vehicle to go shopping or get to appointments. ZipCar was first introduced in Boston in 2000 with ZipCar vehicles located near transit stops. ZipCar has expanded to many other large cities, establishing itself as the leader in this nascent market. The company went public in 2011 and was acquired by Avis in 2013. Chase went on to establish other car-sharing platforms internationally, such as Buzzcar, a peer-to-peer car rental service.

Local Activists Seek to Limit Greenhouse Gas Emissions

One contemporary example of individuals and local groups promoting change in their communities is ICLEI—Local Governments for Sustainability. Founded in 1990 as the International Council for Local Environmental Initiatives, ICLEI, as it was then known, focused on local initiatives to reduce greenhouse gas emissions. As Greene (Chapter 12) has noted, the United States contributes significantly to global GHG emissions; the transportation sector accounts for about one-fifth of anthropogenic GHG emissions worldwide, and one-third of those emissions come from transportation in the United States. Reducing those emissions is an essential component of addressing the challenge of climate change.

In 1991 ICLEI launched its Cities for Climate Protection (CCP) campaign, in which citizens and leaders of cities around the globe worked to reduce their city's fossil fuel consumption in order to limit GHG emissions. Renamed in 2003 to reflect the organization's broadened mission of sustainability, ICLEI now involves several hundred communities in the United States and a global network of more than 1,000 cities and communities representing more than 8 million people around the world. The original CCP campaign has morphed into a suite of programs to facilitate communities' sustainability planning. One major element now is a sophisticated ICLEI software package, Clear Path, which not only helps cities monitor and track local emissions, but also develops forecasts for future local emissions scenarios and calculates the costs and benefits of various emissions reduction measures. ICLEI will also provide guidance and training to local groups in the implementation of Clear Path. In 2014, more than 350 local governments in the United States were using Clear Path as part of their emissions-reductions strategies (ICLEI, 2014).

ICLEI first emerged in the absence of strong action at the national level, especially in the United States, to address the threat of climate change. Instead of lobbying for effective federal policies, ICLEI adopted a bottom-up approach by establishing its network of local groups, each of which set local GHG emissions targets and developed local plans and strategies for meeting those targets. The overall message of ICLEI has been that global change starts in local places and that, moreover, global change (in this case, in reducing GHG emissions) requires coordinated change at the local level. This local focus and the Clear Path tool fit well with the increasingly local focus of transportation planning and decision making. Recognizing the importance of the transportation sector to creating sustainable cities, ICLEI is deeply involved in promoting nonmotorized transportation as a major sustainability strategy.

One might argue that such local initiatives have been influential in persuading national leaders that responding to climate change is not only necessary but possible. In December 2015, 195 countries, including the United States, reached a milestone agreement at the Paris climate talks to take specific steps to limit GHG. Among the actions taken in the Paris Climate Agreement were to keep the global average temperature to "well below 2 degrees Centigrade above pre-industrial levels and to pursue efforts to limit the temperature increase to 1.5 degrees C above pre-industrial levels" (p. 22) and to establish "an enhanced transparency framework for action and support" (p, 28) (United Nations, 2015). While disparaged by some observers as inadequate, these measures represent a significant change in the global approach to addressing climate change. Preceding the Paris agreement, decades of work by ICLEI and other organizations—focused in thousands of local communities and linked into global networks—doubtless helped pave the way to Paris.

CONCLUSION

From the outset, the authors of this book have emphasized that transportation is essential to everyday life and that transportation investment and policy decisions have shaped the structure of today's urban areas. Ongoing processes, particularly globalization and information technology, are changing the supply of and demand for urban transportation. Because transportation is at the heart of planning for more livable cities, there is considerable interest among citizens and policymakers to find ways to make urban transportation more sustainable.

The transportation planning process is a direct descendent of highway planning in that the analytical tools of transportation planning were initially developed to guide regional highway investment. Fortunately, planning tools have improved so that they now enable consideration of land use and induced-demand impacts, the incorporation of all modes, and representation and analysis via GIS. Although these improvements enhance the potential of planning, transportation investment decisions remain inherently political. Who stands to gain and who stands to lose from a proposed transportation system change is often an intensely geographic question.

Throughout this book, authors have stressed the many policy problems that urban transportation poses, including energy consumption, air pollution and other environmental impacts, and inequities in access to transportation. Authors have also discussed specific aspects of transportation, such as land use impacts and transportation finance, that should help you to understand the geography of urban transportation and think through potential solutions to policy problems. Although the United States has made considerable progress in some of these areas, such as environmental quality, we have not made much progress in other areas, such as energy consumption and transportation inequalities. Much remains to be done if we are to achieve the goals of sustainable transportation systems and livable cities.

In this chapter, we have explored a variety of approaches to solving urban transportation problems. We have outlined eight proposals for improving urban transportation and for creating more livable, and more sustainable, cities and communities. Clearly, urban transportation presents society, and especially U.S. society, with many difficult challenges. Our hope is that this book has provided you with the background and the analytical tools needed to take up these challenges and to help create badly needed change in our urban transportation systems. This change is likely to come slowly and incrementally and to be initiated by citizen activists. As the examples in this chapter have shown, you can make a difference!

ACKNOWLEDGMENT

Valuable and greatly appreciated assistance in gathering data and information for this chapter was provided by Arnold Valdez, Master of Planning graduate, 2016, Price School of Public Policy, University of Southern California.

NOTES

1. Calculated by the authors from *EU Transport in Figures Statistical Pocketbook 2015*, accessible at *http://ec.europa.eu/transport/facts-fundings/statistics/pocketbook-2015_en.htm*.

2. Source: *www-fars.nhtsa.dot.gov/Main/index.aspx*.

3. See *www.uber.com/cities*.

4. See *www.smartgrowthamerica.org/complete-streets/complete-streets-fundamentals/complete-streets-faq*.

5. See *www.rita.dot.gov/bts/sites/rita.dot.gov.bts/files/publications/national_transportation_statistics/html/table_01_21.html*.

6. It is important to note that emerging rideshare and carshare services do not solve the societal cost problem of the personal vehicle. While these services will increase the efficiency of using personal vehicles, users are not paying for environmental or congestion costs.

7. Source: *www.ncsl.org/research/transportation/toll-facilities-in-the-united-states.aspx*.

8. See *www.metro.net/news/facts-glance*.

9. Source: *www.afdc.energy.gov/data*.

10. Source: *http://cleantechnica.com/2014/02/03/top-electric-car-states-highest-percentage-electric-cars.*

11. Source: *http://evobsession.com/world-electrified-vehicle-sales-2013.*

12. Citizen opposition to a proposed widening and straightening of Connecticut's Merritt Parkway in the early 1990s exemplifies the value that many people place on such non-Interstate Highway facilities (see Carlson, Wormser, & Ulberg, 1995, pp. 32–39).

13. Sources: *www.arb.ca.gov/html/brochure/history.htm* and *www.c2es.org/us-states-regions/key-legislation/california-cap-trade.*

REFERENCES

American Public Transportation Association. (2015). *2015 public transportation fact book, Appendix A: Historical tables.* Washington, DC: Author.

Brown, J., Hess, D., & Shoup, D. (2003). Fare-free public transit at universities: An evaluation. *Journal of Planning Education and Research, 23,* 69–82.

Carlson, D., with Wormser, L., & Ulberg, C. (1995). *At road's end: Transportation and land use choices for communities.* Washington, DC: Island Press.

Conway, A., Cheng, J., Peters, D., & Lownes, N. (2013). Characteristics of multimodal conflicts in urban on-street bicycle lanes. *Transportation Research Record,* No. 2308, 93–101.

Furth, P. G. (2012). Bicycling infrastructure for mass cycling: A transatlantic comparison. In J. Pucher & R. Buehler (Eds.), *City cycling* (pp. 105–140). Cambridge, MA: MIT Press.

Giuliano, G. (2007). The changing landscape of transportation decision-making. *Transportation Research Record,* No. 2036, 5–12.

Grescoe, T. (2012) *Straphanger: Saving our cities and ourselves from the automobile.* New York: Macmillan.

Hampshire, R., & Gaites, C. (2011). Peer-to-peer carsharing: Market analysis and potential growth. *Transportation Research Record,* No. 2217, 119–126.

Hanemann, W. M. (2007) *How California came to pass AB 32, the Global Warming Solutions Act of 2006* (Working Paper No. 1040). Berkeley: University of California, Berkeley, Department of Agricultural and Resource Economics and Policy. Retrieved November 15, 2016, from *www.hcd.ca.gov/nationaldisaster/docs/uc_berkeley-how_california_came_to_pass_ab_32-global_warming_solutions_act_of_2006.pdf.*

Helling, A., & Mohktarian, P. (2001). Worker telecommunication and mobility in transition: Consequences for planning. *Journal of Planning Literature, 15*(4), 511–525.

ICLEI—Local Governments for Sustainability. (2014). *ICLEI USA 2014 annual report.* Oakland, CA: ICLEI USA. Retrieved February 3, 2016, from *http://icleiusa.org/publications.*

Impacts Modeling Group. (2003). *Impacts Monitoring – First Annual Report.* London: Transport for London. Available at *http://content.tfl.gov.uk/impacts-monitoring-report1.pdf.*

Loukaitou-Sideris, A. (2003, November). *Transportation, land use, and physical activity: Safety and security considerations.* Paper presented at the workshop of the Committee on Physical Activity, Transportation, and Land Use, Washington, DC.

Mitchell, W. J. (2010) *Reinventing the automobile: Personal urban mobility for the 21st century.* Cambridge, MA: MIT Press.

National Safety Council. (2015a). Live to ride another day. Available at *www.nsc.org/learn/safety-knowledge/Pages/news-and-resources-safe-bicycling.aspx.*

National Safety Council. (2015b) Take steps to avoid injury or death while walking. Available at *www.nsc.org/learn/safety-knowledge/Pages/news-and-resources-pedestrian-safety.aspx.*

Nuworsoo, C., & Cooper, E. (2013). Considerations for integrating bicycling and walking facilities into urban infrastructure. *Transportation Research Record,* No. 2393, 124–133.

Pucher, J., & Buehler, R. (2012). Promoting cycling for daily travel. In J. Pucher & R. Buehler (Eds.), *City cycling* (pp. 347–364). Cambridge, MA: MIT Press.

Schmidt, C. (2007). Environment: California out in front. *Environmental Health Perspectives, 115*(3), A144–A147.

Sciara, G.-C., & Wachs, M. (2007). Metropolitan transportation funding: Prospects, progress, and practical considerations. *Public Works Management and Policy, 12*(1), 378–394.

Shaheen, S. (2015). Trends and future of shared-use mobility. Presentation at Transportation Research Board Annual Meeting, Workshop 138. Available at *http://innovativemobility.org/wp-content/uploads/2015/02/Susan-Shaheen-Shared-Use-Mobility-Trends.pdf.*

Shoup, D., (2011). *The high cost of free parking* (updated ed.). Chicago: American Planning Association.

Speck, J. (2013). *Walkable city: How downtown can save America, one step at a time.* New York: Macmillan.

United Nations. (2015). Framework Convention on Climate Change, Conference of the Parties, Adoption of the Paris Agreement, Draft Decision-/CP.21. Retrieved February 3, 2016, from *www.documentcloud.org/documents/2646274-Updated-l09r01.html.*

Winters, M. G., Davidson, D., Kao, K., & Teschke, K. (2011). Motivations and deterrents of bicycling: Comparing influences on decisions to ride. *Transportation, 38,* 153–168.

Index

Italicized page numbers denote a figure or a table

About the Editors

Genevieve Giuliano, PhD, is the Margaret and John Ferraro Chair in Effective Local Government in the Sol Price School of Public Policy, University of Southern California (USC), and Director of the METRANS Transportation Center, a joint partnership of USC and California State University Long Beach. Her research and over 180 publications focus on relationships between land use and transportation; transportation policy analysis; and information technology applications in transportation. Dr. Giuliano is a past chair of the Executive Committee of the Transportation Research Board (TRB), within the National Academies of Sciences, Engineering, and Medicine, and of the Council of University Transportation Centers (CUTC). She is a recipient of honors including the Thomas B. Deen Distinguished Lectureship Award from the TRB, the Distinguished Transportation Researcher Award from the Transportation Research Forum, and the Distinguished Contribution to University Transportation and Research Award from the CUTC. Dr. Giuliano is a frequent participant in National Research Council policy studies and a member of several advisory boards, most recently the National Freight Advisory Committee.

Susan Hanson, PhD, is Distinguished University Professor Emerita in the Graduate School of Geography at Clark University. She is an urban geographer with longstanding interests in gender and economy, the geography of everyday life, and sustainability. Dr. Hanson has been editor of *Urban Geography*, *Economic Geography*, the *Annals of the Association of American Geographers*, and *The Professional Geographer*. She currently serves on the editorial board of the *Proceedings of the National Academy of Sciences* and is division chair for the TRB. Dr. Hanson is a member of the National Academy of Sciences, a Fellow of the American Association for the Advancement of Science and of the American Academy of Arts and Sciences, and a past president of the Association of American Geographers.

Contributors

Ajay Agarwal, PhD, Department of Geography and Planning, Queen's University, Kingston, Ontario, Canada

Evelyn Blumenberg, PhD, Luskin School of Public Affairs, University of California, Los Angeles, Los Angeles, California

Marlon G. Boarnet, PhD, Department of Urban Planning and Spatial Analysis, Sol Price School of Public Policy, University of Southern California, Los Angeles, California

Giovanni Circella, PhD, School of Civil and Environmental Engineering, Georgia Institute of Technology, Atlanta, Georgia

Laetitia Dablanc, PhD, French Institute of Science and Technology for Transport, Development and Networks, Champs-sur-Marne, France

Genevieve Giuliano, PhD, Sol Price School of Public Policy, University of Southern California, Los Angeles, California

David L. Greene, PhD, Department of Civil and Environmental Engineering, University of Tennessee, Knoxville, Knoxville, Tennessee

Susan Handy, PhD, Department of Environmental Science and Policy, University of California, Davis, Davis, California

Susan Hanson, PhD, School of Geography, Clark University, Worcester, Massachusetts

Scott Le Vine, PhD, Department of Geography, State University of New York at New Paltz, New Paltz, New York; Centre for Transport Studies, Imperial College London, London, United Kingdom

Martin Lee-Gosselin, PhD, Graduate School of Planning, Laval University, Quebec City, Quebec, Canada; Centre for Transport Studies, Imperial College London, London, United Kingdom

Harvey J. Miller, PhD, Department of Geography, The Ohio State University, Columbus, Ohio

Patricia L. Mokhtarian, PhD, School of Civil and Environmental Engineering, Georgia Institute of Technology, Atlanta, Georgia

Peter O. Muller, PhD, Department of Geography and Regional Studies, University of Miami, Coral Gables, Florida

Jean-Paul Rodrigue, PhD, Department of Global Studies and Geography, Hofstra University, Hempstead, New York

Gian-Claudia Sciara, PhD, School of Architecture, University of Texas at Austin, Austin, Texas

Lisa Schweitzer, PhD, Sol Price School of Public Policy, University of Southern California, Los Angeles, California

Brian D. Taylor, PhD, Institute of Transportation Studies, University of California, Los Angeles, Los Angeles, California